The Endangered Species Act and Federalism

To my grandfather, by whose side I flushed my first covey of quail and have been in love with wildlife ever since.
– Kaush Arha

To Eliza, Maggie, and Hobie, and all future generations.
– Buzz Thompson

The Endangered Species Act and Federalism

Effective Conservation through Greater State Commitment

EDITED BY

Kaush Arha and Barton H. Thompson, Jr.

earthscan
from Routledge

The findings, interpretations, and conclusions offered in RFF Press publications are those of the authors. They do not necessarily represent the views of Resources for the Future, its directors, or its officers. Similarly, any geographic boundaries and titles depicted in RFF Press publications do not imply any judgment or opinion about the legal status of a territory on the part of Resources for the Future.

First published in paperback 2024

First published 2011 by RFF Press

For a full list of publications please contact:
Earthscan
4 Park Square, Milton Park, Abingdon, Oxon OX14 4RN

by Earthscan
605 Third Avenue, New York, NY 10158

Earthscan is an imprint of the Taylor & Francis Group, an informa business

Publisher's Note
The publisher has gone to great lengths to ensure the quality of this reprint but points out that some imperfections in the original copies may be apparent.

British Library Cataloguing in Publication Data
A catalogue record for this book is available from the British Library

Library of Congress Cataloging in Publication Data
The Endangered Species Act and federalism : effective conservation through greater state commitment / edited by Kaush Arha and Barton H. Thompson, Jr.
p. cm.
Includes bibliographical references and index.
ISBN 978-1-933115-94-8 (hb : alk. paper) 1. United States. Endangered Species Act of 1973. 2. Endangered species—Law and legislation—United States. 3. Endangered species—Government policy—United States—States. 4. Wildlife conservation—Law and legislation—United States. 5. Wildlife conservation—Government policy—United States—States. 6. Federal government—United States. 7. Intergovernmental cooperation—United States. I. Arha, Kaush. II. Thompson, Barton H.
KF5640.E4816 2011
346.7304'69522—dc22

2011000739

ISBN: 978-1-933115-94-8 (hbk)
ISBN: 978-1-03-292132-7 (pbk)
ISBN: 978-1-936331-85-7 (ebk)

DOI: 10.4324/9781936331857

Typeset in Myriad Pro by JS Typesetting Ltd, Porthcawl, Mid Glamorgan
Cover design by Chris Phillips

About Resources for the Future *and* RFF Press

Resources for the Future (RFF) improves environmental and natural resource policymaking worldwide through independent social science research of the highest caliber. Founded in 1952, RFF pioneered the application of economics as a tool for developing more effective policy about the use and conservation of natural resources. Its scholars continue to employ social science methods to analyze critical issues concerning pollution control, energy policy, land and water use, hazardous waste, climate change, biodiversity, and the environmental challenges of developing countries.

RFF Press supports the mission of RFF by publishing book-length works that present a broad range of approaches to the study of natural resources and the environment. Its authors and editors include RFF staff, researchers from the larger academic and policy communities, and journalists. Audiences for publications by RFF Press include all of the participants in the policymaking process—scholars, the media, advocacy groups, NGOs, professionals in business and government, and the public.

Contents

Contributors ix
Preface xiii

Part I. Introduction

1. Federalism under the Endangered Species Act 3
Kaush Arha and Barton H. Thompson, Jr.

Part II. Frameworks

2. An Economic Assessment of Environmental Federalism: The Optimal Locus of Endangered Species Authority 21
Terry L. Anderson and Lawrence Reed Watson

3. Cooperative Federalism and the Endangered Species Act: A Comparative Assessment and Call for Change 35
J. B. Ruhl

Part III. Opportunities for State and Local Involvement under the ESA

4. Listing Decisions, Conservation Agreements, and State–Federal Collaboration: A Litigation Perspective 55
Eileen Sobeck and Paul S. Weiland

5. The Evolution of Federalism under Section 6 of the Endangered Species Act 89
Robert P. Davison

6. California's Natural Community Conservation Planning Program: Saving Species Habitat amid Rising Development 115
Gail L. Presley

Part IV. Case Studies

7. The Karner Blue Butterfly: Wisconsin's Statewide Habitat Conservation Plan 147
David R. Lentz and Jimmy S. Christenson

8. The Red-cockaded Woodpecker: Conservation through Statewide Safe Harbor Agreements 165
Ralph Costa and Richard Gooch

9. Saving *Salmo*: Federalism and the Conservation of Maine's Atlantic Salmon 187
Alison Rieser

10. Oregon Coastal Coho Restoration and the Endangered Species Act 221
Gail L. Achterman and Julia Doermann

11. Grizzly Bear Conservation in the Greater Yellowstone Ecosystem: A Case Study in the Endangered Species Act and Federalism 251
Kaush Arha and John Emmerich

Part V. Concluding Thoughts

12. Toward Greater State and Local Commitment 307
Kaush Arha and Barton H. Thompson, Jr.

Index 323

Contributors

Gail L. Achterman (J.D. and M.S.) has been the director of the Institute for Natural Resources at Oregon State University since 2003. Since 2000, she has been a member of the Oregon Transportation Commission, which she has chaired since 2007. She also serves on the Board of the Oregon Wave Energy Trust and the Advisory Board of the Klamath Basin Rangeland Trust. Among her publications are "The State and Regional Role in Developing Ecosystem Service Markets," published in the *Duke Environmental Law and Policy Forum*.

Terry Anderson (Ph.D.) is the executive director of the Property and Environment Research Center (PERC); senior fellow at the Hoover Institution, Stanford University; and professor emeritus at Montana State University. His work helped launch the idea of "free market environmentalism" with the publication of a book by that title. Anderson's other published books include *Greener than Thou, The Not So Wild, Wild West* and over 30 others. He has been widely published in the press, including the *Wall Street Journal* and the *Christian Science Monitor*.

Kaush Arha (J.D. and Ph.D.) is the lead advisor on agricultural development and governance with the U.S. Embassy in Kabul, Afghanistan. Prior to that he served as the deputy assistant secretary for Fish, Wildlife, and Parks and associate solicitor general at the Department of the Interior. Before his federal service, Dr. Arha practiced natural resource and corporate law in Washington, D.C. He is an editor of *U.S. Agricultural Policy and the 2007 Farm Bill*, and co-author of *Conserving Ecosystem Services Across Agrarian Landscapes*.

Jimmy S. Christenson (J.D.) represented the Wisconsin Department of Natural Resources from 1974 to 2006 in all matters relating to conservation and natural resources law. He served as legal counsel to Wisconsin's statewide Karner Blue Butterfly Habitat Conservation Plan and Incidental Permit from 1991 to 2006 and continues to provide advice. He co-authored the article "Steps for Success in Rare Species Conservation: The Wisconsin Statewide Karner Blue Butterfly HCP," which appeared in *A Monumental Event: Proceedings of the Society of American Foresters 2000 National Convention*.

Ralph Costa (M.S.) has been adjunct assistant professor with the Department of Forestry and Natural Resources at Clemson University, South Carolina, since 1992. He is also the sole proprietor of *Ralph Costa's Woodpecker Outfit* (RCWO), LLC. Previously, he served as a wildlife biologist and coordinator of the Red-cockaded Woodpecker Recovery Project (1991–2007) through the U.S. Fish and Wildlife Service. Mr. Costa is the author of numerous books and professional publications in symposium and conference proceedings. In 2008, he was awarded the U.S. Department of Interior Distinguished Service Award.

Robert Davison (Ph.D.) is a courtesy faculty member in the Department of Fisheries and Wildlife at Oregon State University. Dr. Davison was deputy assistant secretary for Fish, Wildlife, and Parks in the Interior Department from 1993 to 1996. He also was responsible for ESA legislation for eight years as a professional staff member on the U.S. Senate Committee on Environment and Public Works. Dr. Davison is a co-author of "The National Wildlife Refuge System" in *The Endangered Species Act at Thirty*, volume 1.

Julia Doermann (M.S.) is a consultant to Applied Solutions, a national organization working to increase water and energy efficiency and decrease greenhouse gas emissions. She has been an instructor at Oregon State University, teaching graduate and professional courses on water governance and conflict transformation. Before this, Ms. Doermann was a policy advisor to Oregon's Governor Kitzhaber for seven years. She is a co-author of "Retooling Western Water Management," in the *Land and Water Law Review—Survey of Wyoming Law* and other works.

John Emmerich (M.S.) is the deputy director of external operations with the Wyoming Game and Fish Department (WGFD). Previously, he was the WGFD Cody Region Wildlife Management Coordinator supervising the wildlife management program for the Bighorn Basin, including grizzly bear recovery efforts. He has authored six articles in state conservation magazines and state management plans for grizzly bears, wolves, and sage grouse. He has been recognized by the WGFD with the Director's Award; USDA Forest Service Certificate of Appreciation; and WGFD Wildlife Division Employee of the Year.

Richard Gooch is the chief of the Division of Conservation Partnerships in the U.S. Fish and Wildlife Service's Atlanta Regional Office, where he has been employed since 1992. Previously, he served as refuge manager in Maryland and New Mexico; wetland scientist at the U.S. Environmental Protection Agency; natural resource manager in Florida; and a nongame regional biologist with the Florida Fish and Wildlife Conservation Commission.

David R. Lentz has worked at the Wisconsin Department of Natural Resources since 1993, first as a fisheries biologist, and since 1995 as the coordinator of the statewide Karner Blue Butterfly Habitat Conservation Plan. He co-authored the paper "Steps for Success in Rare Species Conservation," which appeared in the proceedings of the North American Wildlife and Natural Resources 2001 National Conference. He is the primary author of the Wisconsin Statewide Karner Blue Butterfly Habitat Conservation Plan and Environmental Impact Statement, 2010.

Gail Presley is executive director of the Georges River Land Trust in Rockland Maine. She was an environmental program manager for the California Department of Fish and Game until 2008, and led the statewide program on Natural Community Conservation Planning from 1998 to 2008. Gail's areas of expertise include natural community conservation planning, habitat conservation planning, conservation and mitigation banking, California and federal Endangered Species Acts, and programs such as safe harbor agreements. Her writings have appeared in California Department of Fish and Game publications.

Alison Rieser (LL.M and J.D.) is Dai Ho Chun Professor of Geography at the University of Hawaii, Honolulu, where she directs the UH@SEA and Graduate Ocean Policy Programs. She is also professor emerita at the University of Maine School of Law in Portland, where she established and directed the Marine Law Institute and taught ocean and coastal law for over 20 years. Ms. Rieser co-authored the leading casebook in coastal and ocean law, "Coastal and Ocean Law," and in 1999 was awarded a Pew Fellowship in Marine Conservation.

J. B. Ruhl (J.D., LL.M., and Ph.D.) is the Matthews & Hawkins Professor of Property at Florida State University College of Law, where he has taught courses on environmental law, land use, and property since 1999. In 2008 he was visiting professor of law at Harvard Law School. He is co-author of *The Law and Policy of Ecosystem Services*, and co-author of two casebooks—*The Law of Biodiversity and Ecosystem Management* and *The Practice and Policy of Environmental Law*.

Eileen Sobeck (J.D.) is deputy assistant secretary for Fish and Wildlife and Parks at the U.S. Department of the Interior. Before that, she was an attorney in the Environment and Natural Resources Division of the U.S. Department of Justice (ENRD), where she served as Deputy Assistant Attorney General from 1999 until 2009. Previously, she worked as a trial attorney and also worked for the Office of General Counsel of the National Oceanic and Atmospheric Administration.

Barton H. "Buzz" Thompson, Jr. (M.B.A. and J.D.) is the Robert E. Paradise Professor in Natural Resources Law at Stanford Law School and the Perry L. McCarty Director of the Woods Institute for the Environment at Stanford University. He also serves as special master for the United States Supreme Court in Montana v. Wyoming and is a member of the Science Advisory Board for the U.S. Environmental Protection Agency. He co-authored *Legal Control of Water Resources*, *Environmental Law and Policy*, and *Property Law*.

Lawrence Reed Watson (M.A. and J.D.) is the coordinator of applied programs and a research fellow at the Property and Environment Research Center (PERC) in Bozeman, Montana. He was a 2008–2009 Associate at the Charles G. Koch Charitable Foundation in Washington, D.C. With Terry Anderson and Brandon Scarborough, Watson co-authored *Water Markets: Increasing the Flow* (2011).

Paul Weiland (J.D. and Ph.D.) is a partner at Nossaman LLP in Irvine, California, and chair of the firm's Environment and Land Use Practice Group. Previously, he served as a trial attorney in the Environment and Natural Resources Division of the U.S. Department of Justice, where he was awarded the Attorney General's Distinguished Service Award. He is co-author of *Managing for the Environment* (1999), which won two National Book Awards. He has also written over 25 articles in the area of environmental law.

Preface

This book represents the first comprehensive examination of the appropriate role of states and local governments in protecting biodiversity in the United States. The book grew out of a major workshop on biodiversity and federalism held at Stanford University in 2005 and sponsored jointly by the Woods Institute for the Environment at Stanford University and by Stanford Law School. National officials (including representatives of the White House, the Department of the Interior, and key Congressional committees), major state officials, environmental representatives, business leaders, and academic experts met for two days to examine the role that states currently play in protecting the nation's imperiled species and how states might play a stronger future role.

The workshop produced wide consensus that states could and should be playing a greater role in implementing the Endangered Species Act (ESA), under the appropriate encouragement and guidance of the national government, and in protecting biodiversity more generally within their borders. Most of the workshop participants agreed that the federal government could foster a larger, more effective role for the states through four mechanisms:

- Promoting trust between state and federal officials involved in biodiversity protection
- Increasing the amount of funding available to the states for biodiversity protection, as well as increasing accountability for that funding
- Contouring opportunities for state involvement under the ESA to each state's particular abilities and interests, and
- Supporting state conservation efforts that prevent the need for listing species under the ESA in the first place.

Many people assisted us in both organizing the workshop and assembling the essays in this volume. Thanks should go first to all of the participants at the workshop, who through their insights and candor ensured a productive examination of the issues. A small committee of experts helped us in organizing and framing the workshop and was instrumental in the workshop's success. The committee included Michael Bean (currently with the Department of the Interior, and with the Environmental Defense Fund at the time of the workshop) and Steve Quarles (with Crowell & Moring).

Greatest thanks go to the authors of the chapters in this book, who have not only provided stimulating analyses but willingly responded to our multiple requests for edits and updates. Tremendous thanks are also due to Susan Carter, who assembled and formatted the final draft, and to Barbara Fahs, who copy-edited and cite-checked the entire manuscript.

K.A.
B.H.T.

PART I

Introduction

CHAPTER 1

Federalism under the Endangered Species Act

Kaush Arha and Barton H. Thompson, Jr.

This book explores the critical role that states can and should play in protecting the nation's vast wealth of biodiversity. Most books and articles on biodiversity protection in the United States have focused on the role of the national government and the national Endangered Species Act (ESA). The steps that states have taken to protect jeopardized species, and the potential contributions that states could play in the future, have received surprisingly little attention. This myopia reflects in part the attention-grabbing headlines that the ESA has received since the snail darter stopped construction of the Tellico Dam in the 1970s. But the myopia ignores the crucial importance of the 50 states plus territories in ensuring the survival of the nation's imperiled species.

States are playing a substantial and growing role in protecting biodiversity. States spend at least as much money as the national government in the protection of wildlife diversity (Richie and Holmes 1999). States also protect a wide variety of species not listed as endangered or threatened under the ESA (George et al. 1998). The United States Fish and Wildlife Service (USFWS), moreover, has increasingly used its administrative discretion under the ESA to give the states a critical role in the implementation of the ESA. States today, for example, often play a major part under the ESA in determining which species get listed and in shaping both recovery plans and private land-use restrictions. As highlighted by many of the chapters in this book, states today are often the central players in the protection and restoration of ESA-listed species.

The central and growing importance of the states in the fight to save imperiled species should not be surprising. States historically played the principal role in protecting fish and game. Indeed, states bear a trust responsibility, sometimes enshrined in their constitutions, to protect fish and wildlife within their borders (Bean and Rowland 1997). States, moreover, often enjoy a variety of advantages over the national government in protecting imperiled species, ranging from greater on-the-ground expertise in the needs of many local species to the ability to enlist local interest and pride in protecting native species.

Even more can be done to encourage effective state protection of imperiled species. States currently play a much smaller role under the ESA than they do under other major federal environmental statutes, such as the Clean Air Act (CAA) and Clean Water Act (CWA), which incorporate strong concepts of cooperative federalism. As implemented, moreover, the ESA today often fails to provide states with a strong incentive to protect listed species and to help in their recovery where imperiled.

The national government must continue to lead the effort to protect endangered and threatened species. Biodiversity protection is a national issue that cannot be left to the states to address by themselves. Many species occupy habitat in more than one state. The nation's interest in protecting particular species, moreover, can sometimes clash with parochial local interests. In addition, national protection of imperiled species can often take advantage of economies of scale and centralized expertise. These factors militate for national oversight and guidance, but they do not undermine the benefits that increased state participation could bring if appropriately encouraged and facilitated by the national government.

Federalism under Today's Endangered Species Act

This book focuses on the opportunities for greater state and local involvement through the ESA. States and local governments already play a significant role in influencing and implementing various aspects of the ESA, and they could play an even more valuable role. The ESA, moreover, provides the national government with its potentially most effective instrument for encouraging states and local governments to commit to stronger efforts to protect biodiversity.

Cooperative federalism was a key concept in the wave of national pollution statutes that Congress passed in the early 1970s. Although the various statutes gave the national Environmental Protection Agency (EPA) the power to establish and maintain the overall environmental standards, the statutes gave willing states a critical role in the implementation of their provisions. As a result, states today play a major role in implementing and enforcing major environmental laws such as the CAA, CWA, and Resource Conservation and Recovery Act. Although the arguments for such cooperative federalism were diverse, Congress believed that states could bring considerable expertise and information to the table and that conditions varied significantly enough among states that local implementation would provide for more effective regulatory regimes.

In passing the 1973 ESA, Congress emphasized the importance of involving states in the protection of the nation's biodiversity too. In section 2 of the ESA, for example, Congress found that encouraging states "to develop and maintain conservation programs" is key to "better safeguarding … the Nation's heritage in fish, wildlife, and plants" (16 U.S.C. § 1531(a)(5)). Reflecting states' historic jurisdiction over water resources, Congress also declared its policy that "Federal agencies shall cooperate with State and local agencies to resolve water resource issues in concert with conservation of endangered species" (16 U.S.C. § 1531(c)(2)).

The substantive provisions of the 1973 ESA, however, appeared to contemplate a more limited role for state governments than the cooperative federalism provisions of contemporaneous pollution statutes such as the CAA and CWA. With the exception of section 6 of the ESA, which provided for cooperative agreements between the national government and the states and explicitly required the national government to "cooperate to the maximum extent practicable with the States" in carrying out the ESA, the ESA gave virtually all of the major responsibilities under the Act to the national government.

In implementing the ESA, the USFWS has arguably undermined the original intent of section 6. At the same time, however, the USFWS has used its discretion under other provisions of the Act to give the states a greater role than Congress may have anticipated.

Listing Decisions

The first step in protecting species under the ESA, deciding whether to list a species as either endangered or threatened, is the exclusive province of the U.S. Departments of Interior and Commerce (16 U.S.C. § 1533(a)).[1] Congress gave a nod to state involvement in the listing process, however, by requiring the national government to take any independent state or local protection efforts into account in deciding whether a species should be listed as endangered or threatened (16 U.S.C. § 1533(b)(1)(A)). Congress also gave a nod to the states in the designation of critical habitat, where the ESA requires Interior and Commerce to consult "as appropriate with affected States" in determining what habitat is "critical" (16 U.S.C. § 1536(a)(2)).

In practice, Interior and Commerce have increasingly collaborated with the states in the listing process, and states have experienced growing influence through their own efforts to protect candidate species. In 1994, the USFWS adopted a formal policy providing for state consultation and advice in pre-listings and listings (George et al. 1998). To help guide determinations of when state and local conservation efforts eliminate the need to list a species, USFWS has also developed a Policy for Evaluation of Conservation Efforts when Making Listing Decisions—or what it calls its "PECE" policy.

The USFWS also has developed a program of candidate conservation agreements in which local landowners—joined in many cases by state agencies (such as fish and wildlife agencies) and local governments and agencies (such as soil conservation districts)—agree to take specific actions to protect a species in order to eliminate the need for a federal listing (and sometimes in return for assurances that, if a species is ultimately listed, the USFWS will not require additional actions of the landowners involved in the agreement)

(Thompson 2005). These agreements provide another opportunity for states to avoid "nationalized protection" of particular species.[2] As of 2003, the USFWS had approved over 100 candidate conservation agreements (of which about two dozen included assurances) (Thompson 2005).

The USFWS's approach, which the language of the ESA itself only inferentially contemplates, in effect permits states and local governments to take responsibility for protecting imperiled species that would normally be subject to national regulatory protection. If states and localities can demonstrate that their efforts are providing adequate protection for an imperiled species, the USFWS will allow them to retain principal authority for the species' protection. If at some future point the USFWS determines that state and local protections are inadequate, the USFWS can "revoke" this delegation by listing the species under the ESA.[3]

Recovery Plans

Interior and Commerce similarly have sole statutory authority to develop and implement recovery plans for listed species (16 U.S.C. § 1533(f)(1)). Although these departments are authorized to "procure the services of appropriate public and private agencies and institutions, and other qualified persons" in developing and implementing recovery plans, the ESA does not provide states with any special, explicit, or ensured role in recovery plans (16 U.S.C. § 1533(f)(2)).

In practice, however, states currently play perhaps their most prominent role under the ESA in developing and carrying out recovery plans for federally listed species. From the very outset of the ESA, the national government has used its authority to procure the services of "appropriate public agencies" in developing and implementing recovery plans to actively seek out the involvement of state agencies. In 1994, the USFWS adopted a policy on the "Role of State Agencies in Endangered Species Activities" that requires the USFWS to seek state participation in both the planning and implementation of recovery plans (George et al. 1998). In the view of some experts, the involvement of state wildlife agencies and their biologists has been critical to the success of recovery programs (Burnham et al. 2005).

Indeed, state agencies may actually play the principal role in developing and implementing recovery plans for many federally listed species. In a recent survey of state biodiversity program managers, researchers asked about the relative role of state and national wildlife managers in field efforts to restore federally listed species. Managers in over two-thirds of the states reported that either the state had primary responsibility for work on federally listed species (true in the majority of states) or the state and national governments shared responsibility. Managers in only two states said that the national government took the lead on field projects. In the average state, biologists in the lead state agency spent approximately a third of their time in efforts to protect and restore federally listed species (Niles and Korth 2005).

The degree of coordination between state and national restoration efforts, however, is less clear. In the survey mentioned above, state biodiversity program managers indicated that a mechanism is lacking for full cooperation between state and national biologists working to restore the same species. Several states indicated that they did not even know what, if any, activities the national USFWS was carrying out in their state, making it impossible to say who was playing the more significant role (Niles and Korth 2005).

Consultations by Federal Agencies

The ESA also gives the departments of Interior and Commerce exclusive jurisdiction over the principal regulatory provisions designed to protect listed species. Federal agencies, for example, must consult with Interior and Commerce, but not states, in evaluating whether a proposed action might jeopardize the continued existence of a listed species or adversely modify its critical habitat (16 U.S.C. § 1536(a)(2)). This contrasts sharply with the Coastal Zone Management Act (CZMA), where Congress not only has authorized states to prepare coastal plans but requires federal agencies to carry out their actions "in a manner which is consistent to the maximum extent practicable with the enforceable policies" of those state plans that have received federal approval under the act (16 U.S.C. § 1456(c)(1)(A)).

Takings Prohibitions

The ESA also gives Interior and Commerce exclusive authority over the enforcement of ESA section 9's prohibition on the "taking" by anyone of an endangered species of fish or wildlife, defined to include harming, hunting, killing, or wounding such a species or, by regulation, adversely modifying its habitat in a way that leads to the death or injury of the species (the "anti-take provision").[4] Interior and Commerce also determine whether and when to issue "incidental-take permits" under section 10 of the ESA exempting the permit holder from the anti-take provision of section 9 when the permittee will only incidentally take a species while engaged in an otherwise lawful pursuit (16 U.S.C. § 1539). Under section 10, Interior can issue an incidental take permit to a private landowner who could harm or kill an endangered species while developing or using his or her property if the landowner has developed an acceptable habitat conservation plan (HCP) (Thompson 1997).[5]

States and local governments often do not become involved in the implementation of section 10 of the ESA even though local land-use decisions can be affected. To date, the vast majority of submitted HCPs and issued permits have involved single properties or development projects (Thompson 2005). States and local governments have seldom been involved in the shaping of these HCPs, which have typically involved lengthy negotiations between private landowners and the USFWS (Lin 1996; Thompson 2005). Property owners must work with local governments to obtain general development and use permits, but the process of negotiating an individual HCP effectively nationalizes issues of species protection.

Because the time and costs involved in this traditional, property-by-property HCP process can significantly impede development and interfere with local land-use planning goals, however, a growing number of local communities throughout the nation have chosen to develop regional HCPs that set out a comprehensive and integrated plan for developing and using property while protecting species that have been listed (and, in a growing number of cases, species that may be listed in the future). As of mid-2003, the USFWS had approved 15 regional HCPs with a terrestrial focus, including HCPs covering major regions of Arizona, Southern California, Nevada, and Utah, and negotiations were ongoing for a sizable number of new regional HCPs (Thompson 1997; Thompson 2005).

State and local governments have played a major role in the negotiation and implementation of regional HCPs, in stark contrast to their minor or nonexistent role in shaping individual HCPs (Beatley 1994; Thompson 1997). They have participated in, and in some cases led, the lengthy and complex negotiations needed to develop plans that are acceptable to all significant stakeholders (e.g., Behan 2005). In 1991, California created its own regulatory framework, known as Natural Community Conservation Planning (NCCP), for developing and structuring regional biodiversity plans (Tarlock 2005). In other cases, states have negotiated or are negotiating state-wide HCPs covering particular species or habitat types (George et al. 1998; Thompson 2005). States and local governments also play a crucial role in implementing the terms of most regional HCPs, which typically dictate where various forms of development and land use can occur and which often provide for the creation and funding of sizable biodiversity reserves (Tarlock 2005).

The national government also has created an opportunity through its "safe harbor" program for state involvement in structuring land-use decisions that help promote biodiversity (50 C.F.R. 17.22(c)). The principal purpose of the safe harbor program is to encourage property owners to enhance, restore, and create habitat on their lands (Salzman and Thompson 2010). Prior to regulatory adoption of the safe harbor program, property owners often feared that if, by restoring or creating habitat, they attracted listed species onto their property, they would lose the ability to make later land-use changes; under safe harbor agreements (SHAs), the national government promises that, in return for the landowners' efforts to promote habitat on their land, a later decision to return the land to its initial conditions will not trigger section 9. Although SHAs can be negotiated for a single property, most have provided for regional protection in order to reduce the overall cost of the program and provide for more coordinated efforts. Such regional agreements, in turn, require an entity that can administer the agreement and enter into subsidiary agreements with individual landowners. States can operate as this entity: Hawaii operates a safe harbor program for the Hawaiian goose, and three states (Georgia, South Carolina, and Texas) run a program for the red-cockaded woodpecker.

Section 6

In contrast to the other provisions of the ESA, section 6 appears to potentially give states a significant role. Section 6 begins by instructing Interior and Commerce to "cooperate to the maximum extent practicable with the States" in implementing the ESA and, more specifically, to consult with a state before acquiring any land or water within the state for conservation purposes (16 U.S.C. § 1535(a)). Section 6 then authorizes Interior and Commerce to enter into two types of agreements with states: (1) "management agreements" in which states can administer and manage conservation areas such as wildlife refuges, and (2) "cooperative agreements" in which states undertake the conservation of listed species within their jurisdiction (16 U.S.C. § 1535(b)–(c)). Of these two opportunities, cooperative agreements appear to offer the stronger potential for cooperative federalism. Section 6(g)(2) of the ESA provides that neither the anti-take provision of section 9(a)(1) nor the provisions of section 4(d), which allow Interior and Commerce to extend the anti-take prohibition to threatened species, applies to the taking of any listed species covered by a cooperative agreement that has been approved by Interior or Commerce (16 U.S.C. § 1535(g)(2)).[6] Like cooperative federalism provisions in other environmental laws, section 6 thus appears to allow states to substitute their own conservation programs for the national regulatory prohibitions of sections 4(d) and 9 where the states have demonstrated an ability to provide adequate, independent protection of the relevant species.

Section 6(c) sets out strict criteria for federal approval of a cooperative agreement. Interior and Commerce can enter into a cooperative agreement only with states that "establish and maintain an adequate and active program for the conservation of endangered species and threatened species" (16 U.S.C. § 1535(c)(1)). The departments must find initially and annually thereafter that the relevant state agency has the authority to "conserve resident species" that the state finds to be endangered or threatened and to "establish programs, including the acquisition of land or aquatic habitat or interests therein, for the conservation of resident endangered or threatened species of fish or wildlife"; the federal agencies also must find that the state agency "has established acceptable conservation programs, consistent with the purposes and policies" of the ESA, for all nationally listed species (16 U.S.C. § 1535(c)(1)).

The legislative history of section 6 suggests that Congress intended to allow states to take over regulation of private threats to federally listed species if the states adopted adequate conservation programs. In reporting on the Senate version of section 6, for example, the Senate Commerce Committee warned that the national government "should not pre-empt efficient [state] programs" that meet minimum federal standards. "Instead, it should encourage these, and aid in the extension or establishment of others, to facilitate management by granting regulatory authority and making available financial assistance to approved schemes" (U.S. Congress 1973c). The Conference Committee for the bill similarly reported that approved cooperative agreements "will of course direct and control the enforcement of endangered and threatened species programs" (U.S. Congress 1973a).

Courts, however, have limited the potential significance of section 6 by refusing to read it to allow Interior and Commerce to permit states to substitute their own conservation programs if those programs do not prohibit the taking of listed species on private lands, including the adverse modification of privately owned habitat. The key to the courts' interpretation is another part of section 6 that provides that any state law or regulation is void to the degree that it may "permit what is prohibited" under the ESA and that any state law "respecting the taking of an endangered species or threatened species may be more restrictive than" sections 4(d) and 9 "but not less restrictive" (16 U.S.C. § 1535(f)). Relying on this provision, two district courts in the early 1990s found that the takings prohibitions of sections 4(d) and 9 preempt any less restrictive state laws, despite the language of section 6(g)(2) which, as the courts recognized, "seems to contemplate less restrictive state law on takings" (*Swan View Coalition, Inc. v. Turner* 1992; *U.S. v. Glenn-Colusa Irrigation Dist.* 1992). Neither case was appealed, so the district court decisions remain the only extant interpretations of section 6 of the ESA.[7]

Under these judicial interpretations, section 6 apparently allows states to take over for the national government through "cooperative agreements" only where relevant state law includes anti-take prohibitions at least as restrictive as the federal prohibitions of sections 4(d) and 9. The decisions thereby turn section 6 into a form of "managerial federalism," in which states can assume responsibility for enforcing the Act's anti-take provisions (and can assume tighter restrictions), but cannot choose a different approach to protecting listed species even if Interior or Commerce determines that a state's approach is equally protective of federally listed species. The legislative history of the ESA is unclear whether Congress intended section 6 to provide only for such narrow managerial federalism or for a broader cooperative federalism.

Federal Funding of State Biodiversity Efforts

The national government can promote state efforts to protect and restore biodiversity not only by allowing states to take over responsibility for key managerial or regulatory functions but also by helping to fund state activities. The ESA expressly authorizes federal cost sharing of the expenses of running cooperative agreements under section 6 (16 U.S.C. § 1535(d),(i)). The USFWS has used this funding to encourage states to enter into cooperative agreements under which states work to protect federally listed species. All 50 states have taken advantage of this opportunity, and many have enhanced their state biodiversity laws in order to meet section 6's qualifying requirements (George et al. 1998). To expand the attractiveness of cooperative agreements, Congress amended the ESA in 1977 to provide for species-specific agreements in addition to the blanket agreements contemplated by the original 1973 legislation (16 U.S.C. § 1535(c)(1)).

Federal funding of state biodiversity efforts has historically represented only a fraction of the costs of running state programs. In recent years, however, Congress has increased the available funding in a sign of increased interest in cooperative federalism under the ESA. Funding under section 6, for example, increased almost sixfold from 1998 to 2003 (Niles and Korth 2005). Over the past decade, moreover, Congress created two new funding programs—the State Wildlife Grant program (SWG) and the Landowner Incentive Program (Niles and Korth 2005). As part of the SWG, Congress required states, as a condition for receiving federal funding, to complete comprehensive wildlife conservation strategies (Shaffer et al. 2005).

State Biodiversity Programs

All but five states now have endangered species acts of their own (up from seventeen rudimentary acts when Congress passed the ESA in 1973). Most of the state acts are far more limited than the national ESA. Although virtually every state act prohibits the killing, wounding, or capturing of imperiled species, most do not regulate habitat destruction or modification, do not require state or local agencies to consult before taking actions that might jeopardize a protected species, and do not provide for express recovery mechanisms (George et al. 1998).

A small but growing number of state statutes, however, have provisions that match and, in a few cases, go beyond approaches in the national ESA. California, for example, applies its regulatory prohibitions to candidate species (unlike the national ESA which provides no protection to candidate species) (Cal. Fish and Game Code § 2080), and eight states provide for emergency listings (George et al. 1998). While the national ESA prohibits, by regulation, habitat modification that leads to the actual death or injury of a listed species, the Massachusetts endangered species act explicitly prohibits disruption to an animal's "nesting, breeding, feeding or migratory activity" or alteration of significant habitat (Mass. Gen. Laws Ann. Ch. 131A, §§ 1–2). Several states also have authorized various forms of agreements, including pre-listing and SHAs, designed to encourage voluntary conservation efforts by landowners (e.g., Haw. Rev. Stat. Ann. § 195D-22; Kan. Stat. Ann. § 32-962(b)(1)). Two states provide for a multi-species, ecological approach to the formulation of recovery plans (e.g., N.M. Stat. Ann. § 17-2-40.1). Several states also provide greater protection of imperiled plants than the national ESA offers (George et al. 1998).

Clear indications of the significant role that states today play in protecting biodiversity are the independent (and voluntarily adopted) programs that states have created to protect and restore imperiled species. Even before Congress' creation of the SWG program, states had begun to create comprehensive state wildlife conservation strategies (Shaffer et al. 2005). For example, approximately 30 states now have active programs for habitat acquisition. In 2003, these states probably spent close to, if not more than, $100 million on such acquisitions (Niles and Korth 2005). Many states, unlike the national government, also actively map imperiled species within their borders (Swain 2005). States are using this information for multiple purposes, including the identification of habitats that are important for the protection of imperiled species (Niles and Korth 2005).

Another indication of the expanded role that states now play in protecting biodiversity is the degree to which they protect imperiled species that are not federally listed. ESA-listed species constitute only a small fraction of those that biologists estimate are actually imperiled; one study estimates that the number of "potentially listed species" is more than 9,000 (far outstripping the 1,200-plus species formally listed under the ESA) (Scott et al. 2005). Forty-four states now have their own endangered species lists (Niles and Korth 2005). While some state lists merely duplicate the national list, many lists include sizable numbers of imperiled species that are not federally listed and thus not protected by the national ESA (George et al. 1998). States often spend significant sums protecting these species (George et al. 1998).

There are limits, however, to state protection of biodiversity. Most state legislatures still have not given their agencies significant regulatory authority for the protection of listed species (Goble et al. 1999). While total state expenditures on imperiled species rival those of the national government, overall state funding of biodiversity efforts from 1998 to 2003 (the dates of the two most recent studies) did not even keep up with the rate of inflation (Niles and Korth 2005). States, in short, are playing a significant role in protecting the nation's biodiversity and, as discussed in the next section of this book, can play a larger role, but national regulation, oversight, and funding are still critical to the success of biodiversity protection.

Schema of this Book

We begin with a pair of assessment frameworks, and then explore local and state opportunities before moving on to case studies. A final section sums up what more can be done at the state and local levels to encourage effective conservation of imperiled species.

Frameworks

The following two chapters are designed to provide frameworks through which to consider the appropriate role of federalism under the ESA. In Chapter 2, "An Economic Assessment of Environmental Federalism: The Optimal Locus of Endangered Species Authority," Terry Anderson and Lawrence Watson provide an economic framework for choosing between state and national implementation of environmental policy. As the authors note, the principal argument for national implementation is the internalization of the benefits and costs of environmental policies with broad geographic impacts. In the case of the ESA, for example, all citizens in the United States have an interest in the protection of the nation's biodiversity, yet individual states might not take the nationwide benefits into account in deciding on the appropriate level of protection. The nationalization of biodiversity protection, however, can generate transaction costs in the form of lost local expertise, agency costs, and "monopoly distortion" in which all citizens are forced to accept a uniform national level of protection. According to the authors, the optimum level of federalism involves a balance between ensuring an appropriate level of internalization and minimizing transaction costs.

In Chapter 3, "Cooperative Federalism and the Endangered Species Act: A Comparative Assessment and Call for Change," J. B. Ruhl provides legal insight into the potential role of states in protecting biodiversity by comparing how the national government has used "cooperative federalism" to implement other environmental policies. Professor Ruhl contrasts the states' role under the ESA with their roles under the CWA and the CZMA. Professor Ruhl finds that the national government has not taken as much advantage of state expertise and resources under the ESA as it has under other statutes. According to Professor Ruhl, this has had a number of disadvantageous effects, including the stunted transformation of state wildlife agencies into effective biodiversity managers and unnecessary hostility from local property owners who feel that the national government is not responsive to their interests and concerns.

Opportunities for State and Local Involvement under the ESA

The next section of the book looks at several provisions and programs under the ESA that might be used to more actively involve the states in biodiversity protection. Chapter 4, "Listing Decisions, Conservation Agreements, and State–Federal Collaboration: A Litigation Perspective," starts by examining the opportunity for states to influence federal listing decisions. Eileen Sobeck and Paul Weiland look at the litigation surrounding decisions by the Department of the Interior not to list species because of existing state efforts to protect the species. As the authors observe, courts have often looked skeptically at such decisions; courts fear that the state efforts have little substance or teeth and that Interior is seeking merely to avoid difficult political decisions. To enable the national government to take legitimate state programs into account in making listing decisions, the authors recommend that Interior use a more regularized process and set of criteria in evaluating state programs and develop better administrative records. The authors also recommend that states launch their protection efforts at an earlier stage so that state efforts do not look merely like means to derail a federal listing decision.

In Chapter 5, "The Evolution of Federalism under Section 6 of the Endangered Species Act," Robert Davison turns to current practice under section 6 of the ESA. In Davison's view, Congress envisioned that section 6 would provide a much greater role for the states than they have enjoyed. The Department of the Interior transformed section 6 into a funding mechanism for specific projects rather than using section 6 to authorize broad cooperative agreements. In 2002, however, the USFWS entered into a broad cooperative agreement with the Arizona Game and Fish Department intended to "facilitate joint participation, communication, coordination, and collaboration." In Davison's view, the Arizona agreement provides a useful model for more robust cooperative agreements in the future between the national government and the states.

Chapter 6, "California's Natural Community Conservation Planning Program: Saving Species Habitat amid Rising Development," examines California's program for NCCP, one of the few instances of a comprehensive state program begun to enhance and improve on the national ESA. According to Gail Presley, the NCCP program has been effective for several reasons. It provides the California Department of Fish and Game with significant authority, including the authority to prohibit the "take" of an endangered species; builds on a strong scientific foundation; enjoys substantial state funding; and provides for significant public participation. In Presley's view, an important question is whether other states could duplicate California's success with the NCCP program. According to Presley, three factors combined to enable California to adopt and develop the NCCP program—California's history of environmental leadership, its rapid growth (which provided significant pressure to create a regulatory system that was more effective than the ESA), and the relative wealth of its development community (which allowed developers to incur the increased expense and time involved in implementing NCCP programs).

Case Studies

The fourth section of the book looks at federalism and the ESA through a series of case studies. Chapter 7, "The Karner Blue Butterfly: Wisconsin's Statewide Habitat Conservation Plan," starts by examining Wisconsin's successful efforts to protect the Karner blue butterfly through a unique statewide HCP that depends to a significant degree on voluntary private action. Jimmy Christensen and David Lentz identify several problems with the ESA that they believe undermine the motivation of states like Wisconsin to undertake significant protection and recovery programs like the Karner blue butterfly HCP.

According to the authors, it is very difficult at the moment to downlist or delist an endangered species, reducing states' incentives to protect a species. Compounding this problem is the national government's uniform treatment of species across their entire territory even if efforts in one state or region reduce the jeopardy the species faces in that area. Further compounding the problem is the failure of the national government to set clear recovery goals that, if met, would allow a state to argue that a species should be removed from the national lists of endangered or threatened species.

In Chapter 8, "The Red-cockaded Woodpecker: Conservation through Statewide Safe Harbor Agreements," Ralph Costa and Richard Gooch examine the partnership between South Carolina and the national government in developing a safe harbor program for the red-cockaded woodpecker. In the authors' view, this safe harbor program is one of the best examples of successful federalism under the ESA and has led to strong cooperation between South Carolina and the national government on biodiversity issues generally. The authors believe that one of the strengths of the red-cockaded woodpecker safe harbor program is its voluntary nature. In their view, moreover, states have a decided advantage over the national government in developing such voluntary programs. Like other authors, however, Costa and Gooch note that limitations in both funding and staff restrict the ability of states to play a more active role in implementing safe harbor programs and other aspects of the ESA.

The next two case studies look at efforts to protect salmon in the United States—and reach very different conclusions. In Chapter 9, "Saving *Salmo*: Federalism and the Conservation of Maine's Atlantic Salmon," Alison Rieser examines efforts to protect Atlantic salmon and concludes that Maine was more an obstacle than a help in trying to protect the species. Professor Rieser tells the story of a state that, concerned about the potential impact of a listing on the local economy, undermined national efforts to list the Atlantic salmon. Echoing Chapter 2 of the book, Rieser's story suggests that states do not always account for the national interest in protecting a species but instead act out of largely parochial concerns.

Chapter 10, "Oregon Coastal Coho Restoration and the Endangered Species Act," tells a very different story about efforts to protect the Oregon Coast coho salmon. According to Gail Achterman and Julia Doermann, it is the national ESA that is broken. Examining efforts to protect Pacific Coast salmon, the authors argue that the national ESA's emphasis on regulation and enforcement has created unhelpful tensions and animosity among major stakeholders and been largely unsuccessful in recovery of the salmon. The authors call for a collaborative, community- and watershed-approach that would emphasize the development of a shared vision, and argue that states and local governments are crucial to the effective implementation of such a vision.

Chapter 11, "Grizzly Bear Conservation in the Greater Yellowstone Ecosystem: A Case Study in the Endangered Species Act and Federalism," concludes with a nuanced story of the role that federalism has played in efforts to protect the grizzly bear in the greater Yellowstone ecosystem. Kaush Arha and John Emmerich suggest that states in the vicinity probably would not have pursued strong conservation efforts if it had not been for the ESA. By incorporating state agencies into recovery efforts, however, the national government helped make the state wildlife agencies into effective biodiversity agencies. The authors, moreover, argue that the national government's strong emphasis on regulation and enforcement has been counterproductive and that the states have played an exceptionally valuable role in promoting and designing more voluntary efforts.

Concluding Thoughts

In Chapter 12, "Toward Greater State and Local Commitment," we provide concluding thoughts on possible ways of increasing state and local involvement under the ESA. The proposals include 1) developing more useful national guidance for when state conservation efforts eliminate the need for federal listing of species, 2) providing states with a greater opportunity to protect and conserve species that are threatened but not endangered, 3) setting out clearer and more certain downlisting and delisting criteria as part of species recovery plans, 4) developing state-based "Species Conservation Committees" that would actively coordinate conservation measures between the national and state governments, and 5) providing needed funding to support effective state conservation efforts.

References

Bean, Michael J. and Melanie J. Rowland. 1997. *The Evolution of National Wildlife Law*. 3rd ed. Westport, Conn.: Praeger.

Beatley, Timothy. 1994. *Habitat Conservation Planning: Endangered Species and Urban Growth*. Austin, TX: University of Texas Press.

Behan, Maeveen. 2005. "County Conservation Planning." In *The Endangered Species Act at Thirty*, vol. 1, Dale D. Goble, J. Michael Scott, and Frank W. Davis eds., 156–160. Washington, D.C.: Island Press.

Burnham, William, Tom J. Cade, Alan Lieberman, J. Peter Jenny, and William R. Heinrich. 2005. "Hands-on Restoration." In *The Endangered Species Act at Thirty*, vol. 1, Dale D. Goble, J. Michael Scott, and Frank W. Davis eds., 237–246. Washington, D.C.: Island Press.

Davison, Steven G. 1995. "Alteration of Wildlife Habitat as Prohibited Taking under the Endangered Species Act." *Journal of Land Use and Environmental Law* 10: 155–195.

George, Susan, William J. Snopes III, and Michael Senatore. 1998. *State Endangered Species Acts: Past, Present, and Future*. Washington, D.C.: Defenders of Wildlife.

Goble, Dale, Susan M. George, Kathryn Mazaika, J. Michael Scott, and Jason Karl 1999. "Local and National Protection of Endangered Species: An Assessment." *Environmental Science and Policy* 2: 43–87.

Lin, Albert C. 1996. "Participants' Experiences with Habitat Conservation Plans and Suggestions for Streamlining the Process." *Ecology Law Quarterly* 23: 369–419.

Niles, Lawrence and Kimberly Korth. 2005. "State Wildlife Diversity Programs." In *The Endangered Species Act at Thirty*, vol. 1, Dale D. Goble, J. Michael Scott, and Frank W. Davis eds., 141–155. Washington, D.C.: Island Press.

Richie, Deborah and Jennifer Holmes. 1999. *State Wildlife Diversity Program Funding: A 1998 Survey*. Washington, D.C.: International Association of Fish and Wildlife Agencies.

Rohlf, Daniel J. 1989. *The Endangered Species Act: A Guide to Its Protection and Implementation*. Palo Alto, CA: Stanford Environmental Law Society.

Salzman, James and Barton H. Thompson, Jr. 2010. *Environmental Law and Policy*, 3rd ed. New York, NY: Foundation Press.

Scott, J. Michael, Dale D. Goble, Leona K. Svancara, and Anna Pidgorna. 2005. "By the Numbers." In *The Endangered Species Act at Thirty*, vol. 1, Dale D. Goble, J. Michael Scott, and Frank W. Davis eds., 16–35. Washington, D.C.: Island Press.

Shaffer, Mark J., Laura Hood Watchman, Sara Vickerman, Frank Casey, Robert Dewey, William J. Snape III, Michael Senatore, and Robert M. Ferris. 2005. "Proactive Habitat Conservation." In *The Endangered Species Act at Thirty*, vol. 1, Dale D. Goble, J. Michael Scott, and Frank W. Davis eds., 286–295. Washington, D.C.: Island Press.

Swain, Hilary. 2005. "A Reality Check from Florida." In *The Endangered Species Act at Thirty*, vol. 1, Dale D. Goble, J. Michael Scott, and Frank W. Davis eds., 137–140. Washington, D.C.: Island Press.

Swan View Coalition, Inc. v. Turner 824 F. Supp. 923 (D. Mont. 1992)

Tarlock, A. Dan. 2005. "The Dynamic Urban Landscape." In *The Endangered Species Act at Thirty*, vol. 1, Dale D. Goble, J. Michael Scott, and Frank W. Davis eds., 127–136. Washington, D.C.: Island Press.

Thompson, Barton H., Jr. 1997. "The Endangered Species Act: A Case Study in Takings and Incentives." *Stanford Law Review* 49: 305–380.

———. 2005. "Managing the Working Landscape." In *The Endangered Species Act at Thirty*, vol. 1, Dale D. Goble, J. Michael Scott, and Frank W. Davis eds., 101–126. Washington, D.C.: Island Press.

U.S. Congress. 1973a. *Congressional Report No. 93-740, 93rd Congress, 1st Session.*

———. 1973b. House of Representatives. *H.R. 37, 93rd Congress, 1st Session.*

———. 1973c. Senate. *Report No. 93-307, 93rd Congress, 1st Session.*

U.S. v. Glenn-Colusa Irrigation Dist. 788 F. Supp 1126 (E.D. Cal. 1992)

Endnotes

1 National jurisdiction over the ESA is split between Interior and Commerce. Commerce has authority over marine and anadromous fish species; Interior has authority over all other species. The USFWS is the executing agency for the Act in the Department of Interior, and the National Marine Fisheries Service (NMFS) is the executing agency for the Act in the Department of Commerce (Rohlf 1989).

2 According to Defenders of Wildlife, states and local regions have used such agreements to "avoid listing threatened and endangered species, such as the Coho salmon in Oregon and Atlantic salmon in Maine, the jaguar in the Southwest, the

Barton Springs salamander in Texas and the flat-tailed horned lizard in Arizona and California" (George et al. 1998).

3 This approach resembles the CAA's delegation of authority to states to implement national ambient air quality standards, where states have the first shot at developing an effective plan for implementing the national standards but EPA can impose its own federal implementation plan if a state fails to act or acts ineffectively (Salzman and Thompson 2010). While the CAA does this by directly delegating administrative responsibility to willing states, the USFWS accomplishes the same goal indirectly under the ESA through its authority to decide which species to list.

4 Section 9(a)(1)(B) of the ESA prohibits the taking of any endangered species of fish or wildlife (16 U.S.C. § 1538(a)(1)(B)). Section 3(19) defines "take" to mean "to harass, harm, pursue, hunt, shoot, wound, kill, trap, capture, or collect, or to attempt to engage in any such conduct" (16 U.S.C. § 1532(19)). Regulations promulgated by Interior and Commerce further define "take" to include the adverse modification of a species' habitat where the modification leads to the actual death or injury of the species (50 C.F.R. § 17.3 (1995)).

5 To be approved under the ESA, an HCP must minimize the impact of the incidental taking "to the maximum extent practicable," ensure that the private land use will not "appreciably reduce the likelihood of the survival and recovery of the species in the wild," and be adequately funded (16 U.S.C. § 1539(a)(2)(B)). Salzman and Thompson (2010) and Thompson (1997) provide overviews of how incidental take permits and habitat conservation plans work in practice.

6 This exemption does not apply to species covered by the Convention on International Trade in Endangered Species (CITES) or any other international treaty or national law. It also does not apply where the taking of the species is itself a violation of state law.

7 Davison (1995) provides a good discussion of the *Swan View* and *Glenn-Colusa* opinions.

PART II

Frameworks

CHAPTER 2

An Economic Assessment of Environmental Federalism: The Optimal Locus of Endangered Species Authority

Terry L. Anderson and Lawrence Reed Watson

Environmental policy in the United States is largely seen as the domain of the national government, but there are good reasons to consider devolving some responsibilities for environmental policy to lower levels of government and even to private stewards. Especially when resource management issues do not involve large areas and when knowledge about the resource is localized, devolution can make sense. For example, depending on the permeability of aquifers, groundwater management could be more or less localized; in cases where aquifers are impermeable, one user's actions might only affect immediate neighbors, making devolution more viable. On the other hand, with air quality management, especially with emissions that travel long distances in the atmosphere, higher levels of governmental regulation may be necessary.

This chapter presents a framework for thinking about the appropriate level of resource management and applies that framework to endangered species and their habitat. The basic argument is that the optimal locus of control is determined by the trade-off between the potential of centralized governmental regulation to take account of wide-ranging spillovers and the higher information and management costs associated with that centralized control.

The debate over the appropriate locus of governmental actions was fundamental to the founding fathers, even though they were not focused on environmental issues. After the failed attempt with the Articles of Confederation, which left the national government with too little power to carry out its functions, the framers of the Constitution debated how much power was appropriate for the national government vis-à-vis the states. The Anti-Federalists contended that a strong national government would lead to infringement on individual rights. The Federalists, on the other hand—although far from proponents of a big national government—argued that without certain powers, the national government would fail, as it had under the Articles. The Constitution created a political system aimed at reserving certain powers and privileges for the states, while giving the national government sufficient power to handle functions such as military protection that clearly required more centralized control.

Environmental issues in general, and endangered species protection in particular, provide a good opportunity to reconsider federalism. We present a transaction-cost framework for determining the optimal locus of control for species conservation. This framework combines several economic approaches to the optimal locus of decisionmaking authority, including collective action in firms (Cheung 1983), governmental action for the environment in general (Haddock 1997a, 1997b), and governmental action for wildlife in particular (Lueck 1995; Anderson 1998, 2001). Combining these theories, the chapter provides several examples that illustrate how the optimal locus of governmental power changes with different transaction costs and, consequently, different endangered species. The chapter concludes by making a case that reform of the Endangered Species Act (ESA), from the perspective of both economic optimality and environmental pragmatism, should include provisions for assigning endangered species authority to different levels of government.

The Economics of Environmental Federalism

A robust body of economic literature discusses the different institutions that affect the principal-agent problem. This literature originally focused on "The Nature of the Firm" (Coase 1937) and has expanded to include detailed discussions of how transaction costs affect our ability to manage resources and our choice of institutions with which to manage them. In this context, the "firm may be as small as a contractual relationship between two input owners or, if the chain of contracts is allowed to spread, as big as the whole economy" (Cheung 1983, 17).

First consider what economists mean by transaction costs, of which there are at least three types—the costs of contracting with other owners of inputs and outputs to incorporate them into production; the costs of measuring and monitoring performance of agents; and the costs created by monopoly power, which makes it difficult to withdraw from transactions.

It is generally acknowledged that the costs of contracting with other owners decline as the size of the governing structure increases from a sole proprietor firm to a totally centrally planned economy. This is because the larger firm has more ability to dictate how inputs are used and therefore needs to negotiate less with other input owners. Specifically for endangered species management, this is the cost of contracting with each habitat owner to acquire control of the necessary inputs and with demanders, who value endangered species, to get them to pay for habitat that is being preserved. As the range of a species increases, it might require habitat owned by many different owners and perhaps even owners in different countries. This will make it relatively cheaper for a central authority in control of habitat to dictate its use rather than contracting with the multiple owners.[1] Moreover, the coercive power of a central government can be used to overcome the "free-rider problem," to the extent that those who value the existence of endangered species can free-ride on the efforts of others who preserve them.

On the other side of the transaction-cost equation are the principal-agent costs. Think of management as a contract between a principal and an agent. In the case of the corporation, the principal is the stockholder, and the agents include the CEO and other managers. In the case of government, the principals are the citizens, and the agents are the politicians and bureaucrats. With all principal-agent relationships, there are costs of measuring and monitoring the actions of the agents to see if they are consistent with the objectives of the principals. However, as the firm grows in size, from single proprietorship to large corporation to centrally planned economy, the agency costs rise for two reasons. First, as size increases, it is more and more difficult for individual principals to measure and monitor agent performance. This is even the case in large corporations where the stock market provides an aggregate estimate of agency performance and a way for principals to exit. In the case of large governments, there is no such aggregate, no transparent measure, and exit (moving to another jurisdiction) can be costly. Second, as the size of the firm increases, there are more and more agents, making it more and more difficult to observe malfeasance and nonfeasance. Again, this problem exists in the corporation, and again the stock market provides some transparency, which does not exist with centralized governments.

Putting agency costs in terms of endangered species habitat production, we can think of a single customer of wildlife contracting for inputs and capturing most of the value of the endangered species that is preserved. For example, Ted Turner owns large ranches that provide some endangered species habitat. Turner faces zero costs in contracting for habitat production inputs if those ranches are sufficiently large, because he owns them. He rarely has to negotiate with other habitat input owners, although he does face agency costs associated with measuring and monitoring the performance of his managers. Furthermore, he captures most of the value of the habitat enhancement.[2] In this case, transaction costs are relatively low.

An example of a larger firm supplying endangered species habitat is The Nature Conservancy (TNC). Unlike Mr. Turner, TNC must contract with habitat owners by acquiring leases, easements, or outright ownership. Higher agency costs associated with such a large firm are indicated by the Conservancy's now-banned practice of selling land to its own trustees and financial supporters at a significant loss (Stephens and Ottaway 2003).

Putting the first two types of transaction costs together helps us understand the optimal locus of endangered species management. As the size of the organization controlling habitat increases, the marginal costs of contracting with other habitat owners decline, while the marginal agency costs rise. As long as the former exceeds the latter, it makes economic sense to assign management authority to higher and higher levels of collective action.

Economist David Haddock (1997a) applies this framework to the "benefits and costs of regulatory centralization." On the benefit side of regulatory centralization is the ability of a centralized regulator to account for costs or benefits for which market transactions do not account, commonly referred to as externalities.[3] In Haddock's words, "because far-flung boundaries for a governing body are more likely to contain externalities fully, centralized regulation confers benefits" (1997a, 45).

The benefits of regulatory centralization, however, must be weighed against three types of transaction costs. First, there are "the costs of obtaining information about field conditions and citizen preferences" (Haddock 1997a, 45). Second, there are agency costs. Haddock explains that these costs are compounded if regulators are appointed. "Just as voters rationally monitor their elected officials poorly, elected officials rationally monitor their appointees poorly. Hence, regulatory behavior can deviate from what the officials would want, a problem that grows with the number of regulatory agencies that officials oversee" (1997a, 47). The alphabet soup of environmental regulatory agencies in Washington, D.C., or in state capitals illustrates how difficult it can be to monitor regulators.

The third cost, known as monopoly distortion, compounds the principal-agent problem. Monopoly distortion refers to the fact that monopolies reduce the ability of people to choose the agents with whom they wish to transact. If a company has monopoly power, it means that consumers do not have good substitutes and therefore must pay the price set by the monopolist or do without the product.

In the case of governmental jurisdictions, monopoly power increases with the geographic size of the jurisdiction because a larger geographic jurisdiction gives citizens less ability to migrate to another jurisdiction where political agents are taking actions more to the liking of the principals. In other words, decentralization to more local levels of government allows citizens to "vote with their feet"[4] by moving to competing jurisdictions. As centralization increases, the cost of voting with one's feet increases, giving centralized regulators more monopoly power to impose their preferences—or that of specialized constituents—on the general citizenry.

Combining all three transaction costs suggests that the optimal locus of any regulatory authority is not necessarily at the national level. Lower organizing costs must be weighed against higher agency and monopoly distortion costs. True, higher levels of centralization can capture more benefits from regulation, but the costs of creating these benefits also rise. The key is to balance the net benefits.

In a much earlier era, John Wesley Powell, the 19th-century explorer of the West's great waterways, recognized the importance of finding the optimal locus of governmental authority for water management. Speaking to the Montana Constitutional Convention in 1889, Powell described what he thought would be the optimal geographical units for organizing county government. "I want to present to you what I believe to be ultimately the political system which you have got to adopt in this country, and which the United States will be compelled sooner or later ultimately to recognize. I think each drainage basin in the arid land must ultimately become the practical unit of organization, and it would be wise if you could immediately adopt a county system which would be convenient with drainage basins" (quoted in Kemmis 2001, 177). Although his suggestions were largely ignored by Montana and other western states, his insights into the connection between the physical characteristics of natural resources and the optimal geographic region for organizing government are as profound today as they were then.

Just as Powell's insights suggest greater or lesser centralized water management, depending on the costs and benefits of collective action, they suggest greater or lesser centralization of endangered species management. To the extent that a species migrates over a larger territory, the costs of coordinating private decisions about habitat over that territory will be higher, and hence the gains from centralized regulation will be higher. Lueck (1995, 14) notes that:

> *Since Geer [an 1896 Supreme Court ruling that articulates the state ownership of wildlife], there have been changes, in two directions, both correlating control with the territory of wild populations. First, where state control was insufficient to cover the entire territory of a species (as for migratory birds), federal or international control has been assigned. Second, where state control was too extensive, ownership changed to private hands or smaller government units.*

Wildlife applications by Lueck (1995) and Anderson (1998, 2001) shed further light on the optimal locus of control of endangered species and their habitat. Although not directly focused on the issues of federalism, Lueck's economic analysis of wildlife management nicely frames the variables relevant to the optimal locus of endangered species regulation (Lueck 1995, 9–11). His analysis considers the trade-offs between constraints of existing landownership patterns, the highest value of land use, the land requirements for wildlife species, and the contracting costs for reorganizing landownership to account for changing wildlife (endangered species) values. Four propositions emerge from his analysis:

1 As the relative value of wildlife increases, whether it is for hunting, viewing, or existing, the greater will be the gains from private contracting with landowners to provide habitat. This explains the abundance of private hunting in states such as Texas.
2 The more productive land is for wildlife, the greater will be the gains from private contracting with landowners to provide habitat. This explains why

organizations such as TNC contract with certain landowners and not others for endangered species habitat.

3 As the optimal size of landholding for other purposes (such as ranching) increases relative to the territory of the species in question, the costs of contracting among landowners will decrease, because fewer landowners will be needed to provide the optimal habitat size. As discussed above, Ted Turner's ranches are sufficiently large that no contracting with other landowners is necessary.

4 As a species' territory increases, contracting costs among landowners will increase, making it less likely that private contracting will occur. International treaties to regulate migratory birds' hunting and habitat preservation can be understood in this context.

Anderson (1998, 2001) builds on Lueck's argument, suggesting that the transaction costs of providing habitat vary with the number of habitat owners, the number of species demanders, and the potential for separating out the contribution of each habitat owner to the preservation of species. Contracting among interested parties will be less costly if there are a number of competing habitat owners, each of whom can ensure survival of the species. Contracting will also be less costly if the demanders of the species are easily recognized, if they are easily excluded from enjoying species preservation if they do not pay, therefore eliminating the free-rider problem,[5] and if the contribution of each habitat owner to the survival of the species can be easily identified or separated from the contributions of other habitat providers.

Let us now pull these various models of transaction costs together into a coherent way of thinking about the optimal locus of regulatory activity to preserve endangered species. Two general categories of costs flow from this analysis. The first is the cost of articulating the demand for endangered species, and the second is the cost of bringing the habitat inputs together to ensure species survival.

The following questions and hypotheses are under the demand category:

1 Can demanders be excluded from receiving endangered species benefits if they do not pay? If they can be more easily excluded, the locus of endangered species provision will be more local or even private.

2 Related to question 1, is the value of endangered species survival largely the result of existence values, and are the existence values spread across a large geographic area? If the existence value is spread across a larger geographic area, the locus of endangered species provision will be more centralized.

3 When agents are hired, can their actions be easily monitored by the principals? If costs of monitoring agents are higher, the farther removed they are from those demanding endangered species survival. In that case, the locus of endangered species management will be more local.

The following questions and hypotheses are under the cost of bringing the habitat inputs together:

1 Is there good scientific information about the requirements for endangered species survival? If the scientific "production function" is well known, the locus of endangered species provision will be more local, because agents with specialized knowledge will not be as necessary.[6]
2 Do scale economies exist in the production of scientific information on species survival? If there are significant scale economies in the production of scientific information, the locus of endangered species provision will be more centralized, because centralization captures these economies.
3 Is the territory of the endangered species large relative to the size of landownership? If the territory is large relative to the optimal size of landownership for other purposes, the locus of endangered species provision will be more centralized.

Applying the Transaction-Cost Approach

From this theory, we can examine how endangered species have been regulated to determine whether regulation has approximated the optimal locus of control. Because nearly all endangered species provision has taken place at the national and state levels, we confine our examples to those levels. At the outset, however, we emphasize that more localized control is not unheard of and can play an important role in endangered species management. Indeed, the private sector can contribute to endangered species protection, as organizations such as TNC illustrate.

National Endangered Species Regulation

The federal government's role in wildlife management stems largely from Congress' regulatory power over interstate commerce, beginning with the passage of the Lacey Act of 1900. The Lacey Act prohibited the transport of wild game killed in violation of state law and the import of certain animals that could threaten agricultural and horticultural assets.

The Migratory Bird Treaty Act, enacted in 1918, relied on a different congressional authority than interstate commerce altogether, one that has greater bearing on the discussion of endangered species management today. Discussing bird migration from Canada, Huffman (1995, 30) notes that this legislation "relied upon the treaty-making, rather than the commerce-regulating, power of Congress, and reflected an assumption that the states are not in a position to effectively regulate wildlife that migrate across international borders. Of course, this same assumption would justify federal regulation of wildlife that migrate across state borders as well." Hence the Migratory Bird Treaty Act considered species-specific migration—not transport or import by human methods—as justification for federal authority.

National regulation of migratory species that cross international borders makes sense in the transaction-cost framework discussed above. The national government covers a broad enough geographic area to reduce the transaction costs of negotiating with all the necessary parties. Imagine the costs of individual states contracting with Canada over migratory waterfowl management. Our third hypothesis under the cost of species inputs suggests that if the territory of a species is large relative to landownership, the locus of endangered species provision will be more centralized. This does not mean that other institutions cannot further reduce transaction costs. The International Association of Fish and Wildlife Agencies (IAFWA), Ducks Unlimited, Delta Waterfowl, and the Peregrine Fund are examples of organizations that have played a role in national and international species regulation.

Effective management of some salmon species requires centralized control because these fish travel from coastal waters far inland and, like migratory birds, fail to recognize state and national borders. The rivers in which these fish spawn, in fact, cross and sometimes represent state boundaries. Since habitat modification in one state can impact the survival of a salmon species in another, the possibility of scale economies in salmon preservation seems to warrant centralized control.

The bald eagle provides perhaps the best example of a species warranting national management authority. Given its importance as a national symbol, the bald eagle produces existence values over a large geographic area.[7] It follows that state endangered species management would be less able to capture such broad benefits compared to national control. And, like salmon and other migratory species, bald eagles range across state and international borders.

Decentralized Endangered Species Regulation

Although national regulation can reduce the costs of negotiating with multiple landowners, it also has more potential for higher agency costs. One of these costs is the potential for creating the "tyranny of the majority." This is the source of much opposition from private landowners to the ESA. A majority of the population might desire endangered species habitat preservation and, under ESA regulatory authority, can force others to provide that habitat without compensation. For example, residents of New York City or Los Angeles could capture the far-flung benefits of northern spotted owl survival, while many or all of the costs associated with their survival fall on private landowners and the timber industry in the Pacific Northwest. As Simmons and Frost (2004, 10) note, "Seventy-five percent of all listed species have portions or all of their habitat on private lands, and landowners are not compensated for their losses from ESA regulations." Because states have less power to "take" habitat without compensation and because such takings are more transparent at lower levels of government, state or local jurisdictions are less likely to force local landowners to bear the costs of providing habitat for the democratic majority.

Rancher opposition to gray wolf reintroduction in Yellowstone National Park can be explained in this context. In the 1980s and early 1990s, proposals to reintroduce the gray wolf into the park spawned fears among area ranchers that the wolves would prey on their livestock and significantly reduce ranch profits. By establishing a private compensation fund to pay for proven livestock losses, Defenders of Wildlife minimized the private costs of wolf reintroduction and, consequently, alleviated much of the local opposition. Categorizing the reintroduced wolves as an experimental population gave local governmental officials the authority to relocate or kill chronic livestock predators, which further reduced local opposition. The locally administered compensation fund and the local management of problem wolves decentralized wolf management authority and reduced the transaction costs of wolf management.[8]

This devolution of endangered species management proved highly successful. According to the United States Fish and Wildlife Service's (USFWS's) 2008 estimate, the Northern Rocky Mountain (NRM) distinct population segment consisted of 1,639 wolves in 95 breeding pairs (U.S. Department of the Interior 2009). Those estimates were approximately five times higher than the minimum population recovery goal and three times higher than the minimum breeding pair recovery goal (U.S. Department of the Interior 2009). Accordingly, on April 2, 2009, the Service delisted the NRM gray wolf populations in Montana and Idaho and turned over wolf management authority to those states.[9] The pendulum swung back to centralized management authority on August 5, 2010, when the U.S. Federal District Court in Missoula, Montana vacated the delisting on the technical point that delisting on a state-by-state basis is prohibited by the ESA.[10] As the resulting political conflict demonstrates (Byron 2011), the delisting decision increased the transaction costs of wolf management without apparent off-setting management benefits.

Viewed through the transaction-cost lens, devolution seems appropriate for another protected species of Montana and the Rocky Mountain West, the bull trout. First, most of the habitat issues are confined within state boundaries; second, it is likely that any existence values for bull trout are more local than national, compared with the national scope of existence values for the bald eagle. Listed as threatened across the coterminous United States, the bull trout is known to inhabit only Idaho, Montana, Nevada, Oregon, and Washington. Among those states, separate and distinct populations vary in both number and health. In some states, bull trout populations are not healthy, but, according to Pat Graham, former Director of the Montana Department of Fish, Wildlife, and Parks, "bull trout populations in the South Fork of the Flathead River and in Swan River/Swan Lake are healthy and at recovery levels" (Montana FWP 1997).

Graham attributes the health of some Montana populations to "state management authority for bull trout," as well as "continued implementation of cooperative habitat restoration efforts and a commitment of funds and resources to implement restoration. So far, four local watershed groups have formed to address the problem in specific drainages" (Montana FWP 1997). The more local nature of the demand for species such as the bull trout makes it easier to organize private groups to partner with state regulators, as evidenced by the number of watershed groups, fishing organizations, and property owners who have risen to the occasion.

Another example of the potential for decentralized endangered species management is the black-tailed prairie dog. Again, the species is not so susceptible to transjurisdictional habitat impacts and is not likely to have huge national existence values. The conservation agreement among western states to protect the black-tailed prairie dog exemplifies the effectiveness of proactive species management at the state and local levels. By "maintaining prairie dog habitat, setting population goals for each state, [instituting] a prairie dog monitoring program, and developing incentives for prairie dog conservation on private lands," the black-tailed prairie dog is no longer a candidate for listing (Montana FWP 2000).

To be sure, the federal government played a role by encouraging state action. Ralph Morgenweck, then director of the USFWS's mountain-prairie region, said with "recent state estimates of occupied black-tailed prairie dog habitat, the Service has determined that the black-tailed prairie dog does not meet the Endangered Species Act's definition of 'threatened'" (USFWS 2004). Specifically, Morgenweck noted:

> *until now, the best scientific and commercial information available to the Service indicated that the impacts of disease, chemical control, and other lesser factors were substantial enough to warrant listing of the black-tailed prairie dog as a threatened species. Since 2002, State agencies, Federal agencies, Tribes, and other parties provided additional information regarding the black-tailed prairie dog, which was considered by the Service in an evaluation of the status of the species.*

Conclusion

The transaction-cost approach described above provides a lens for examining ESA reform. This approach suggests that the optimal locus of authority is not necessarily at the national level. As Haddock (1997b, 16) notes, "pointing to the benefits while ignoring concurrent costs is inappropriate, for ideal regulation would maximize net rather than gross benefits. Many of the gross benefits could be preserved through properly devolved regulation, while substantial costs could be avoided." Allowing more authority for endangered species management below the national level could improve the efficiency and effectiveness of the ESA and reduce the potential for it to foster a "tyranny of the majority." Pat Graham, former director of Montana Fish, Wildlife, and Parks, describes the gains from devolution (Graham 1996):

> *While working with the U.S. Fish and Wildlife Service and many public and private groups to recover species, the State of Montana is also working to reform the Endangered Species Act to provide more incentives to landowners, to place more emphasis on preventing species from becoming threatened, and to provide the opportunity for state and local efforts to replace the need for federal action.*

This transaction-cost approach might prove useful as the reader considers the various analyses in subsequent chapters. Several of these chapters—particularly the case studies—present specific advantages in terms of the effectiveness and efficiency of local species conservation—conservation where the locus of authority more local than the nation's capital. The transaction-cost method allows the reader to evaluate different approaches to endangered species conservation and arrive at his or her own conclusion as to where the optimal locus of authority of species conservation may rest.

References

Anderson, Terry L. 1998. "Viewing Wildlife Through Coase-Colored Glasses." In *Who Owns the Environment?* Peter J. Hill and Roger E. Meiners eds., 259–282. Lanham, MD: Rowman & Littlefield Publishers.

———. 2001. "The ESA through Coase-Colored Glasses." In *Protecting Endangered Species in the United States: Biological Needs, Political Realities, Economic Choices*, Jason F. Shogren and John Tschirhart eds., 226–249. New York: Cambridge University Press.

———. 2004. "Viewing Land Conservation through Coase-Colored Glasses." *Natural Resources Journal* 44 (Spring): 361–381.

Anderson, Terry L., and Donald R. Leal. 1997. *Enviro-Capitalists: Doing Good While Doing Well.* Lanham, MD: Rowman & Littlefield Publishers.

———. 2001. *Free Market Environmentalism*, revised edition. New York: Palgrave, St. Martin's Press.

Boudreaux, Donald J., and Roger E. Meiners. 1998. "Existence Value and Other of Life's Ills." In *Who Owns the Environment*, Peter J. Hill and Roger E. Meiners eds., 153–186. Lanham, MD: Rowman & Littlefield Publishers.

Byron, Eve. 2011. "Gov Calls for Wolves to Be Killed." *Helena Independent Record*, February 17.

Cheung, Steven N. S. 1983. "The Contractual Nature of the Firm." *Journal of Law and Economics* 26 (April): 1–22.

Coase, Ronald H. 1937. "The Nature of the Firm." *Economica* 4(3): 386–405.

Fischer, Hank. 1995. *Wolf Wars: The Remarkable Inside Story of the Restoration of Wolves to Yellowstone.* Helena and Billings, MT: Falcon Press Publishing.

Graham, Patrick J. 1996. "Our Point of View." *Montana Outdoors.* March–April.

Haddock, David D. 1997a. "Must Water Regulation Be Centralized?" In *Water Marketing—The Next Generation*, Terry L. Anderson and Peter J. Hill eds., 43–61. Lanham, MD: Rowman & Littlefield Publishers.

———. 1997b. "Sizing Up Sovereigns: Federal Systems, Their Origins, Their Decline, Their Prospects." In *Environmental Federalism*, Terry L. Anderson and Peter J. Hill eds., 1–21. Lanham, MD: Rowman & Littlefield Publishers.

Huffman, James L. 1995. "In the Interests of Wildlife: Overcoming the Tradition of Public Rights." In *Wildlife in the Marketplace*, Terry L. Anderson and Peter J. Hill eds., 25–42. Lanham, MD: Rowman & Littlefield.

Kemmis, Daniel. 2001. *This Sovereign Land: A New Vision for Governing the West.* Washington, D.C.: Island Press.

Lueck, Dean L. 1995. "The Economic Organization of Wildlife Institutions." In *Wildlife in the Marketplace*, Terry L. Anderson and Peter J. Hill eds., 1–24. Lanham, MD: Rowman & Littlefield Publishers.

Montana FWP (Montana Department of Fish, Wildlife, and Parks). 1997. "Montana Seeks Federal Conservation Agreements in Face of Bull Trout Listing." News release, October 17. http://fwpiis.mt.gov/news/article_751.aspx (accessed Sept. 8, 2010).

———. 2000. "States' Conservation Plan Plays Role in Fed's Prairie Dog Decision." News release, February 4. http://fwpiis.mt.gov/news/article_1661.aspx (accessed Sept. 8, 2010).

Simmons, Randy T., and Kimberly Frost. 2004. *Accounting for Species: The True Cost of the Endangered Species Act.* Bozeman, MT: PERC.

Stephens, Joe, and David B. Ottaway. 2003. "Nonprofit Sells Scenic Acreage to Allies at a Loss." *Washington Post*, May 6, A1.

Thompson, Jr., Barton H. 1997. "The Endangered Species Act: A Case Study in Takings and Incentives." *Stanford Law Review* 49: 305–380.

Tiebout, Charles M. 1956. "A Pure Theory of Local Expenditures." *Journal of Political Economy* 64 (October): 416–424.

U.S. Department of the Interior. 2009. "Final Rule to Identify the Northern Rocky Mountain Population of Gray Wolf as a Distinct Population Segment and to Revise the List of Endangered and Threatened Wildlife." *Federal Register* 74(62, 2 April): 15123–15188.

U.S. FWS (U.S. Fish and Wildlife Service). 2004. "Black-tailed Prairie Dog Removed from Candidate Species List." News release, August 12. www.fws.gov/mountain-prairie/pressrel/04-57.htm (accessed Sept. 8, 2010).

Endnotes

1 The central government's ability to dictate habitat provision at low cost is bound, of course, by the constitutional rights of private property owners and their ability to assert those rights in court. For a more complete discussion, see Thompson, Jr. (1997).

2 For a discussion of Ted Turner's land holdings in this context, see Anderson and Leal (1997 and 2001).

3 Elsewhere Anderson (2004) argues that the term "externality" is vague and reflects a lack of clearly defined property rights.

4 This is described as the Tiebout effect (see Tiebout 1956).

5 This would be the case if the species in question was mainly demanded by hunters who only get value from the hunting experience and can be excluded by the habitat owner from enjoying that experience if they do not pay.

6 For additional discussion of the public decisionmaking balance between science and politics, see Huffman (1995, 31).

7 For a discussion of the pros and cons of existence value arguments, see Boudreaux and Meiners (1998).

8 For a discussion of the "wolf wars," see Fischer (1995).

9 Though the USFWS has approved wolf management plans in Montana and Idaho, the service concluded that the state law and management plan in Wyoming were insufficient to protect Wyoming's portion of the NRM gray wolf population.

10 *Defenders of Wildlife et al. v. Salazar, Secretary of the Interior*, case no. CV-09-77-M-DWM; *Greater Yellowstone Coalition v.* Salazar, case no. CV 09-82-M-DWM.

CHAPTER 3

Cooperative Federalism and the Endangered Species Act: A Comparative Assessment and Call for Change

J. B. Ruhl

The Endangered Species Act (ESA) is an oddball among federal environmental laws in more ways than one, but among its most prominent distinctions is the approach it takes to federal–state relations. Many federal environmental laws are held out as examples of "cooperative federalism," under which the federal legislation provides many opportunities for states to implement national goals and standards through state-run programs. Under this model of federal–state relations, if a state program satisfies prescribed federal policy criteria, the state receives some institutional benefit in return, ranging anywhere from funding to virtually complete responsibility for implementation and enforcement of a federal statute. Although debate rages over the best recipe, there is widespread agreement that the successes of federal environmental law over the past 40 years owe much to its strong taste for cooperative federalism. Despite appearances, however, and notwithstanding its strong influence over state and local land-use decisions, the ESA is not an example of this approach.

To be sure, with section 6 of the statute devoted in its entirety to "cooperation with the states," the ESA boldly proclaims in that provision that the federal government "shall cooperate to the maximum extent practicable with the States." But there is little in that section or elsewhere in the statute to back up this lofty promise. So-called "cooperative agreements" established between the federal and state governments demand very little from the states and offer the same in return. A few provisions of the ESA require the two federal agencies chiefly responsible for implementing the statute—the Department of the Interior's United States Fish and Wildlife Service (USFWS) and the Department of Commerce's National Oceanic and Atmospheric Administration Fisheries Division (NOAA-Fisheries)—to provide notice to states before taking actions that might impose regulatory burdens. The sum total of "cooperation" demanded in the ESA, however, goes little further than this and a few other ministerial procedures. By and large, the states are just like any other regulated entity under the ESA, rather than being partners with the federal government in policy development, implementation, and enforcement. In short, compared to its counterparts in federal environmental law, the ESA barely qualifies as an example of cooperative federalism at work.

This chapter uses a comparative assessment to examine the structure and efficacy of the ESA's weak form of cooperative federalism. The first section provides background on the general principles and instruments of cooperative federalism. The second describes the federalism structure of the ESA and its implementation history. The third compares the ESA's record with two other federal environmental laws, representing different models of cooperative federalism: the Clean Water Act (CWA), which is an example of the delegation model under which state agencies run federal programs, and the Coastal Zone Management Act (CZMA), which represents the consistency model under which federal agencies adjust their actions to state policies.

Drawing on this foundation, the final section of this chapter then examines the costs that the ESA's weak form of cooperative federalism has imposed in terms of federal–state relations and the status of the ESA at large. The strong federal presence in endangered species conservation throughout the states—with no meaningful integration of the states into federal program policy development, implementation, and enforcement—has crowded out state initiative. In general, the result has been that state endangered species conservation programs on the whole are weak, lacking in regulatory teeth and policy innovation. Moreover, because it is the exclusive source of regulatory authority, the federal program has taken the brunt of criticism from property rights and land development interests for the ESA's regulatory impact. This feature has led, in many states, to friction between state government and federal agencies, as well as to broad and often intense public hostility to the ESA and "The Feds." Finally, in an ironic twist, the lack of federal resources for ESA enforcement has led some environmental protection interests to scapegoat the states, holding them vicariously liable for private actions that violate the ESA, situations that can only further erode the sentiment some states have for the ESA.

Hence, there is good reason to explore the potential for enhancing the place of cooperative federalism in the ESA. The federal program appears to have reached its limits. It is not getting any stronger, richer, or more popular under current political conditions. Short of a truly significant shift in the halls of Congress, the most promising outlet for enhancing our nation's endangered species conservation effort is through the states. A new approach to cooperative environmental federalism and the ESA thus provides a potentially huge opportunity to substantially improve species conservation efforts under the continuing evolution of the ESA.

Framework: Tools of Cooperative Federalism

As national policy with respect to environmental goals is formulated, the federal government must consider implementation options that range quite broadly with respect to federal–state relations. At one extreme, the federal government might have no power at all, failing to match the policy with an enumerated authority over federal public lands, interstate commerce, or other federal domains. At the other extreme, federal power is clearly present and pervasive, and the national interest so overriding of particular state concerns as to lead the federal government to preempt all state action. Options between these two extremes represent increasing dominance of policy outcomes by the federal government. So, for example, in some contexts the federal government would have authority to act, but instead it relies on state action to address the issue. Or, in the once-prevailing model known as "dual federalism," the federal government might take action to regulate the field but take no position with respect to the exercise of concurrent authority by the states, other than to preempt state actions repugnant to the federal program. Under the model prevailing since the New Deal era, known as "cooperative federalism," the federal government uses its authority to act in a particular field, and its power to preempt state authority, as leverage to enlist states in implementing state programs consistent with the federal objectives, without commandeering them.

The shift from dual federalism to cooperative federalism tracks the rise of the modern regulatory state as well as the increasing proliferation of social issues that transcend precise lines between federal and state interests. In the environmental policy context, where both phenomena are keenly represented, the federal government has given life to cooperative federalism through a wide range of tools. At one extreme, the federal government can provide a strong impetus to states to pursue national goals by agreeing to subordinate federal policy to state decisions—such as by incorporating state law on a particular matter as federal law or policy ("incorporation") or by binding federal agencies to act consistently with state program requirements ("consistency"). A less constricting approach places federal agencies in an advice-seeking role by requiring them to consider state positions as an important factor in making federal action decisions ("consideration"), or at least to provide states heightened opportunities to receive notice of and provide comments about proposed federal actions ("notice and comment"). Incentive-based approaches to cooperative federalism include the provision of federal financial and staff support of state planning efforts designed to initiate state programs that satisfy national goals ("planning support"), as well as broader unconditional federal financial and staff support for implementing state programs that facilitate national policies ("unconditional support"). The federal government can pursue more persuasive strategies by making financial and staff support contingent on meeting federally prescribed criteria ("conditional support"); offering states a package of benefits if they agree to opt in to a program the federal government implements within the state ("opt in"); or allowing a state to work as a judicial partner with the federal government in implementation decisions ("partnership"). Under the most coercive of the cooperative environmental federalism options, known as the "delegation" model, the federal government threatens to enforce a comprehensive federal regulatory program in a state unless the state uses its authority to regulate the field in a manner that satisfies federal program criteria.

Overall, then, cooperative federalism fits into the larger context of federal–state relations options, as shown in Figure 3-1. The federal government's decision to employ cooperative federalism uses pervasive federal authority as leverage to influence states to act in line with federal policy choices. The tools of cooperative federalism, however, range from "soft" strategies that place the state in the driver's seat for policy development, to "firm" arrangements, under which the federal government prescribes national goals but enters partnership relationships with states for implementation, to "hard" prescriptive approaches that dictate regulatory details and leave the states little latitude. Combinations of these different tools also can be used within one comprehensive federal program, such as control of water pollution—making cooperative federalism potentially quite complex as a system for implementing national environmental policy objectives.

Focus: Federalism and the Endangered Species Act

Environmental law has a rich tradition of employing the tools of cooperative federalism. For example, as shown in more detail below, the CWA primarily uses a coercive delegation model for its pollution permit program. The CZMA, by contrast, relies on federal consistency commitments to entice states to compile coastal land management plans, offering in return to bind federal agencies to the plan provisions. In the case of the ESA, however, cooperative federalism of any variety has not materialized as a significant factor in program implementation. To be sure, USFWS and NOAA-Fisheries administer several core ESA programs that are prime candidates for employing cooperative federalism:

- Section 4 authorizes USFWS and NOAA-Fisheries to identify "endangered" and "threatened" species, known as the "listing" function, and then to designate their "critical habitat."
- Section 4 also requires USFWS and NOAA-Fisheries, after the agency has listed a species, to develop a "recovery plan" for the species.
- Section 7 requires all federal agencies to ensure that actions they carry out, fund, or authorize do not "jeopardize" the continued existence of listed species or "adversely modify" their critical habitat.

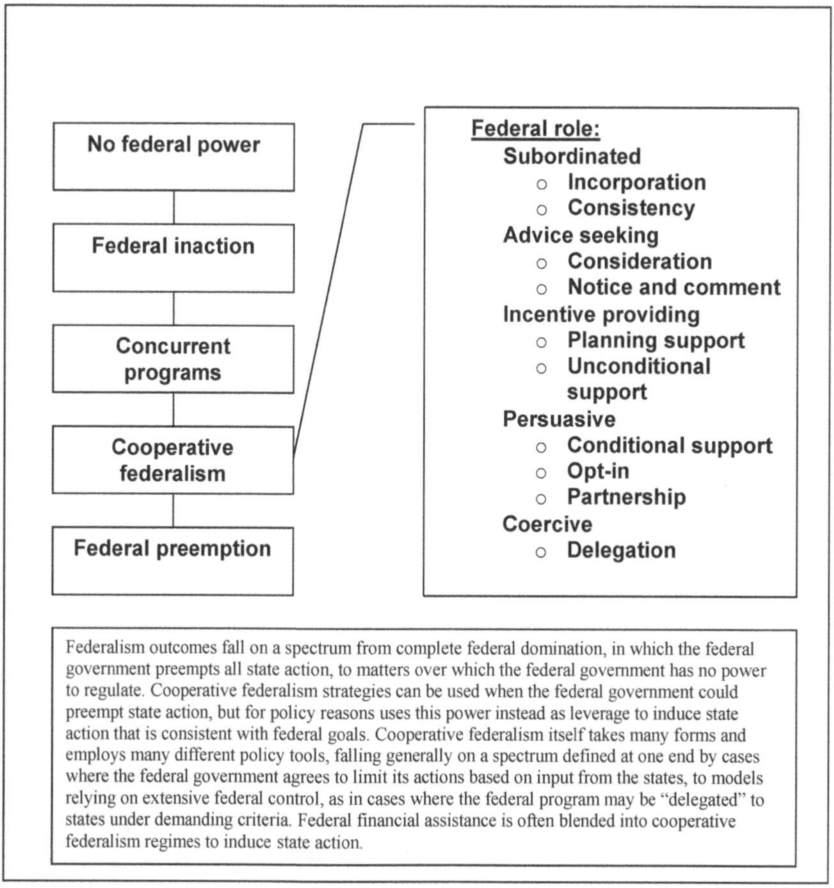

FIGURE 3-1. The Anatomy of Cooperative Federalism

- Section 9 requires that all persons, including all private and public entities subject to federal jurisdiction, avoid committing "take" of listed species of fish and wildlife.
- Section 7 (for federal actions) and section 10 (for actions not subject to section 7) establish a procedure and criteria for USFWS and NOAA-Fisheries to approve "incidental take" of listed species.

The opportunity for using cooperative federalism as a principal mechanism for implementing these core ESA programs surely was there in terms of constitutional setting. Few fields in environmental policy present a more appropriate opportunity for using cooperative federalism approaches than does the objective of conserving imperiled species. State and federal interests alike are strong in the field, and both the states and the federal government have significant authorities with long historical roots from which to draw for policy implementation. State authority over endangered species conservation derives from the extensive interests and authorities the states hold with respect to wildlife management, as held in *Greer v. Connecticut* (161 U.S. 519 [1920]), and land-use regulation, as observed in *Solid Waste Agency of Northern Cook County v. Army Corps of Engineers* (531 U.S. 159 [2001]). Federal authority over endangered species conservation derives from a host of federal powers:

- Treaty Power: In *Missouri v. Holland* (252 U.S. 416 [(1920]), the Court upheld the Migratory Bird Treaty Act of 1918 as an exercise of the Treaty Power.
- Property Clause: In *Kleppe v. New Mexico* (426 U.S. 529 [1976]), the Court described the plenary federal power over federal public lands.
- Commerce Clause: Well after confirming its basis in the Treaty Power, in *Andrus v. Allard* (444 U.S. 51 [1979]), the Court upheld the Migratory Bird Treaty Act as an implementation of Congress' power to regulate commerce between the states.
- Spending Clause: As the Court explained in *New York v. United States* (505 U.S. 144 [1992]), the federal government can condition financial assistance to the states as long as any conditions relate to the nature of the program.
- Self-Regulation: Where an enumerated power exists to support federal action, the federal government can place restrictions and other conditions on the federal action, such as a restriction regarding endangered species.

Notwithstanding this promising constitutional medium for implementing the ESA through cooperative environmental federalism, the statutory framework is virtually passive in this regard. True enough, section 2(c)(2) announces "the policy of Congress that Federal agencies shall cooperate with State and local agencies to resolve water resource issues in concert with conservation of engendered species." Even more pointedly, section 6(a) proclaims that USFWS and NOAA "in carrying out the program authorized by this chapter ... shall cooperate to the maximum extent practicable with the states." And Congress repeatedly has loaded praise for cooperative federalism into the legislative record on the ESA (Fischman 2004). The legislative history of the original enactment emphasized the importance of state involvement in program implementation (House Conference Report 1973), and Congress later proclaimed that it "intended that a cooperative, federal–state relationship be initiated and sustained." (House Conference Report 1982, 25–26). But the statute simply does not follow through on these promises.

The states are not enlisted in the statute to participate meaningfully in the core ESA programs, but rather are treated primarily as regulated entities. Although states receive some financial assistance to conserve species and their habitat and enjoy special notice and comment status for some federal agency decisions, they are for the most part treated in the statute the same as any other regulated entity.

In short, compared with other environmental laws, the ESA is remarkably devoid of creative strategies for putting cooperative federalism into play, especially given the importance of the statute's policy objectives to both national and state interests. In particular, the engines of ESA policy—listing of species (section 4), prohibition of take (section 9), federal agency consultation (section 7), and permitting (section 10)—are virtually empty of cooperative federalism spirit, much less substantive provisions. Appendix 3-1 to this chapter presents a survey of all significant substantive and procedural provisions of the ESA. It identifies those in which the states play a role or enjoy some special status in the statutory implementation—revealing the paucity of cooperative federalism tools at work and the vague parameters articulated in the few provisions that profess to adopt cooperative federalism tools.

Of these provisions, section 6 of the ESA, titled "Cooperation with the States," is the provision one reasonably would expect to contain the core architecture of the ESA's cooperative federalism. Its convoluted structure, however, has prevented it from gaining any traction in this respect. Section 6 has three main components: 1) a "cooperative agreements" authority; 2) an exception to federal enforcement of the section 9 take prohibition; and 3) a federal preemption clause.

Cooperative Agreements

Section 6(c) establishes the authority for USFWS and NOAA to enter into agreements for the purpose of assisting the states in implementing state endangered species conservation programs. Very few criteria are specified for state programs to qualify for this "cooperation." They must demonstrate, for example, that "authority resides in the State agency to conserve resident species" and that "the State agency has established acceptable conservation programs." No details beyond these are prescribed, and the benefits states derive from entering into cooperative agreements are limited primarily to financial assistance. All states have entered into section 6(c) agreements, but almost no state programs exercise as much regulatory authority as is found in sections 7 and 9 of the ESA. The result is that the subject matter of, and financial assistance provided under, most section 6(c) cooperative agreements relates to the listing and monitoring of species and to implementation of nonregulatory species conservation efforts. At best, this is a weak example of the conditional financial assistance model of cooperative federalism.

Enforcement Limitation

Section 6(g) provides that in states that have entered section 6(c) cooperative agreements, the section 9(a)(1)(B) prohibition against taking of endangered species—and, pursuant to section 4(d), of threatened species, when USFWS or NOAA provide by rule—does not apply in the state unless the taking would violate state law or the state requests federal enforcement of section 9. Nothing in this passage requires that a state program that forms the basis of a section 6(c) cooperative agreement must contain a take prohibition that is congruent with the ESA section 9 take prohibition. Indeed, most state programs do not. The ESA extends its take prohibition to cover "harm" of species, which USFWS and NOAA have interpreted to include habitat modification that leads indirectly to the actual death of or injury to a listed species. Most state programs limit take to include only direct actions such as hunting and killing. Section 6(g) appears to provide, therefore, for one of two options in cooperative agreement states: either 1) ESA section 9 applies to its fullest extent only in states that have regulated to the same extent, or 2) in states that have not, only if the state so requests. Arguably, the idea behind such an approach would be to arrange to withdraw federal regulatory clout as the incentive for states to adopt meaningful nonregulatory conservation programs. This is not, however, how section 6(g) has been interpreted or implemented, given the presence of a strong preemption statement in section 6(f).

Federal Preemption

Section 6(g) is, to be fair, rather convoluted and lacking a clear framework for cooperative federalism. Accordingly, courts have declined to immobilize the full extent of the ESA section 9 take prohibition in cooperative agreement with states that do not extend the state take prohibition to cover habitat modification. The reasoning has been that section 6(f) provides that "any state law or regulation respecting the taking of an endangered species or threatened species could be more restrictive than the exemptions or permits provided for in this chapter or in any regulation which implements this chapter but not less restrictive than the prohibitions so defined." Hence, regardless of what section 6(g) appears to anticipate and allow in terms of less restrictive state take regulations, the argument can be made that section 6(f) appears to preempt less restrictive state take regulations. So ruled the court in *Swan View Coalition v. Turner* (824 F. Supp. 923 [D. Mont. 1992]), although the court in that case recognized that section 6(g) does, on its own, appear to suggest otherwise. The *Swan View* court resolved the apparent conflict based on "the overwhelming priority Congress has given to the preservation of threatened and endangered species." The result, however, is that there is no meaningful *regulatory* incentive for states to enter cooperative agreements; states cannot beef up nonregulatory conservation programs to get out from under the regulatory impact of ESA section 9, so why bother?

Notwithstanding the statute's uncoordinated and incoherent approach to cooperative federalism and the cramped reading the courts have given to what little there is in the statute on which to hang any hope of a stronger role for states, there might have been room for USFWS and NOAA-Fisheries to take the lead in advancing a more substantive role for the states in policy implementation and enforcement. Yet this also has not happened. Although the Department of the Interior, under Bruce Babbitt's leadership, accomplished many notable ESA reforms (Ruhl 2004), forging innovative federal–state relations was not one of them. In the midst of its aggressive administrative reform effort, USFWS, with NOAA-Fisheries in tow, issued a feeble policy statement on the role of state agencies in ESA activities (FWS and NOAA 1994). Basically, the agencies committed to doing no more than to "utilize the expertise and solicit the information of State agencies" with respect to a long list of ESA functions, from listing of species to incidental take permitting. The policy thus offered nothing the states would not already have in the way of participation through statutory provisions and cooperative agreements. The Bush Administration went no further, focusing instead on providing financial support to state, local, and private interests for voluntary habitat conservation programs, but failing to reformulate in any meaningful way the role states play in implementation and enforcement decisionmaking. The Obama Administration, thus far, has initiated no new perspectives on the topic.

Comparison: The Clean Water Act and Coastal Zone Management Act

Most federal environmental laws, particularly those with substantial land-use impacts, lean heavily on cooperative federalism to set the theme of federal–state relations. Two examples from opposite ends of the cooperative federalism toolbox are section 301 of the CWA water quality program, known as the National Pollutant Discharge Elimination System (NPDES), and the CZMA. Like it did for the ESA, Congress presented both programs as being constructed around the ideal of cooperative federalism. Section 101(b) of the CWA, for example, announces "the policy of Congress to recognize, preserve, and protect the primary responsibilities and rights of States to prevent, reduce, and eliminate pollution." For the CZMA, Congress based the statute on a finding in section 302(m) that "coastal states have substantial and significant interests in the protection, management, and development of the exclusive economic zone that can only be served by the active participation of coastal states in all Federal programs affecting such resources." It thus expressed the statute's primary purpose, in section 303(2), as being "to encourage and assist the states to exercise effectively their responsibilities in the coastal zone." From there, however, the similarities end.

The CWA, for example, contains a sprawling regulatory program administered by the Environmental Protection Agency (EPA) at the federal level, but which runs wide and deep with cooperative federalism (Craig 2004). The core of the statute is the NPDES permitting program, which Congress intended states to administer through delegation of federal authority, conditioned on state programs meeting a litany of prescribed standards. One of the primary inputs to the program, moreover, is the water quality standards determined by the states with little oversight from EPA. The statute also contains a consistency requirement limiting federal agencies from interfering with the standards the states adopt. The CZMA is more narrowly designed than this, focusing its federal–state interface on the consistency tool. Nevertheless, it represents a far more developed sense of cooperative federalism than does the ESA. Under section 307 of the CZMA, states that adopt coastal management plans and related measures that satisfy broadly articulated national goals, according to the Department of Commerce, enjoy the effect of a provision requiring federal agencies to ensure that their actions are carried out in a manner consistent, to the maximum extent practicable, with the enforceable policies of approved state management programs. When laid out side-by-side, as shown in Appendix 3-2, the CWA and CZMA illustrate just how puny the ESA is by comparison on the matter of cooperative federalism.

Consequences: The Costs of Incoherent Federalism

Unlike the cooperative federalism found in most federal environmental laws—particularly pollution control statutes—it is clear from the foregoing examination that the ESA does not employ anything close to the program delegation or consistency models. Indeed, given the puzzling relation between sections 6(g) and 6(f) and the lack of any clear criteria in section 6(c) for cooperative agreements, the ESA comes off as looking remarkably incoherent on the topic of cooperative federalism. This has proven costly to the ESA in four respects: 1) it has stunted the development of strong state endangered species conservation programs; 2) it has fueled much public resentment of the ESA and federal program in general; 3) it has led to friction between the USFWS and NOAA, on the one hand, and states on the other; and 4) it has facilitated a strategy of blaming the states for failures in the enforcement of endangered species regulations that are attributable, in fact, to USFWS and NOAA.

Weak State Programs

With California as a significant exception, the vast majority of states have adopted a weak version of the ESA as their state program model. The most glaring omission in this respect is the failure to regulate habitat modification as an indirect source of take (Defenders of Wildlife 1998) or to provide some similar level of habitat conservation through nonregulatory programs such as landowner incentives and local empowerment. Regulation of habitat modification under the ESA—accomplished through regulatory interpretation of the statutory term defining take—is unquestionably the engine of ESA authority, defining a scope far broader than would be the case were take limited to actions that only directly kill or injure protected species (Glen and Douglas 2001; Liebesman and Petersen 2010). But if the federal program goes that far, one might ask, why worry that the states do not? The answer is obvious to anyone who has followed the world of wetlands conservation law—that is, the scope of the federal program might not remain intact. For a variety of reasons, few states had sought delegation of the federal CWA section 404 program regulating fill of wetlands—which limits the scope of delegation to waters that are not navigable in use (33 U.S.C. 1344 [g])—and only a handful had developed state wetlands programs as extensive as section 404. As was illustrated in *Solid Waste Agency of Northern Cook County v. Army Corps of Engineers* (531 U.S. 159 [(2001]) (*SWANCC*), in such circumstances any withdrawal of federal authority—in *SWANNC*, by the Court's narrow interpretation of geographic jurisdiction—leaves a potentially huge gap in regulatory authority, which might or might not be filled by new state initiative.

That danger was a very real threat for the ESA in *Babbitt v. Sweet Home Chapter of Communities for a Greater Oregon* (151 U.S. 687 [1995]), where the Court considered a facial challenge to the habitat modification rule USFWS and NOAA promulgated to implement the take prohibition. Had *Sweet Home* turned out as Justice Scalia wished—limiting the take prohibition to actions that directly take protected species—an enormous gap in coverage would have opened for species conservation, given that states also do not generally prevent injurious habitat modification or aggressively promote affirmative habitat conservation. Moreover, the Court has considered, but thus far rejected, *certiorari* petitions raising Commerce Clause challenges to the constitutionality of the application of the ESA to intrastate species, which comprise the majority of listed species. There is no certainty that the Court in the future will continue to refrain from opening the topic. The ESA has failed to promote a culture of comprehensive species protection in the states; any rollback of the federal program, whether by administrative reform or judicial interpretation, threatens to open a gaping hole in species conservation.

Public Hostility to the ESA and "The Feds"

The ESA is primarily federal in context, as implemented, and it often has significantly adverse social and economic effects. In many states, as a result, the ESA has attracted extensive criticism—not only of the program, but also of the federal agencies responsible for it. The ESA is seen by many affected interests as a land-use regulation program administered by "the Feds" from afar, with little attention to local interests. This sentiment is by no means confined to the West, where local interests have long traditions of resenting the federal government for reasons extending well beyond endangered species protection policy. For example, federal protection of manatees in Florida under the ESA and related laws has engendered sharp attacks on the statute and USFWS officials by many of the large number of boating enthusiasts in Florida. This and other pockets of public hostility to the ESA are neither trivial nor isolated.

Frictional Federal–state Relations

Over time, political hostility to the ESA has increased with the public's hostility. In many parts of the nation, lauding one's plans for cooperative efforts with USFWS or NOAA would not make for a successful bid for local or state office. When the ESA is blamed for any woes, the federal government has put itself in the position of taking all the heat and of allowing state officials to make that perfectly clear to the electorate. For example, the Western Governors' Association (WGA) held a two-day summit on the ESA in 2004, which resulted in their loud call for comprehensive reform. The primary theme, which the WGA has been harping on for several years, is a stronger role for the states (Western Governors' Association 2004).

Scapegoating of the States

Overall, the ESA's incoherent approach to cooperative environmental federalism has done little to contribute to the development of state endangered species programs; quite the opposite, it has fostered hostile reactions by the public and state and local politicians to the ESA and "the Feds" who implement it. In a final slap in the face for states, a wave of litigation has struck, in which states are accused of violating the ESA because they failed to adequately restrict their private citizens and businesses from violating the statute (Ruhl 2001). Never mind that this is a *federal* program that invites little state participation, and never mind that the *federal* agencies are failing to enforce the *federal* law against the private violators; under this theory of vicarious liability, the states shoulder all the responsibility. Surely this can do nothing but further alienate state interests from the ESA.

Conclusion

It is time to get creative and serious about cooperative environmental federalism and the ESA. Absent congressional action on this front, the most effective strategy for doing so is through a reconceived approach to the existing statutory tools available to USFWS and NOAA. But cooperative regulatory programs will not be accomplished by administrative fiat of the federal agencies. Simply put, there has to be a quid pro quo approach, under which the states receive some substantial institutional benefit for pursuing endangered species protection to levels desired by national policy objectives. There are several potential sources of authority for doing so.

Section 4(d) of the ESA provides that when USFWS or NOAA lists a species as threatened, the section 9 take prohibition applies only to the extent the agency prescribes by regulation; however, for states that have entered cooperative agreements, "such regulations shall apply … only to the extent that such regulations have also been adopted by such state." Fischman and Hall-Rivera (2002) have focused on this provision to lay out a promising blueprint for enhancing cooperative environmental federalism—under which states and the federal listing agency (USFWS or NOAA) would work cooperatively to develop appropriate take regulations for a species, on the assumption that the species would be listed as threatened.

Section 6 does not permit as straightforward an approach. As long as the *Swan View* interpretation of section 6 persists, states have no incentive to provide aggressive species conservation programs unless they are also willing to adopt a state equivalent to the federal take prohibition, which could undercut the incentive to provide aggressive species conservation programs. USFWS and NOAA could solve the section 6 puzzle by developing an official administrative interpretation that limits the preemptive effects of section 6(f) and which could be used to sway courts to open their minds about the force of section 6 in general. Section 6(f) is a general federal preemption clause that makes no reference to cooperative agreements. Section 6(g)—which prevents application of the section 9 take prohibition in states that have entered cooperative agreements—and section 4(d), discussed in the previous section, specifically trump the take prohibition in cooperative agreement states. The specific should rule the general. Of course, following this approach would make cooperative agreements more important. It should lead USFWS and NOAA to demand more of states, in terms of demonstrating that enforcement of the take prohibition is not necessary, to its fullest extent, in order for the state program to qualify for a cooperative agreement that does not rely on that regulatory instrument.

Re-envisioning these provisions of the statute, while intriguing, might not be necessary to gain some traction for cooperative federalism. For example, section 10(a) habitat conservation plans have been used to authorize incidental take on a regional basis (Liebesman and Petersen 2010; Stanford Environmental Law Society 2001). There is no reason why section 10(a) could not be used to permit a broad variety of actions on a statewide basis pursuant to the state's agreement to implement a statewide program of species conservation (including perhaps its own permitting program). And as an extension of the section 10(a) program, USFWS and NOAA could also use interagency memoranda to specify conditions under which compliance with state endangered species programs would cause the federal agency to conclude that no take of a species will occur, and that no adverse effects to the species would be present that would require a section 7 consultation.

So there are, after all, some dusty corners of the ESA that could be investigated as new means of energizing cooperative federalism in our nation's endangered species protection policy. Short of that, only congressional commitment to a new approach to cooperative federalism in the ESA will rescue the statute from its incoherence on federal–state relations. With the statute gaining no ground in popularity among the states, it seems time, through one of these avenues, to take cooperative federalism seriously for the ESA.

APPENDIX 3-1. Searching for Cooperative Federalism in the ESA

ESA Provision	*Program Function of Provision*	*Role/Status of States in Program Function*	*Type of Cooperative Federalism Instrument*
4(a)(1)	Criteria for identifying endangered species and threatened species ("E/T species")	None	None
4(a)(3)	Criteria for identifying critical habitat of E/T species	None	None
4(b)(1)	Methodology for identifying E/T species	USFWS and NOAA-Fisheries must consider efforts of states to protect species	Advice seeking: consideration
4(b)(2)	Methodology for identifying critical habitat	None	None
4(b)(3)–(7) and 4(i)	Procedures for making decisions under sections 4(a)(1) and (3)	USFWS and NOAA-Fisheries must provide notice to affected states of proposed decisions and submit written justification for decisions that conflict with state positions	Advice seeking: notice and comment
4(d)	Prescribing the discretion associated with identifying a species as threatened (as opposed to the more serious endangered status)	Regulations USFWS and NOAA-Fisheries develop for threatened fish and wildlife E/T species apply in a state that has entered into a section 6 "cooperative agreement" only to the extent the state adopts the regulations	Subordinated: incorporation
4(f)	Standards and procedures for developing recovery plans for E/T species	None	None
4(g)	Requires USFWS and NOAA-Fisheries to monitor status of species determined to have recovered from E/T status	Agencies must implement a monitoring system "in cooperation with the States"	Persuasive: partnership with unspecified terms
5	Land acquisition	None	None
6(a)	Requires that USFWS and NOAA-Fisheries "shall cooperate to the maximum extent practicable with the States"	Cooperating party with unspecified status	Persuasive: partnership with unspecified terms

6(b)	Allows USFWS and NOAA-Fisheries to enter "management agreements" with states for administration and management of any area established for E/T species conservation	Party to agreement	Depends on terms of agreement
6(c)–(e)	Standards and procedures for "cooperative agreements" between USFWS/NOAA-Fisheries and states	State establishes program for conservation of E/T species and seeks federal approval	Unclear [see Appendix 3-2]
6(f)	Preemption	Preempted by federal law or regulation	None
6(g)	Describes effect of cooperative agreement on section 9 prohibitions within the state	Party to cooperative agreement	Unclear [see Appendix 3-2]
7(a)–(d)	Consultation between federal agencies and USFWS/NOAA-Fisheries to evaluate impacts of federal agency actions and impose necessary conditions	None	None
7(e)–(p)	Criteria and procedures for exemption from conditions imposed through consultation under sections 7(a)–(d)	Federal agency provides notice to affected states of exemption application; president appoints representative from each affected state to serve on exemption decision committee	Advice seeking: notice and comment; and persuasive: partnership
9	Specifies prohibitions that apply to all persons with respect to take, trade, collection, and other actions affecting identified species	None with respect to animal species; prohibitions apply to plant species on nonfederal property only to the extent of state law prohibitions	None, other than incorporation for plant species
10	Standards and procedures for issuing permits and other exceptions from section 9 prohibitions	None	None
11	Enforcement authorities	None	None

APPENDIX 3-2. Comparing Cooperative Federalism Across Federal Resource Statutes

Instrument	*ESA*	*CWA-NPDES*	*CZMA*
Subordinated: incorporation	Limited: section 4(d) extends federal rules regarding threatened species to states with cooperative agreements only if the state adopts similar rules; section 9 take prohibition incorporates state law with respect to plant species only	Extensive: section 303 assigns to states the responsibility of developing water quality standards and related determinations which are used in other substantive provisions of the CWA, which EPA must adopt unless limited criteria are not met	None
Subordinated: consistency	None	Extensive: section 401 requires that any applicant for a federal agency permit or license to construct or operate facilities that could result in water pollution provide the agency a certificate from the state confirming the facility will not violate state-adopted water quality standards and related criteria	Extensive: section 307 requires federal agency actions to be carried out consistent with state coastal plans adopted by states and approved by the Secretary of Commerce pursuant to section 306
Advice seeking: consideration	Limited: USFWS and NOAA-Fisheries must consider state efforts when deciding whether listing of a species is necessary	None	None
Advice seeking: notice and comment	Limited: USFWS and NOAA-Fisheries must provide direct notice of proposed species listings to affected states and provide reasons why final decisions depart from state comments	Extensive: numerous provisions of the CWA requiring EPA to develop nationwide standards for control of water pollutants require that EPA consult with the states	None

Incentive providing: planning support	None	None	Extensive: sections 309 and 310 provide federal grants for the development of state plans required under the CZMA as a basis for imposing the consistency requirement on federal agencies
Incentive providing: unconditional support	None	None	None
Incentive providing: conditional support	Limited: section 6 funding is provided through the terms of cooperative agreements	Extensive: section 1281(g) et seq. establish a grants program for state construction of wastewater treatment facilities subject to specified conditions	Extensive: section 306 establishes a program for administrative grants to states that adopt coastal plans that satisfy CZMA criteria; section 306A establishes a grants program for restoration and preservation of coastal resources meeting specified criteria
Persuasive: opt in	None	None	None
Persuasive: partnership	Limited: section 4(g) requires USFWS and NOAA-Fisheries to monitor delisted species through a system implemented in cooperation with the states	None	None
Coercive: delegation	Unclear: section 6 suggests some form of delegation is possible, but has been interpreted more narrowly as a preemption provision	Extensive: section 306 delegates to states the power to enforce federal standards for new sources of water pollution; section 402 delegates primary responsibility for implementing and enforcing the NPDES permit program to the states	None

Note: ESA = the Endangered Species Act; CWA-NPDES = the National Pollutant Discharge Elimination System of the Clean Water Act; CZMA = Coastal Zone Management Act.

References

Craig, R.K. 2004. *The Clean Water Act and the Constitution: Legal Structure and the Public's Right to a Clean and Healthy Environment*. Washington, D.C.: ELI Press.

Defenders of Wildlife. 1998. *State Endangered Species Acts: Past, Present, and Future*. Washington, D.C.: Defenders of Wildlife. www.defenders.org.

Fischman, R. L. 2004. "Predictions and Prescriptions for the Endangered Species Act." *Environmental Law* 34(2): 451–482.

Fischman, R. L. and J. Hall-Rivera. 2002. "A Lesson for Conservation from Pollution Control Law: Cooperative Federalism for Recovery under the Endangered Species Act." *Columbia Journal of Environmental Law* 27(1): 45–172.

FWS and NOAA (United States Fish and Wildlife Service and Department of Commerce's National Oceanic and Atmospheric Administration Fisheries Division). 1994. *Federal Register* 59: 34275, July 1.

Glen, A. M. and C. M. Douglas. 2001. "Taking Species: Difficult Questions of Proximity and Degree." *Natural Resources & Environment* 16(2): 65–69, 132.

House Conference Report. 1973. H.R. Conf. Rep. 93-307.

———. 1982. H.R. Conf. Rep. No. 97-835.

Liebesman, L. R. and R. Petersen. 2010. *Endangered Species Deskbook*, 2nd ed. Washington, D.C.: Environmental Law Institute.

Ruhl, J. B. 2001. "State and Local Vicarious Liability Under the ESA." *Natural Resources and Environment* 16(2): 70–77.

———. 2004. "Endangered Species Act Innovations in the Post-Babbittonian Era—Are There Any?" *Duke Environmental Law & Policy Forum* 14(2): 419–49.

Stanford Environmental Law Society. 2001. *The Endangered Species Act*. Stanford, CA: Stanford University Press.

Western Governors' Association. 2004. *Endangered Species Act Executive Summit*. www.westgov.org/wga/initiatives/esa/panel3.pdf (accessed Sept. 9, 2010).

PART III

Opportunities for State and Local Involvement under the ESA

CHAPTER 4

Listing Decisions, Conservation Agreements, and State–Federal Collaboration: A Litigation Perspective

EILEEN SOBECK AND PAUL S. WEILAND

The focus of this chapter is the role of state and federal collaboration under the Endangered Species Act (ESA), particularly as elucidated by litigation in federal courts. ESA cases that touch directly on the state–federal relationship are relatively rare. The area where litigation has most contributed to the understanding of the appropriate nature of the state–federal relationship has been with respect to section 4 of the Act and whether the United States Fish and Wildlife Service (USFWS) and the National Marine Fisheries Service (NMFS) may rely on conservation agreements (or analogous measures) in making determinations as to whether to list a species. Herein, we review the relevant section 4 litigation, identifying lessons that can be learned from that litigation and then applied to state and federal efforts to utilize conservation agreements in the future.

In the following section of the chapter, we provide a brief overview of the major litigation during the 35 years that have passed since the enactment of the ESA, including litigation under the major sections of the Act. The subsequent section focuses on state–federal cooperation under section 4 of the ESA, while the fourth examines the relevant case law regarding the state–federal relationship and listing. The response of the federal wildlife agencies to litigation, through the development of specific policy guidance, is examined next. We then make some observations about what lessons can be learned from past litigation and what opportunities for state–federal cooperation might exist. Concluding thoughts are set forth in the final part of the chapter.

Overview of ESA Litigation

The course and pattern of litigation over the last 35 years has profoundly influenced the interpretation and implementation of the ESA, especially since the

Act itself has been sparingly amended.[1] In this section, we will provide a brief overview of ESA litigation trends, before turning to a more detailed analysis of litigation regarding section 4, specifically conservation agreements.

Litigation Concerning Section 7 of the ESA

Perhaps the most important litigation under the Act has arisen under section 7. It became obvious shortly after passage of the ESA and was confirmed by the Supreme Court in *Tennessee Valley Authority v. Hill* in 1978 that lack of compliance with section 7 of the ESA could result in far-reaching injunctive relief, halting massive federal projects based upon the likelihood of jeopardy to even a single, humble listed species of fish.[2] Section 7 of the Act applies to all actions authorized, funded, or carried out by federal agencies and requires that each agency ensure, in consultation with the USFWS or NMFS, that its actions are not likely to jeopardize the continued existence of any listed species or result in the destruction or adverse modification of the critical habitat of such species. The far-reaching scope of section 7 has been reinforced and elaborated upon in a number of major cases issued by the Supreme Court and various courts of appeals.

Agency action subject to section 7 requirements has been broadly defined to include issuance of a permit or regulation,[3] contract renewals,[4] Forest Service Land Resource Management Plans,[5] registration of pesticides by the Environmental Protection Agency (EPA),[6] and administration of the National Flood Insurance Program.[7] USFWS and NMFS promulgated regulations that require agencies to consider the indirect, as well as the direct, effects of their actions.[8] The obligation to consult applies only to discretionary actions,[9] and litigation has often been required to determine the limits of agency discretion, and, thus, what actions are or are not subject to the consultation requirement.[10] Initially, the duty to consult under section 7 focused on the potential of jeopardy to listed species, but more recently courts have considered the section 7 requirements as they apply to critical habitat. Two circuit courts have concluded that consultation is required even if a federal action affects the prospects for recovery alone (rather than the likelihood of jeopardy to the species' continued existence) through the alteration of critical habitat.[11]

Who may sue to enforce section 7, and under what circumstances, has been an important area of judicial interpretation. In *Bennett v. Spear*,[12] the Supreme Court parsed the citizen suit provisions of the Act, explaining which agency actions may be challenged under those provisions and which are subject to review under the Administrative Procedure Act (APA). In *Bennett*, the court also analyzed Article III standing, holding that the ESA citizen suit provision negates the traditional zone of interests test.[13]

The Supreme Court also has mandated that jurisdiction under the APA is only granted for review of specific agency action, not to obtain broad programmatic relief.[14] While under the *Lujan* decision plaintiffs may not stop a federal program by bringing a single case, plaintiffs have, throughout the life of the Act, initiated clusters of litigation focused on particular agency programs or geographical areas of interest to the plaintiff groups. For example, in the 1980s there were numerous challenges to the exploration, drilling, and production of oil under the Outer Continental Shelf and Lands Act.[15] The use and allocation of federal water in California and Arizona has been entangled in ESA litigation for years.[16] In the early days of the Act, the activities of the Forest Service were of little interest to litigants. That soon changed, however, and an onslaught of cases challenging timber harvesting practices on federal lands ensued.[17] For a number of years now, there has been substantial litigation regarding management of large river systems like the Missouri and Columbia Rivers by virtue of federal control of infrastructure, including dams and navigation, as well as fisheries management.[18] These cases demonstrate that plaintiffs and courts have ensured that the net of section 7 has been widely, if not always evenly, cast. For every case litigated to a point where a judicial opinion is issued, many more are settled. While these settled cases add little to the ESA jurisprudence, they do have significant effects on the operations of federal programs.

Litigation Concerning Sections 9 and 10 of the ESA

Section 7 applies only to federal actions and is concerned with the threats to a species' continued existence or its critical habitat, but section 9 of the Act applies to all persons, including private individuals, and its prohibitions apply to effects on individual members of a species.[19] The prohibitions of section 9 cover a number of actions, including the import or export of an endangered species, but it is the Act's prohibition on the "take" of a protected species that has been the focus of most litigation under that section. The ESA defines take to mean "harass, harm, pursue, hunt, shoot, wound, kill, trap, capture, or collect, or to attempt to engage in any such conduct."[20] USFWS regulations further define "harm" to mean an act that not only actually kills or injures fish or wildlife, but one that might include significant habitat modification or degradation where it actually kills or injures wildlife by significantly impairing essential behavioral patterns, including breeding, feeding, or sheltering.[21] The Supreme Court has upheld the habitat modification portion of the regulatory definition of harm.[22] Generally, the United States may prosecute violators of the section 9 prohibitions criminally.[23] Occasionally, the government seeks an injunction against prospective violations.[24] More frequently, section 9 cases arise as a result of civil enforcement pursued by individuals or organizations under the citizen suit provisions of the Act.[25]

Not all takes are prohibited by the Act. Incidental take statements, issued as part of consultation under section 7, may authorize take otherwise prohibited by section 9 of the Act.[26] In addition, any person may apply for an incidental take permit under section 10 of the Act.[27] An applicant for such a permit must submit a conservation plan (commonly referred to as a habitat conservation plan or HCP) outlining, among other things, the likely impacts of the activities covered by the plan on the species covered by the plan and the steps to be taken to minimize and mitigate those impacts.[28] Litigation regarding section 10 permits is rare relative to the body of case law that has developed regarding section 7 consultations.[29] This is, in part, because of the fact that until the 1990s only a handful of section 10 permits, with their attendant HCPs, were issued.[30] But it also may be due, in part, to the efforts of both Congress (which enacted the 1982 ESA amendments) and the executive branch (which adopted various administrative reforms initiated by Interior Secretary Bruce Babbitt) to devise an incentive-based conservation planning process. In 1998, USFWS promulgated its "No Surprises" regulation, which provides regulatory assurances to HCP permittees that no additional land-use restrictions or financial compensation will be required of them in the event of unforeseen (as contrasted with changed) circumstances.[31] In 2007, a district court issued a decision upholding the rule.[32]

Litigation Concerning Section 4 of the ESA

Sections 7 and 9 impose significant burdens and restrictions after a species is listed. Section 4, which governs the listing process itself, generated little litigation, but much dissatisfaction, in the first decade of the Act's existence. In the 1982 amendments to the Act, Congress completely rewrote Section 4(b), which governs the listing process. The amendments required the Secretary of the Interior to make listing determinations "solely" on the basis of the "best scientific and commercial data available" and, among other changes, streamlined the listing process by reducing the time periods for rulemaking.[33]

There then followed a trickle of cases challenging the merits of decisions of USFWS and NMFS not to place a species on the list—notably a successful challenge by the Sierra Club on the merits of a USFWS decision not to list the Northern spotted owl.[34] These challenges frequently involved claims that the agencies had not employed the best scientific information available in their decisions. But more often, they involved process issues, such as missed statutory deadlines and the effects of a congressionally imposed moratorium on funding for listing-related activities.

For many years, USFWS struggled to cope with a backlog of species waiting to be considered for listing. Species remained on the candidate list for long periods of time. Critical habitat was habitually found to be "not determinable" or "not prudent" at the time of listing, so that, even for species that were evaluated, critical habitat was rarely designated. Eventually, dissatisfied with the pace of progress, plaintiff groups opted to utilize litigation to advance their interests, and the section 4 litigation trickle became an avalanche. Early attempts to manage the litigation included a settlement with a timetable for moving species off the candidate species list.[35] Challenges continued, focusing on statutory deadlines routinely missed by USFWS. In 1995, Congress enacted an appropriations bill with a rider preventing the USFWS from making any final listing decisions.[36] This contributed to the backlog of pending listing determinations. After the moratorium was lifted in 1996, USFWS promulgated its Listing Priority Guidance, setting out its strategy for reducing the existing listing backlog, based on available appropriated resources.[37]

Again, for a short time, the USFWS made progress in reducing the listing backlog. At about this time, however, there was a huge increase in the number of critical habitat determinations that USFWS had to consider. This was in large part due to the decision in *Natural Resources Defense Council v. Department of the Interior*,[38] overturning the USFWS finding that designating critical habitat for the California gnatcatcher was "not prudent." The USFWS provided two bases for that finding: 1) designation of critical habitat could make a species more vulnerable to habitat modification and/or destruction and 2) whereas most populations of the species are found on private lands, the additional protection afforded by designation occurs during the section 7 consultation process that is triggered only by federal actions.[39] The Ninth Circuit's decision rejecting the position of USFWS led to a series of cases requiring USFWS to revisit many of its prior "not prudent" determinations and to stop its practice of relying on this approach to defer critical habitat determinations indefinitely.

USFWS has been unable to dig itself out from under the twin burdens of backlogged listing and critical habitat determinations. As a result, through court orders and settlements, plaintiffs and the judiciary, rather than USFWS, are determining how the limited funds appropriated by Congress to USFWS for listing activities are spent. The seemingly endless stream of missed deadline cases filed in different districts and circuits is nearly impossible to manage given scarce resources provided to the agency. Any change in the schedule for completing consideration of one species in one court can have a cascading effect on other species being considered but which are subject to disparate and sometimes competing and contradictory court orders. Unless the Act is amended or USFWS obtains the resources necessary to address its statutorily mandated duties, both USFWS and the Department of Justice have no choice but to spend staggering amounts of time and money on litigation, which cannot solve the underlying problem.[40]

Against this litigation backdrop, the focus of this chapter is a handful of section 4 cases in which USFWS or NMFS have attempted to justify a decision not to list a species based, at least in part, on the conservation actions of states. Despite the hundreds of cases litigated over the more than 35-year life span of the Act, very few such cases have focused, or even shed much light on, the relationship, under the ESA, between the federal government and state and local entities, either directly or indirectly. In this chapter we will examine the scant body of case law relating to this topic.

The Role of State–Federal Cooperation under ESA Section 4

A review of the role of state–federal relations outlined in section 4 of the ESA itself is useful before turning to the case law. When Congress enacted the ESA in 1973, the structure of the Act itself provided evidence of congressional intent that the federal government would have principal, and in most cases total, responsibility for determining which species would be listed, the bounds of critical habitat, what actions would be prohibited (in the case of threatened species), whether permits would be issued, and whether federal actions would result in jeopardy.[41]

Under section 4, the ESA requires the Secretary to determine whether a species is an endangered or threatened species because of any of the following five factors:

1 The present or threatened destruction, modification, or curtailment of its habitat or range;
2 Overutilization for commercial, recreational, scientific, or educational purposes;
3 Disease or predation;
4 The inadequacy of existing regulatory mechanisms;[42] or
5 Other natural or manmade factors affecting its continued existence.[43]

Section 4(b) of the Act requires that the Secretary make a listing determination "*solely on the basis of the best scientific and commercial data* available to him[.]"[44] The Act goes on to identify a specific role for states in the pre-listing evaluation, and provides that this determination shall be made:

> *after conducting a review of the status of the species and after taking into account those efforts, if any, being made by any State or foreign nation, to protect such species, whether by predator control, protection of habitat and food supply, or other conservation practices, within any area under its jurisdiction, or on the high seas.*[45]

Further, the Secretary is required to notify and invite the comments of any affected states when considering whether to list a species.[46]

The statutory authority set out in these provisions gives states and their subdivisions the opportunity to avoid the federal listing of a species by pointing to or adopting sufficient state or local legislation, regulations, or programs or entering into conservation agreements, Memoranda of Agreements, or other arrangements sufficient to convince the listing agency (that is, either USFWS or NMFS) that federal listing criteria are not triggered, and federal protection is not required.

State–Federal Relationship in Section 4 Cases

Since the inception of the Act, the listing agencies have sought a successful strategy that would both benefit species and also avoid the need for such species to be listed by developing and then relying on state statutes, regulations, or other conservation measures, or the promise of such actions, in evaluating the listing criteria of section 4. In a handful of instances, USFWS and NMFS based determinations not to list a species in whole or in part on such conservation agreements, management plans, or similar endeavors.[47] Environmental groups challenged a number of these determinations in court. The results of the litigation are mixed. A few agency decisions not to list have been upheld. But in a number of other cases, the agency decision not to list a species was struck down because of deficiencies in the conservation agreements relied upon by the agencies. As the discussion of the cases below demonstrates, courts are concerned about whether such agreements contain measures that are sufficiently likely to conserve species and whether the proposals are likely to come to pass. In cases where agency decisions were struck down, the courts have frequently found that agency explanations and discussions in the administrative record regarding these factors were inadequate. This group of cases led USFWS and NMFS to rethink how, and to what extent, they could rely on the conservation efforts of states and other entities, as well as how to evaluate such efforts in their decisionmaking. This resulted in a change in agency policy in 2004, discussed below.

Winter-Run Chinook Salmon[48]

The first instance in which an agency decided not to list a species based on the adequacy of existing management measures occurred in 1987, when NMFS determined that listing the Sacramento winter run of Chinook salmon was not warranted. That determination was promptly challenged by the American Fisheries Society, which had petitioned for the species to be listed as threatened in 1985.[49] NMFS made a preliminary 90-day finding that listing "may be warranted."[50] The agency subsequently conducted a status review of the species and evaluated it under the five listing criteria set out under section 4(a) of the Act.

During the status review, NMFS identified a number of threats to the winter run chinook salmon.[51] The agency concluded, based principally on the actions state and federal agencies had adopted and were implementing, that a proposed listing of the winter-run chinook was not warranted.[52] In assessing the adequacy of existing regulatory mechanisms, NMFS referred to a number of state and federal laws, finding that the available laws and regulations provided adequate mechanisms for restoring the winter run in the Sacramento River.[53] The agency noted the central dilemma that arises in most of these cases: if the existing laws and regulations are adequate, why is the species in such trouble? Here, NMFS said, "the precipitous decline in the size of the run since the late 1960s indicates that these regulatory mechanisms *have not been applied effectively* with respect to winter-run chinook .... The resource agencies are now applying existing authorities to the specific restoration of winter-run chinook salmon."[54] The notice of determination provided a limited description and discussion of federal and state agency efforts that would benefit the fish, including a habitat restoration project sponsored by the California Department of Fish and Game (CDFG), California Water Quality Control Board[55] standards for releases of polluted water into the Sacramento River, and the CDFG's in-river fishing regulations, designed to prevent salmon from being harvested on their way to the spawning grounds.[56] These and other measures deemed by NMFS to be the best means to ensure restoration of the winter-run chinook were included in a "Ten Point Plan." Subsequently, three federal agencies—NMFS, the Bureau of Reclamation (BOR), and USFWS—and one state agency, CDFG, formalized the Ten Point Plan by entering into a cooperative agreement.[57] As the court noted, the cooperating agencies agreed to take *extraordinary* measures on a *voluntary* basis.[58]

In ruling on the adequacy of the determination by NMFS, the court made some initial legal findings regarding section 4—holding that, in making a decision as to whether to list, the Secretary considers his conclusions regarding the five factors in section 4(a)(1) in combination with the information on existing federal and state programs.[59] The court also held that the Secretary could consider other federal programs, as well as those of states, in making his determination. Finally, the court found (with limited analysis) that, in this case, when the activities of other agencies reflected in the cooperative agreement were examined and their effectiveness considered, most, if not all, of the threats to the species were addressed. The court concluded, therefore, that the agency's decision not to list should be upheld.[60]

This early challenge came at a time when the wildlife agencies rarely decided not to list petitioned species and when such determinations generally were not challenged. The Ten Point Plan and cooperative agreement came late in the listing determination process. Nevertheless, the court, without any elaboration, found that "the agencies undertaking [cooperative agreement] activities have exhibited substantial commitments to their promised duties."[61] While the court noted that the measures to be taken were "extraordinary" and "voluntary," it did not find these factors to be problematic.[62]

Barton Springs Salamander[63]

A second case involved a rare aquatic amphibian found only in the waters of Barton Springs, which is located in a park in Austin, Texas.[64] In this case, the court found that the Secretary of the Interior violated the ESA in withdrawing its proposal to list the Barton Springs salamander as endangered in reliance on a conservation agreement between USFWS and agencies of the state of Texas.[65]

On February 17, 1994, the Secretary issued a proposed rule to list the salamander.[66] The Service is normally required under the Act to make a final listing determination within one year of a proposal to do so.[67] The Secretary extended that deadline in the case of the salamander for six months pursuant to section 4(b)(6)(B).[68] After a challenge by plaintiffs, the court eventually ordered the Secretary to make a final listing by August 30, 1996.[69] On June 24, 1996, USFWS reopened the comment period on the proposed rule to determine whether the state was offering any additional regulatory protection for the species. On that same day, representatives of the state of Texas met with USFWS officials regarding this matter.[70]

The court closed the comment period on July 10, 1996.[71] On August 13, 1996, USFWS and several Texas state agencies entered into the "Barton Springs Salamander Conservation Agreement and Strategy."[72] The Secretary took final action, withdrawing the proposed rule to list the Barton Springs salamander as endangered, on September 4, 1996.[73] There was no opportunity for public comment regarding the provisions of the conservation agreement or the decision to withdraw the proposed listing.

As part of its original evaluation of listing criteria under section 4(a)(1), the USFWS specifically, and in some detail, evaluated existing state and local regulatory mechanisms, detailing evidence of the inadequacies of these programs.[74] The provisions of these regulatory mechanisms did not change between 1994, when first analyzed by USFWS, and the time of the final decision to withdraw the proposed listing in 1996. As a result, the court concluded that the decision not to list must have been based on the conservation agreement. The agreement contained measures designed to reduce significantly the threats to the salamander, including risks from catastrophic events, and to establish a captive breeding program.[75]

As an initial matter, the court found that the Secretary had violated the ESA on three separate occasions by missing mandatory deadlines specified under section 4 of the Act.[76] The court also held that the Secretary's determination was deficient in a number of ways. The court noted that the agreement was entered into after the close of the public comment period.[77] When it turned to the merits of the Secretary's decision not to list, the court held that the Secretary failed to follow proper procedures under the APA and ESA, that he failed to allow comment on issues that were fundamental to his decision, and that he considered factors other than those contemplated by the ESA. The court therefore determined that the Secretary's decision violated the APA and ESA.[78]

The court articulated a number of concerns with the Secretary's reliance on the conservation agreement as the rationale for declining to list the species. First, the court pointed to the analysis of the agency itself at the time it proposed listing the salamander, in which the agency enumerated a number of inadequacies in the existing regulatory mechanisms.[79] In the Federal Register notice withdrawing the proposed rule, the Secretary set out an explanation of why a new development, the conservation agreement, was sufficient to offset the deficiencies described in the proposed rule.[80] However, the court identified eight concerns, previously articulated by the agency in its 1994 proposed rule, that it asserts were not addressed or remedied by the conservation agreement.[81] In addition, the court identified 28 actions that were to be implemented "immediately," but found that none would significantly reduce the immediate threats to the species. Finally, because the agreement was so recently concluded, the court stated that it had no way of determining whether or not the actions have been implemented. The court skeptically noted, however, that "the terms and phrases 'identify,' 'evaluate,' 'review,' 'develop,' 'work with' and 'establish' used in the conservation agreement suggest to this Court that even if the activities have been fully implemented, they do not take any tangible steps to reduce the immediate threat to the species."[82] The court therefore concluded that the measures in the conservation agreement were speculative in that "[t]here are no assurances that the measures will be carried out, when they will be carried out, nor whether they will be effective in eliminating the threats to the species."[83] Further, the court stated that, "[w]hen the Secretary permitted an agreement with no proven track record for effectiveness in protecting the species, to play the pivotal role in his listing decision," he acted in an arbitrary and capricious manner. "[T]he Secretary placed the continued existence of a species, found only one place in the natural world, in the hands of State agencies and a conservation agreement with no proven track record for success."[84]

In addition to the missed statutory deadlines and the finding that the Secretary's decision was arbitrary and capricious, the court found that two serious process violations had occurred, which contributed to the death knell of the "no list" decision. First, there had been no opportunity for prior public comment on the conservation agreement, which the court found to be a flaw in the validity of the agency's decision.[85] Second, the court found that political measures had been taken to influence the listing decision, and that strong political pressure was applied to the Secretary to withdraw the proposed listing of the salamander, resulting in the Secretary considering improper factors in making his decision not to list the salamander as endangered.[86] Less than two months after the district court issued its decision, USFWS listed the Barton Springs salamander as endangered.[87]

Oregon Coastal Coho Salmon[88]

By the early 1990s, it was apparent that many runs of salmon from California to Alaska were in steep decline, and that curtailing ocean and in-river harvests alone would not stem the decline of these native fish. In 1993, several environmental organizations filed a petition to list West Coast populations of coho salmon.[89] In response to the petition, NMFS conducted a detailed status review of the species, and, in so doing, it violated the one-year ESA deadline for making such a determination.[90] On July 25, 1995, NMFS issued a proposed regulation establishing that West Coast populations of coho should be divided into six evolutionarily significant units (ESUs), each to be considered separately, and that three of those ESUs, including the Oregon Coast ESU, should be listed as threatened.[91] In response, in October 1995, Oregon launched a state plan, the Oregon Coastal Salmon Restoration Initiative (OCSRI), designed to halt or limit the decline of Oregon coastal coho and other salmonids.[92] On October 31, 1996, NMFS extended the deadline to list by six months for two of the ESUs, including Oregon Coast coho, based on a determination that there was substantial scientific disagreement regarding the available data.[93] Oregon issued the final version of the OCSRI in March 1997.[94]

On March 28, 1997, NMFS issued an updated status review that evaluated the adequacy of all existing mechanisms, including the OCSRI, and found that despite the significant protections offered by the state's initiative, NMFS could not conclude that these measures would be adequate to meet the criteria to forestall listing. NMFS was especially concerned about the adequacy of habitat measures set forth in the OCSRI. In April 1997, NMFS entered into a memorandum of agreement with Governor John Kitzhaber, under which NMFS agreed to provide guidance to Oregon regarding the specific measures it considered necessary for Oregon coastal coho habitat protection, and the state agreed to implement these or equivalent measures within two years. Otherwise, the state argued that NMFS would change the status of the ESU.[95] On May 6, 1997, NMFS published a final rule with its determination that the Oregon Coast ESU did not warrant listing at that time. The agency concluded that the Oregon Coast coho is not likely to become endangered in the interval between this decision and the adoption of the improved habitat measures by the state of Oregon.[96]

The court invalidated the final no-list determination. As an initial matter, the court held that NMFS had applied the wrong legal standard by not applying the statutory standard of determining whether a species is threatened if it is "likely to become an endangered species within the foreseeable future... ."[97] The court found that NMFS had determined only that the Oregon Coast coho would not become endangered within the next two years, "falling far short of any reasonable definition of the 'foreseeable future.'"[98] The court found that the ESA "requires a determination as to the likelihood—rather than merely the prospect—that a species will become endangered in the foreseeable future ... based on the current regulatory structure."[99]

The court then went on to consider whether NMFS could properly consider future or voluntary protective efforts in applying the listing criteria, and concluded that it could not. The court found that "with respect to future actions ... [e]ven the broad language of § 1533(b)(1)(B) cannot reasonably be interpreted to include future efforts, whether regulatory or nonregulatory."[100] The court determined that, in accordance with the Barton Springs salamander case, the Secretary may not rely on "plans for future actions to reduce threats," instead, it may only rely on conservation efforts that are currently operational.[101] Further, the court held that the agency may not rely on unenforceable (that is, voluntary) efforts since they are, in the words of the court, "necessarily speculative." As a result of these legal conclusions, the court concluded that the final NMFS rule improperly relied on future and voluntary measures contained in the OCSRI.[102]

After making its legal finding of what factors the listing agency may take into account in making its determination, the court then reviewed the final decision not to list the Oregon coastal coho and the administrative record supporting it to determine whether NMFS had relied on any improper factors. The court found that the rule itself acknowledges its reliance on the OCSRI and the memorandum of agreement. At the time that the agency proposed listing, the Northwest Forest Plan (NFP) and the Oregon Forest Practices Act were already in existence,[103] but NMFS had determined that they did not provide adequate habitat protections for the coho. Therefore, the court concluded that NMFS must have relied on the OCSRI and the memorandum of agreement to remedy the previously identified inadequacies articulated in the proposal and to justify a no-list determination. The court held that both the OCSRI and the memorandum of agreement contain promises for future action and voluntary measures. In addition, the court noted several similar NMFS decisions where the agencies concluded that the OCSRI or similar measures were not sufficient to avoid a listing—for example, the NMFS decision to list the Transboundary ESU and the Umpqua River cutthroat trout.[104] In addition, the court noted that the record itself seemingly illustrated numerous ways in which the OCSRI is insufficient to protect the habitat needs of the species.

The Court concluded that:

> *NMFS's ultimate conclusion not to list the Oregon Coast ESU flies in the face of its express concerns regarding the OCSRI. The wait-and-see stance of the NMFS has no support in the ESA or case law. Instead of placing the risk on the future and voluntary conservation measures proposed by the OCSRI, the NMFS unlawfully placed the risk of failure squarely on the species.*[105]

Despite this ruling, NMFS made a subsequent determination not to list Oregon Coast coho. That determination was struck down, in part because of improper reliance by NMFS on future rather than existing state regulatory mechanisms.[106] A final rule listing the Oregon Coast coho ESU was published in the Federal Register on February 11, 2008.[107]

Flat-tailed Horned Lizard[108]

In 1993 USFWS published a proposed rule to list the flat-tailed horned lizard as threatened.[109] Subsequently, USFWS constituted the Rangewide Strategy Working Group that was tasked to design a plan to mitigate threats to the lizard. The Working Group included representatives of several federal agencies, as well as state representatives from the California Department of Parks and Recreation, the CDFG, the Arizona Game and Fish Department, and a number of local representatives from Imperial County, California, and Yuma County, Arizona. Several years later, the Working Group drafted a Rangewide Management Strategy. The proposed rule was reopened for additional public comments in March 1997 and extended again in May 1997.[110]

Conservation plaintiffs finally filed suit to force USFWS to make a final determination, and a federal district court in Arizona ordered USFWS to make such a final decision by July 1997. The final Management Strategy was completed in May 1997, and a conservation agreement, governing implementation of the Strategy, was entered into by state and federal agencies in June 1997. As the Ninth Circuit noted in its disposition of an appeal from a lower court challenge to the final determination by USFWS, the designation of five management areas on public lands—to be protected by restrictive measures, some of which had been in effect for some time and others of which were new—was central to the implementation of the conservation agreement.[111] On July 15, 1997, USFWS made its final "no-list" decision regarding the lizard and withdrew the proposed rule.[112]

In December 1997, Defenders of Wildlife and others filed suit, challenging the USFWS decision not to list the lizard as threatened. Defenders raised three arguments: first, that the agency's analysis of the lizards' population trends was faulty; second, that the agency's conclusion, based on its reliance on the conservation agreement, was arbitrary and capricious because the agreement was specifically designed to reduce threats only to that portion of the lizard's habitat comprised of public lands; and third, that the reliance of USFWS on the untested conservation agreement and the Rangewide Management Strategy was improper and insufficient to justify withdrawing the proposed rule.[113]

With respect to the first claim, the district court found that USFWS properly used and relied upon the appropriate scientific information regarding population trends.[114] Further, the court concluded that, although USFWS acknowledged that the conservation agreement did not impose restrictions on private lands, the administrative record before the agency adequately supported its conclusion that threats to the lizard's habitat had diminished sufficiently to justify the no-list determination. Finally, the court held that USFWS's reliance on the conservation agreement was proper.[115]

The court disposed of plaintiffs' arguments by distinguishing other cases—for instance, noting that unlike in the Barton Springs salamander case, USFWS reopened the public comment period for 60 days to allow for public input regarding the Management Strategy[116]—or simply disagreeing with the prior decisions. The court, in this case, was not troubled by the short time that the Management Strategy had been in effect, disagreeing with the federal district court in Oregon that had considered the case of the Oregon Coast coho salmon. The court found that the conservation agreement concerning the lizard had been finalized and executed a month prior to the final withdrawal decision and was operational at the time of the USFWS decision. Therefore, the court concluded, it was properly considered by the agency.[117]

The court noted:

> *The Rangewide Strategy Working Group, composed of various state and federal agencies, spent nearly three years developing a resource management strategy for the FTHL between 1993 and 1997. The group included experts and scientists in lizard ecology, herpetology, and conservation biology who had conducted significant FTHL research since 1978. This highly qualified group developed the proposed policies of the Conservation Agreement.*[118]

Further, the court had no qualms about the propriety of USFWS taking the newly minted efforts of the states into account in making its determination.

> *To require USFWS to not consider state conservation strategies and agreements simply because the strategy is "newly implemented" would discourage states from engaging in any conservation efforts at all. … Though the ESA regulations may represent many species' last chance at survival, Congress surely did not intend to make it the only chance at survival.*[119]

The court upheld the USFWS decision to withdraw the proposed listing of the lizard, finding that the decision was supported by the record and was not arbitrary and capricious.[120]

On appeal, the Ninth Circuit reversed, concluding that the USFWS decision to withdraw the proposed rule recommending listing the lizard was arbitrary and capricious.[121] The Court of Appeals, however, focused on two legal arguments barely addressed by the district court. The court examined the question of whether the Secretary's determination that the protections afforded to the lizard on public lands were sufficient to remove the threat of extinction of the species "throughout all or a significant portion of its range."[122] It carefully analyzed the meaning of that statutory phrase, concluding that a species can be extinct "throughout all or a significant portion of its range" if there are major geographical areas in which it is no longer viable but once was. The court then found that in the case of the lizard, where the area in which it is expected to survive is much smaller than its historical range, the agency must explain its conclusion that the area in which the species can no longer live is not a "significant" portion of its range. The court found that the agency simply had not expressly considered this issue at all. It, therefore, found that this omission, with respect to a significant legal issue, was a sufficient basis for remanding the case. In addition, the court found that the agency was required to address the lizard's viability in a site-specific manner as to the alleged benefits of the conservation agreement, which it had not done.[123]

Klamath Mountains ESU of Steelhead[124]

This case involved a challenge to a final rule issued by NMFS in which the agency decided not to list the Klamath Mountains ESU of steelhead as a threatened species. The facts surrounding the steelhead decision are similar to those in the Oregon Coast coho decision not to list. The plaintiffs in this case petitioned NMFS to list the steelhead in 1992. After conducting a number of status reports, NMFS issued a proposed rule to list the Klamath Mountains Province ESU as threatened. NMFS specifically noted that numerous efforts to halt the decline in the population of the fish had been inadequate to do so.[125] As was the case with Oregon Coast coho, at the time that it issued the proposed rule the agency concluded—after a review considering existing regulatory and other conservation programs—that the steelhead should be listed. But before making its final decision, NMFS entered into a memorandum of agreement with California because "the State's habitat protection measures … are not presently adequate to secure properly functioning habitat conditions over the long-term."[126] At the same time in Oregon, NMFS entered into a "Steelhead Supplement" to the Oregon Plan and the 1997 memorandum of agreement between NMFS and Oregon developed for Oregon Coast coho salmon.[127] On March 19, 1998, NMFS issued its final determination not to list the Klamath Mountains ESU as threatened.[128]

The court held that the NMFS rule was arbitrary and capricious because it relied on future and voluntary state actions in making its determinations.[129] The issue in the case was whether, pursuant to ESA section 4(b)(1), NMFS could consider and rely on existing regulatory measures, here including the memorandum of agreement, that include elements to be implemented in the future. The court found that in light of the findings of the biological review team during the steelhead status reviews, the NMFS decision not to list relied heavily on state conservation efforts in both Oregon and Washington, and the agency's determination that these efforts were sufficient to ameliorate the threat to the fish.[130] The court found that since NMFS had changed its position from a proposal to list the steelhead to a decision not to list, the agency's decision had to be based on "newly discovered or newly implemented conservation efforts," that is, those discovered or implemented after the date of the proposed rule.[131] The court found that most of the state plans that NMFS relied on were, in fact, only proposals for future action, and the rest were voluntary measures. With respect to the voluntary measures, the court noted: "[a]lthough it was appropriate for NMFS to consider such measures, it was arbitrary and capricious for NMFS to rely, in effect, exclusively on voluntary actions despite its finding in the Proposed Rule that past state conservation efforts were inadequate."[132] Finally, the court found that it was proper for NMFS to rely on existing federal regulatory schemes (here, the National Forest Plan promulgated by the Forest Service).[133] But the court concluded that these federal measures, without the problematic state measures, were not enough to support the final decision.[134]

Short-leaved Dudleya[135]

The short-leaved Dudleya is a small, slow-growing succulent plant, located mainly on bluffs in southern California.[136] USFWS proposed to list the Dudleya as an endangered species in 1993.[137] At that time, USFWS included an analysis of existing regulatory mechanisms. The State of California had listed the Dudleya under the California Endangered Species Act and the California Native Plant Protection Act. USFWS concluded, however, that these protections were not adequate, in that the California laws contained too many exemptions and were not adequately enforced.[138]

In 1995, the Southwest Center for Biological Diversity sued to force USFWS to make a final determination regarding the Dudleya listing. A court ordered USFWS to make its final decision by September 30, 1996.[139] In response, USFWS withdrew its proposed listing, based on new information, principally that the Dudleya was a species protected under the Multiple Species Conservation Program (MSCP) for southern San Diego County. The MSCP was scheduled to go into effect within a few months of the final USFWS decision.[140] Plaintiffs brought a new challenge to the final decision and USFWS's reliance on the MSCP, which was not in existence at the time of its decision, but which USFWS argued "was reasonably certain to be enacted and implemented in the near future."[141] The court held that in reading sections 1533(a)(1)(D) and (b)(1)(A) together, it is plain that:

> *the Service may consider the imminent adoption of proposed regulations as a factor in assessing the adequacy of an existing regulatory mechanism in the interim. Simply because a proposed regulation has not been adopted does not mean that the proposed regulation must be ignored.*[142]

Plaintiffs also argued that the agency's decision was arbitrary and capricious because San Diego's existing regulatory mechanism, the San Diego County Resource Protection Ordinance (RPO), was not deemed by USFWS to be sufficient at the time of its proposed listing. The court noted that USFWS, in its withdrawal of the listing, stated that the RPO would provide "substantial interim protection until the newly developed MSCP was passed, and that passage was expected shortly."[143] The court found that it was not arbitrary and capricious for USFWS to have found, in 1993, that the RPO by itself was insufficient, but that, in 1996, after further regulatory mechanisms were developed, the RPO provided adequate short-term protection until the imminent adoption of proposed long-term protection.[144]

Agency Development of PECE and CCAA Policy

In the context of the litigation described above, USFWS and NMFS have promulgated two policies that are relevant to the species conservation efforts of states (and their subunits) in cooperation with the federal government. The more recent is the Policy for Evaluation of Conservation Efforts When Making Listing Decisions (PECE), finalized in 2003. The latter is the Candidate Conservation Agreement with Assurances Policy (CCAA Policy), issued in 1999.

PECE

In large measure due to their relative lack of success in the cases discussed above, USFWS and NMFS decided to promulgate a policy describing how the agencies would evaluate conservation efforts identified in conservation agreements, management plans, or similar documents developed by federal agencies, state, local or tribal governments, and/or other individuals or organizations. The two agencies published a draft of the policy on June 13, 2000.[145] They issued the final policy on March 28, 2003.[146]

The goal of the policy is to provide "direction" to the listing agencies regarding how to consider a conservation agreement when making a decision on whether a species should be listed or not. In the Federal Register publication of the final notice, the agencies also noted that the policy would provide useful information to the entities working to develop agreements or plans. In addition, the agencies acknowledge that they hope that the policy will help address deficiencies found by courts in past litigation by establishing a set of consistent standards.[147]

The policy applies to formalized conservation agreements and similar programs designed to benefit a species.[148] The policy specifies that, in determining whether such conservation efforts contribute to making the listing of a species unnecessary, the agencies will evaluate two key factors:

1 For those efforts yet to be implemented, the certainty that the conservation effort will be implemented; and
2 For those efforts that have not yet demonstrated effectiveness, the certainty that the conservation effort will be effective. [149]

Criteria for evaluating each of the two factors were also provided. With reference to the certainty that the conservation effort will be implemented, the relevant factors are included in Table 4-1. With respect to the certainty that the conservation effort will be effective, the agencies will consider the factors included in Table 4-2.

CCAA Policy

In 1999, USFWS and NMFS issued a policy and implemented regulatory changes that offer assurances as an incentive for citizens, states, local governments, tribes, businesses, organizations, and other nonfederal property owners to implement conservation measures for species that are proposed for

TABLE 4-1. Factors Used to Determine Whether Conservation Efforts Will Be Implemented[150]

1	The conservation effort, the party(ies) to the agreement or plan that will implement the effort, and the staffing, funding level, funding source, and other resources necessary to implement the effort are identified.
2	The legal authority of the party(ies) to the agreement or plan to implement the formalized conservation effort, and the commitment to proceed with the conservation effort are described.
3	The legal procedural requirements (for example, environmental review) necessary to implement the effort are described, and information is provided indicating that fulfillment of these requirements does not preclude commitment to the effort.
4	Authorizations (for example, permits, landowner permission) necessary to implement the conservation effort are identified, and a high level of certainty is provided that the party(ies) to the agreement or plan that will implement the effort will obtain these authorizations.
5	The type and level of voluntary participation (for example, number of landowners allowing entry to their land, or number of participants agreeing to change timber management practices and acreage involved) necessary to implement the conservation effort are identified, and a high level of certainty is provided that the party(ies) to the agreement or plan that will implement the conservation effort will obtain that level of voluntary participation (for example, an explanation of how incentives to be provided will result in the necessary level of voluntary participation).
6	Regulatory mechanisms (for example, laws, regulations, ordinances) necessary to implement the conservation effort are in place.
7	A high level of certainty is provided that the party(ies) to the agreement or plan that will implement the conservation effort will obtain the necessary funding.
8	An implementation schedule (including incremental completion dates) for the conservation effort is provided.
9	The conservation agreement or plan that includes the conservation effort is approved by all parties to the agreement or plan.

listing under the ESA (that is, candidate species), as well as species that are likely to become candidates or proposed species in the near future.[152] The idea behind the policy is to remove the need to list a species through early conservation efforts, thereby allowing property owners maximum land-use flexibility. The policy, with its assurances, provides that nonfederal property owners who commit to implement specifically approved conservation measures will receive "assurances" from the Services that additional conservation measures will not be imposed in the event the species is listed in the future. Such assurances will be evidenced in the property owner's agreement and in any permits issued under section 10 of the Act.[153]

Litigation under the New Policies

There has been a trickle of cases touching on evaluating agency implementation of the new policies. Despite the potential of the policies in strengthening

TABLE 4-2. Factors Used to Determine Whether Conservation Efforts Will Be Effective151

1	The nature and extent of threats being addressed by the conservation effort are described, and how the conservation effort reduces the threats is described.
2	Explicit incremental objectives for the conservation effort and dates for achieving them are stated.
3	The steps necessary to implement the conservation effort are identified in detail.
4	Quantifiable, scientifically valid parameters that will demonstrate achievement of objectives, and standards for these parameters by which progress will be measured, are identified.
5	Provisions for monitoring and reporting progress on implementation (based on compliance with the implementation schedule) and effectiveness (based on evaluation of quantifiable parameters) of the conservation effort are provided.
6	Principles of adaptive management are incorporated.

the ability of states to devise plans to stave off a placement of a species on the federal endangered species list, the cases underscore that courts will still rigorously examine the effectiveness of state regulation to determine whether the criteria to avoid listing have, in fact, been met.

The first judicial examination of the PECE and CCAA policies arose in a challenge brought by environmental plaintiffs to a USFWS decision not to list the slickspot peppergrass.[154] The slickspot peppergrass is a small annual or biennial plant in the mustard family that occurs only in southwestern Idaho. In 2001, the Service received an emergency petition to list the species. Later that year, petitioners filed an action seeking to compel USFWS to list the species. In April 2002, the litigation was settled; the Service agreed to issue a proposal to list in July 2002 and a final determination in July 2003. In July 2003, the Service announced a six-month extension of the deadline for a final listing determination.[155] In the meantime, in October 2003, the Service published a notice of availability of a Candidate Conservation Agreement (CCA) involving the slickspot peppergrass. For a number of years preceding the publication of the notice of availability, state and federal agencies intermittently coordinated conservation efforts, but work on the CCA occurred primarily from July to December 2003. The CCA was completed in December 2003, and the following month, in January 2004, the Service made a no-list decision.[156]

The Service cited the efforts to protect the slickspot peppergrass as "an outstanding example that listing a species and designating critical habitat is not necessarily the only means to achieve conservation."[157] In making a final decision regarding the proposed listing of the slickspot peppergrass, USFWS evaluated the CCA and additional conservation efforts contained in the recently updated Air Force Integrated Natural Resource Management Plan (INRMP) under the newly adopted PECE policy and determined that they met the PECE criteria. Therefore, USFWS took these conservation efforts into account when making its decision. USFWS identified wildfire, livestock grazing management, military training activities, residential and agricultural development, gravel or cinder mining, and recreational use as threats to the slickspot peppergrass, but it concluded that the CCA and INRMP reduced these threats to a level such that listing was not required. It therefore decided to withdraw the proposed rule and not to list the slickspot peppergrass as threatened or endangered.

Environmental plaintiffs challenged the no-list decision on a number of grounds, including the application of the PECE policy in this case,[158] arguing that USFWS improperly applied the PECE policy. The court, however, did not base its ruling on the PECE policy claim, since it had already determined that USFWS had violated the ESA by misapplying the term "foreseeable future" and improperly creating a new standard, "high risk of extinction."[159] The court nevertheless felt compelled to provide nonbinding commentary on the PECE policy and the manner of its implementation.

> *... the Court does not want to discourage or undermine conservation agreements undertaken outside the auspices of the ESA. The Court commends the efforts and hard work undertaken by the many different parties involved in developing the Candidate Conservation Agreement for the slickspot peppergrass. Such efforts are laudable and can be very effective in arresting the decline of a species in need of monitoring. At a minimum, it would seem that an appropriate time for the implementation of such agreements would be when a species is first designated as a category 2 candidate species, which warrants careful monitoring, or even sooner. As this case demonstrates, formalized conservation efforts should be implemented before a species stands so near the precipice of extinction. The Court hopes the federal and state agencies and various private groups continue to work cooperatively to conserve the slickspot peppergrass within the mandate of the ESA, and the CCA can serve as a solid foundation upon which to build additional protection provided by the ESA.*[160]

The court remanded the matter to USFWS.[161]

The PECE policy was also considered, albeit rather incidentally, in a case involving a challenge to a USFWS "not warranted" 12-month finding on a petition to list the Greater sage grouse as threatened or endangered.[162] In that case, the Court found the USFWS decisionmaking process flawed in three ways: 1) by not allowing the experts it convened to opine regarding the listing determination; 2) by failing to create an adequate record of the experts' opinions of the USFWS team; and 3) by ignoring those parts of the experts' opinions that were preserved in the record. The court noted that at the time of the 90-day finding, USFWS had concluded that there were no state regulations that conserve sage grouse habitat or encourage such efforts on private land, and that for the most part, federal programs did not exist.[163]

Nevertheless, in making its 12-month determination USFWS, in evaluating the adequacy of existing regulatory mechanisms pursuant to 16 U.S.C. 1533(a)(1)(D), seemingly ignored this conclusion and significant information gaps, instead concluding that conservation efforts were likely to moderate the loss of grouse habitat. As a result, the court held that FWS had failed to "coherently consider" the adequacy of existing regulatory mechanisms, and its decision was thus arbitrary and capricious. With regard to the agency's consideration of the adequacy of existing regulatory mechanisms, the court stated:

> *Lacking data on existing programs, and deciding not to review the PECE conservation programs, the Director must have been relying on his assumptions about future conservation efforts; assumptions that "cannot be relied upon in an agency's decision not to list." See Trout Unlimited v. Lohn, 2007 WL 2973568 at 25 (D. Or. July 13, 2007).*[164]

How to Assess Litigation Risk in the Post-PECE World

The cases described above demonstrate that efforts to avoid listing yielded some modest—and, in some cases, transient—successes, but that there were also significant failures. The listing agencies have, in the formulation and issuance of PECE, spent some time carefully considering and addressing, in the abstract, the problems courts have identified with state conservation agreements and how to best avoid these problems in future negotiations with states regarding conservation agreements that could have a bearing on new listing decisions. The challenge now is to formulate and implement specific cooperative conservation efforts that are adequate to protect at-risk species and thereby avoid listing. Both the federal government and the states must focus greater attention and resources on the challenges that they are likely to face if they want to significantly increase the likelihood that no-list decisions, based on conservation agreements, will be upheld by the courts. If they enter into this exercise seriously, it could well provide an avenue for fruitful future partnership. Despite past adverse decisions in a number of district courts, there have been no appellate decisions giving a binding interpretation of law to the agencies in this area. Now, with a formal agency policy to which courts should give some degree of deference, states and the federal wildlife agencies are better positioned to develop and defend conservation agreements and no-list decisions.

The value of a clean litigation slate before a challenge on the merits of a no-list decision cannot be underestimated. Most of the cases described above began with court actions in which plaintiffs brought one or more successful actions alleging that USFWS had missed mandatory ESA section 4 deadlines. While perhaps it should not matter as an abstract legal matter, failure of a federal agency to meet mandatory federal statutory deadlines cannot help but set the stage in a negative manner for further judicial review of future or voluntary measures embarked upon by the same states and federal agencies acting in concert. Courts might be able to overcome disappointment in an agency that fails to meet a deadline, but generally, failure to meet a deadline provides courts with a basis (whether fair or not) to scrutinize an agency's other actions.

In addition, as in many other contexts, last-minute deals are likely to elicit skepticism from interested stakeholders and the judiciary. Greater scrutiny has been given to conservation agreements formulated and implemented immediately before a final decision not to list than to those formulated in advance of the submission of a petition to list. Further, last minute deals are less likely to be adequately supported by the administrative record.

Now that environmental plaintiffs are not willing to let missed section 4 deadlines slip for years, or even months, it becomes increasingly difficult for states and federal agencies to sew up substantive conservation agreements before the proposed or final listing deadline. This timing issue was discussed in the comments to and agency responses to the PECE publication.[165] States and federal agencies should review candidate lists early and focus resources and set deadlines before ESA requirements are triggered to avoid the substantial challenges associated with last-minute implementation.[166] This, in turn, will require discipline, but it will greatly enhance the likelihood of successfully defending a challenge to a no-list decision. Regardless of the timing of formulation of a conservation agreement, such agreement is more difficult to defend if a proposal to list has already been made. The agency's burden then becomes rebutting its own proposal rather than merely supporting an affirmative proposal. This underscores the importance of timing. Conservation agreements, state regulatory changes, and cooperative state–federal measures are better entered into before a species is proposed for listing than after.

Additionally, regardless of when a final decision not to list a species is made, it is imperative to build an adequate administrative record of such decision. The record should reflect deliberation as to why state actions are a sufficient basis not to list as opposed to conclusory assertions lacking in analysis. It will be important to address the relevant factors and criteria for evaluating these factors specifically set out in PECE. An agency decision that includes a full discussion of the PECE factors should be accorded deference by a court; an agency decision that does not will be open to attack by plaintiffs and scrutiny by the judiciary. Criticism of a conservation measure or a no-list decision contained in the administrative record, whether raised by plaintiffs or agency personnel, must be rebutted in order to maximize the government's ability to defend a decision. If ignored or left until the last minute, or disputed without a reasonable analysis, it will be difficult to expect courts to accept a no-list decision. Given sufficient time and a willingness to candidly take on a difficult task, such criticism often can be rebutted. Courts frequently give agencies the benefit of the doubt under the APA when questions raised during the administrative process are squarely addressed by the agency (whereas perceived evasiveness by agencies leads to heightened scrutiny).

Not all potential listings can be adequately addressed by state or local conservation efforts. USFWS, NMFS, and the states must consider carefully where their scarce resources might be best deployed when working on conservation agreements. Perhaps the best use of resources is with respect to an important, geographically extensive species, like the bull trout or grizzly bear. On the other hand, perhaps species with a small geographic footprint, such as various narrowly endemic plant species, are the most sensible targets of a listing avoidance strategy.

Finally, soliciting participation and input from stakeholders in the course of formulating conservation agreements is essential and should serve to increase the likelihood that concerns raised by such stakeholders are addressed and that all stakeholders will support a no-list decision by the Services.

Opportunities for State–Federal Cooperation under Other Sections of the ESA

This chapter focuses on litigation arising under the listing provisions of section 4 that has contributed to defining the parameters of state–federal cooperative efforts. Without going into the same level of detail as with respect to the section 4 listing provisions, we next will note in passing a few cases that illustrate other parts of the ESA where the cooperative efforts of federal agencies and the states have met the requirements and advanced the goals of the Act and have been recognized and upheld by the courts.

Many similar cases have arisen under section 7 of the Act. While section 7 involves the duty of federal agencies to consult and ensure that their actions are not likely to threaten the continued existence of a listed species, in some circumstances the actions of a state are relevant to the section 7 exercise. For example, in *Selkirk Conservation Alliance v. Forsgren*, the Ninth Circuit held that USFWS may rely on final conservation agreements in determining whether an agency was complying with its section 7 obligations.[167] In *Ramsey v. Kantor*, the Ninth Circuit found that issuance of an incidental take statement under section 7 may permit parties that are neither federal agencies nor applicants (in that case, states) to engage in incidental takes without applying for section 10 permits.[168] Similarly, in *Friends of the Wild Swan*, the court upheld the validity of an incidental take statement issued for a voluntary conservation agreement entered by nonfederal parties.[169] To the extent that states and federal agencies are engaged in joint or closely related projects involving already listed species, cooperating to ensure that the scope of a section 7 consultation covers relevant state conduct could be beneficial. In such circumstances, states can play an active role in the consultation process by providing accurate and timely information to the action agency and the Service.

Finally, in evaluating reasonable and prudent alternatives formulated under section 7 to ensure that federal actions are not likely to jeopardize a listed species, courts have been willing to recognize the benefits of cooperative state–federal efforts. In *Southwest Center for Biological Diversity v. Bureau of Reclamation*, the court held that although the Secretary must rely on "best available scientific and commercial data" in formulating a reasonable and

APPENDIX 4-1. Summary of Cases

Case name	*Species*	*Citation*	*Federal/state action*	*Holding*
American Fisheries Society v. Verity	Winter-run chinook salmon	Unpublished, Civ. No. 88-0174 RAR-JFM (E.D. Cal. Feb. 24, 1989)	NMFS determined that listing was not warranted due to: (1) the existence and effective application of state and federal laws that provide adequate mechanisms to restore the species and (2) the adoption of a cooperative agreement between one state agency and three federal agencies.	The court upheld the agency's decision while finding that the cooperative agreement adopted specified that the agencies would take measures that were voluntary and extraordinary.
Save our Springs v. Babbitt	Barton Springs salamander	27 F. Supp. 2d 739 (W.D. Tex. 1997)	USFWS withdrew a proposed rule to list based, at least in part, on a conservation agreement entered into between several state agencies and USFWS.	The court struck down the agency's decision holding as follows: • On three separate occasions, the USFWS missed mandatory deadlines specified under the ESA. • The conservation agreement was entered into after the close of the comment period for the listing decision. • The conservation agreement does not include tangible steps to reduce threats to the species. • The conservation agreement was too recently concluded to determine whether it would be implemented and, if implemented, whether it would be effective.

Oregon Natural Resources Council v. Daley	Oregon coastal coho salmon	6 F. Supp. 2d 1139 (D. Or. 1998)	NMFS determined that listing was not warranted in light of the adoption of a recovery initiative by the state and the state's commitment to improve the habitat measures as part of that initiative.	NMFS cannot consider *voluntary* measures to protect a species when deciding whether to list the species.
Defenders of Wildlife v. Norton	Flat-tailed horned lizard	Unpublished, No. 97-CV-2230 W(LSP) (S.D. Cal. June 14, 1999)	USFWS withdrew a proposed rule to list following the development of a Rangewide Management Strategy and a conservation agreement among a number of federal and state agencies.	USFWS properly considered a conservation agreement executed shortly before the no-list decision. The Ninth Circuit reversed on primarily legal grounds.
Federation of Fly Fishers v. Daley	Klamath mountains ESU of steelhead	131 F. Supp. 2d 1138 (N.D. Cal. 2000)	NMFS withdrew a proposed rule to list after amending a plan to protect the species and entering into agreements with California and Oregon.	The court struck down the agency's decision holding as follows: • NMFS cannot rely on future, voluntary measures as a basis for a no-list decision. • NMFS can properly rely on federal regulatory schema as a basis for a no-list decision, but here those schema are insufficient to justify the no-list decision.
Southwest Center for Biological Diversity v. Babbitt	Short-leaved Dudleya	Unpublished, No. CV 9506471 (C.D. Cal. July 30, 1996)	USFWS withdrew a proposed rule to list based primarily on the development of a Multiple Species Conservation Plan covering a portion of San Diego County not yet in effect as of the time of the decision.	The court upheld the agency's decision holding as follows: USFWS may consider the imminent adoption of a proposed regulation as a fact in assessing the adequacy of existing regulatory mechanisms.

prudent alternative, the ESA does not explicitly limit USFWS's analysis to apolitical considerations.[170] It went on to state that if two proposed alternatives would avoid jeopardy, USFWS is permitted to choose the one that best suits all of its interests, including political or business interests.

Conclusion

Perhaps because the ESA evokes strong emotional responses from many stakeholders, efforts to amend the Act over the past quarter century have, for the most part, failed. Yet, eventually it will be revisited, and any efforts to change the Act are certain to involve revisions to the listing process. Given the uncertainties in the judicial interpretation of conservation agreements, Congress would be wise to provide guidance regarding the circumstances under which such agreements can be taken into account under a revised section 4. The PECE provides a good starting point for Congress to build upon. In the meantime, the policy will now form the backdrop for judicial evaluation of decisions not to list that are, at least in part, based on the existence and effectiveness of cooperative state–federal measures. As the court noted in the Oregon Coast coho case, "[t]he whole purpose of listing a species as 'threatened' or 'endangered' is not simply to memorialize species that are on the path to extinction, but also to compel those changes needed to save these species from extinction."[171] The more documentation state and federal agencies can acquire as to the effectiveness of how their efforts conserve and protect species, the more effective legislative or policy changes are likely to be, and the greater the likelihood courts will defer to agency actions promised in such agreements. Appendix 4-1 lists a summary of the cases.

Endnotes

1 Congress has significantly amended the ESA on only three occasions since 1973. Each time the amendments included provisions aimed at improving the listing process and balancing more equitably the conflicts between species protection and economic development. See, e.g., Act of Nov. 10, 1978, Pub. L. No. 95-632, 92 Stat. 3751; Act of Dec. 28, 1979, Pub. L. No. 96-159, 93 Stat. 1225; Act of Oct. 13, 1982, Pub. L. No. 97-304, 96 Stat. 1411.

2 437 U.S. 153 (1978).

3 *Connor v. Burford*, 848 F.2d 1441 (9th Cir. 1988), cert. denied, 489 U.S. 1012 (1989).

4 *Natural Resources Defense Council v. Houston*, 146 F.3d 1118 (9th Cir. 1998), cert. denied 526 U.S. 111 (1999).

5 *Pacific Rivers Council v. Thomas*, 30 F.3d 1050 (9th Cir. 1994), aff'g *Pacific Rivers Council v. Robertson*, 854 F. Supp. 713 (D. Or. 1993) (Forest Service Land Resource Management Plans for timber sales, range activities, and road building projects are agency actions requiring consultation).

6 *Washington Toxics Coalition v. EPA*, 413 F.3d 1024 (9th Cir. 2005), cert. denied 126 S. Ct. 1024 (2006).
7 *Florida Key Deer v. Paulison*, 522 F.3d 1133 (11th Cir. 2008); *National Wildlife Federation v. Federal Emergency Management Administration*, 345 F. Supp. 2d 1141 (W.D. Wash. 2004).
8 50 C.F.R. § 402.02 (definition of "effects of the action"); see also *National Wildlife Federation v. Coleman*, 529 F.2d 359 (5th Cir. 1976), reh'g denied, 532 F.2d 1375, cert. denied, 429 U.S. 979 (1976).
9 50 C.F. R. §§ 402.03, 402.16; see also *National Association of Home Builders v. Defenders of Wildlife*, 551 U.S. 644 (2007).
10 *Defenders of Wildlife v. Norton*, 257 F. Supp. 2d 53 (D.D.C. 2003). Bureau of Reclamation did not have a duty to consult because it did not have the discretion to release water into Mexico due to a Supreme Court injunction, the Law of the River, and international treaty.
11 *Gifford Pinchot Task Force v. Fish and Wildlife Service*, 378 F.3d 1059 (9th Cir. 2004); *Sierra Club v. Fish and Wildlife Service*, 245 F.3d 434 (5th Cir. 2001).
12 520 U.S. 154 (1997).
13 Bennett, 520 U.S. at 164; see also *Lujan v. Defenders of Wildlife*, 504 U.S. 555 (1992).
14 *Lujan v. National Wildlife Federation*, 497 U.S. 871 (1990).
15 See, e.g., *Village of False Pass v. Clark*, 733 F.2d 605 (9th Cir. 1984); *Conservation Law Foundation v. Watt*, 716 F.2d 946 (1st Cir. 1983); *North Slope Borough v. Andrus* 642 F.2d 589 (D.C. Cir. 1980).
16 See, e.g., Delta Smelt Consolidated Cases, 693 F. Supp. 2d 1145 (E.D. Cal. 2010); *Pacific Coast Federation of Fishermen's Associations v. Gutierrez*, 606 F. Supp. 2d 1122 (E.D. Cal. 2008); *Pacific Coast Federation of Fishermen's Associations v. Bureau of Reclamation*, 426 F.3d 1082 (9th Cir. 2005); *Southwest Center for Biological Diversity v. Bureau of Reclamation*, 143 F.3d 515 (9th Cir. 1998).
17 See, e.g., *Lane County Audubon Society v. Jamison*, 958 F.2d 290 (9th Cir. 1992), holding that an announcement of Bureau of Land Management (BLM) timber harvest levels for fiscal years 1991 and 1992 in Oregon is an "action" for ESA section 7 purposes; *Kentucky Heartwood, Inc. v. Worthington*, 20 F. Supp. 2d 1076 (E.D. Ky 1998), holding that a district court has jurisdiction to review ESA challenges for failure to consult on forest plans; *Pacific Rivers Council v. Thomas*, 30 F.3d 1050 (9th Cir. 1994), aff'g *Pacific Rivers Council v. Robertson*, 854 F. Supp. 713 (D. Or. 1993), holding that Forest Service Land Resource Management Plans for timber sales, range activities, and road building projects are agency actions requiring consultation.
18 See, e.g., In re: Operation of Missouri River System Litigation, 421 F.3d 618 (8th Cir. 2005) (Missouri River); *National Wildlife Federation v. National Marine Fisheries Service*, 524 F.3d 917 (9th Cir. 2008) (Columbia River). Even management of more modest river systems was, in some cases, the basis for acrimonious litigation: *Pacific Coast Federation of Fishermen's Associations v. Bureau of Reclamation*, 426 F.3d 1082 (9th Cir. 2005) (Klamath River).
19 16 U.S.C. § 1532(19).
20 50 C.F.R. § 17.3.
21 *Sweet Home Chapter of Communities for a Greater Oregon v. Babbitt*, 515 U.S. 687 (1995); see also *Palila v. Hawaii Department of Land and Natural Resources*, 649 F. Supp. 1070 (D. Haw. 1986), aff'd, 852 F.2d 1106 (9th Cir. 1988) (Palila II).

22 *United States v. Billie*, 667 F. Supp. 1485 (S.D. Fla. 1987).

23 *E.g., United States v. Town of Plymouth*, 6 F. Supp. 2d 81 (D. Mass. 1998).

24 *E.g., Strahan v. Coxe*, 127 F.3d 155 (1st Cir. 1997) (affirming a lower court decision that a State's permitting regime for commercial fishing violated the ESA's take prohibition because fishing activities resulted in the take of listed Northern Right whales); *Marbled Murrelet v. Pacific Lumber Co.*, 880 F. Supp. 1343 (N.D. Cal. 1995), aff'd as amended, 83 F.3d 1060 (9th Cir. 1996), cert. denied, 519 U.S. 1108 (1997) (holding that defendant's implementation of timber harvest plan would "harm" the endangered species as defined in 50 C.F.R. § 17.3, and thereby cause a "take" of the species in violation of 16 U.S.C. § 1538(a)(1)(B)); *Forest Conservation Council v. Rosboro Lumber Co.*, 50 F.3d 781 (9th Cir. 1995) (In action to enjoin lumber company from clear-cutting timber on private lands plaintiff showed that defendant's proposed harvest was reasonably certain to injure the endangered owls by significantly impairing their essential behavioral patterns, including breeding, feeding, and sheltering. The court held that injury to wildlife that occurs, in the past, present, or future, would satisfy the injury requirement of the Secretary's definition of "harm," including "actual injury.").

25 16 U.S.C. § 1539.

26 16 U.S.C. § 1539(a)(2)(A).

27 But see *Southwest Center for Biological Diversity v. Bartel*, 470 F. Supp. 2d 1118 (S.D. Cal. 2006); *National Wildlife Federation v. Norton*, 306 F. Supp. 2d 920 (E.D. Ca. 2004); *Sierra Club v. Babbit*, 15 F. Supp. 2d 1274 (S.D. Ala. 1998); *Friends of Endangered Species, Inc. v. Jantzen*, 760 F.2d 976 (9th Cir. 1985).

28 See 64 *Federal Register* 11,485 (Mar. 9, 1999).

29 63 *Federal Register* 8,859 (Feb. 23, 1988).

30 See *Spirit of the Sage Council v. Kempthorne*, 511 F. Supp. 2d 31 (D.D.C. 2007).

31 Pub. L. No. 97-304 § 2(a), 96 Stat. 1411 (1982).

32 *Northern Spotted Owl v. Hodel*, 716 F. Supp. 479 (W.D. Wash. 1988).

33 *Fund for Animals v. Lujan*, No. 92-800 (D.D.C. Dec. 15, 1992) (Settlement).

34 Pub. L. No. 104-6, 109 Stat. 7386 (1995).

35 61 *Federal Register* 24,722 (May 16, 1996). The guidance was last updated in 1999.

36 113 F.3d 1121 (9th Cir. 1997).

37 58 *Federal Register* 16,742, 16,756 (March 30, 1993).

38 In missed deadline cases, the government typically files a lengthy declaration from a USFWS official setting out other deadlines, including court-ordered dates, that may constrain agency action. See, e.g., *Sierra Club v. Norton*, Civ. No. CV-03-0377 (S.D. Ala. Dec. 20, 2004) (declaration of Elizabeth Stevens).

39 It should be noted that while this chapter focuses on section 4 of the ESA, sections 2 and 6 of the Act also reference federal–state cooperation. In section 2, Congress encourages the states, through federal financial assistance and other incentives, to "develop and maintain conservation programs" (16 U.S.C. § 1531[a][5]). Further, in an odd turn of the phrase, Congress states that "federal agencies shall cooperate with State and local agencies to resolve water resource issues in concert with conservation of endangered species" (16 U.S.C. § 1531[c][2]).

Section 6 deals specifically with cooperation between the states and the federal government, focusing, however, on consultation and funding regarding land acquired by the federal government or managed by a state for the conservation of endangered or threatened species (16 U.S.C. § 1535). The Secretary is authorized to enter into cooperative agreements with states regarding conservation programs that meet certain criteria, and further to provide financial assistance to states with whom it has such agreements. Section 2(f) of the Act generally provides for resolution of conflicts between state and federal law in this area. There has been little litigation under this section, and what few cases there have been are significant principally in defining the role of preemption of state authority over listed species. See, e.g., H. J. *Justin & Sons, Inc. v. Brown* 702 F.2d 758 (9th Cir. 1983), cert. denied, *H. J. Justin & Sons, Inc. v. Deukmejian*, 464 U.S. 823 (1983); *Swan View Coalition, Inc. v. Turner*, 824 F. Supp. 923 (D. Mont. 1992). If anything, this litigation has generally reinforced the primacy of the federal government under the ESA.

40 At least one court has specifically agreed with USFWS that, with respect to this factor, the inquiry is not whether existing regulatory mechanisms are adequate, but whether, alone or in conjunction with the other listing factors, they are so inadequate as to lead to the conclusion that the species is threatened or endangered. See *Center for Biological Diversity v. FWS*, 402 F. Supp. 2d 1198 (D. Ore. 2005), aff'd in part and rev'd in part on other grounds 274 Fed. Appx. 542 (9th Cir. 2008).

41 16 U.S.C. § 1533(a)(1).

42 16 U.S.C. § 1533(b)(1)(A) (emphasis added).

43 Id.

44 16 U.S.C. § 1533(b)(5)(A)(ii).

45 The agencies estimated that between 1994 and 2000, USFWS had entered into approximately 60 conservation agreements, and about 14 of the agreements had contributed, in some degree, to a determination that listing a species was unnecessary. Similarly, between 1997 and 2000, NMFS had entered into three conservation agreements, all of which it determined had contributed to making it unnecessary to list covered species. 65 *Federal Register* 36,102, 37,106-07 (June 13, 2000).

46 *American Fisheries Society v. Verity*, Civ. No. 88-0174 RAR-JFM (E.D. Cal. Feb. 24, 1989).

47 51 *Federal Register* 5,391 (Feb. 13, 1986).

48 Id.

49 In its decision that listing was not warranted, NMFS concluded that the most serious problem for the winter run was the barrier posed by the Red Bluff Diversion Dam to the salmon reaching their spawning grounds. Other threats included the operations of other dams and irrigation facilities, predation, harvesting, and pollution from abandoned mines. 52 *Federal Register* 6,041, 6,042-45 (Feb. 27, 1987).

50 Id. at 6,041.

51 Id. at 6,406.

52 Id. (emphasis added).

53 It is not possible to discern from the NMFS decision whether the agency is referring to the State Water Resources Control Board or the Central Valley Regional Water Quality Control Board, but we believe the reference is to the latter of the two.

54 Id.

55 *American Fisheries Society v. Verity*, Civ. No. 88-0174 RAR-JFM, slip op. (E.D. Cal. Feb. 24, 1989).

56 Under the cooperative agreement, CDFG agreed to assist in gravel replenishment to rehabilitate spawning habitat, to regulate the size of in-river harvesting of winter-run chinook, to participate in the efforts of a Restoration Team, and to participate in a number of cooperative studies (slip op, at 4).

57 Id. at 6.

58 *American Fisheries Society v. Verity*, Civ. No. 88-0174 RAR-JFM, slip. op. at 7 (E.D. Cal. Feb. 24, 1989).

59 Id.

60 NMFS was obliged to revisit this determination in 1994, and to list the Sacramento winter-run chinook as endangered. See 59 *Federal Register* 440 (Jan. 4, 1994).

61 *Save Our Springs, et al. v. Babbitt*, 27 F. Supp. 2d 739 (W.D. Tex. 1997).

62 61 *Federal Register* 46,608, 46,615 (Sept. 4, 1996).

63 Id. at 748.

64 In this case, USFWS had missed its statutory deadline for making a warranted but precluded finding by nearly one year (27 F. Supp. 2d at 739, 741). The agency missed its statutory deadline for publishing a proposed regulation by more than one year (Id).

65 16 U.S.C. § 1533(b)(6)(A).

66 16 U.S.C. § 1533(b)(6)(B).

67 27 F. Supp. 2d at 739, 741.

68 27 F. Supp. 2d at 741.

69 Id.

70 61 *Federal Register* 46, 608, 46,611 (Sept. 4, 1996).

71 Id. at 46,608.

72 59 *Federal Register* 7,968, 7,975-76 (Feb. 17, 1994).

73 27 F. Supp. 2d at 739, 743.

74 These violations included failure to make a timely 90-day finding under section 4(b)(3)(A); failure to make a determination as to whether listing was warranted within 12 months of the initial finding as required by section 4(b)(3)(B); and failure to meet the nondiscretionary 12-month deadline to determine either to list, withdraw the proposed listing, or to extent the period for an additional six months (27 F. Supp. 2d at 747).

75 Id. at 744.

76 Id. at 748.

77 27 F. Supp. 2d at 743.

78 61 *Federal Register* 46,608, 46,611-15 (Sept. 4, 1996).

79 These include failure to list the species under the state analog of the ESA, the ineffectiveness of state water quality rules, the absence of a legally enforceable commitment for the City of Austin to protect the salamander, and the absence of regulatory authority over a substantial percentage of groundwater pumping, as well as limited enforcement authority over groundwater pumping subject to regulation (27 F. Supp. 2d at 743).

80 Id. at 744.

81 27 F. Supp. 2d at 744.

82 Id. at 747. The court also stated that "any listing decision that considers the conservation agreement will be deemed by the Court to be arbitrary and capricious until sufficient time has elapsed to permit the Secretary to determine its effectiveness

in protecting the species. This Court considers a sufficient track record to be two years" (Id. at 747-748). The Court provides no explanation as to why two years as opposed to some other period would be sufficient, but clearly Judge Bunton believed that the lack of any track record was problematic.

83 Id. 747.

84 Id. at 744, 748.

85 62 *Federal Register* 23,377 (April 30, 1997).

86 *Oregon Natural Resources Council v. Daley*, 6 F. Supp. 2d. 1139 (D. Or. 1998).

87 Id. at 1143.

88 In June 1995, the Oregon Natural Resources Council (ONRC) filed a complaint in the Northern District of California regarding the missed deadline. Id.

89 60 *Federal Register* 38,011 (July 25, 1995).

90 6 F. Supp. 2d at 1147.

91 61 *Federal Register* 56,138, 56,211 (Oct. 31, 1996).

92 6 F. Supp. 2d at 1147.

93 62 *Federal Register* 24,588, 24,605-06 (May 6, 1997).

94 Id.

95 6 F. Supp. 2d at 1150-51, citing 16 U.S.C. § 1532(20).

96 Id. at 1151.

97 6 F. Supp. 2d at 1151.

98 Id. at 1153.

99 Id. at 1154.

100 Id. at 1154-55 (voluntary measures included in the OCSRI and mentioned by the court include the establishment of Watershed Councils).

101 The court also concluded it was proper for the Secretary to rely on existing federal regulations. Id. at 1156-57.

102 Id. at 1158-59.

103 Id. at 1160-61.

104 *Trout Unlimited v. Lohn*, 2007 WL 2973568 at 25 (D. Or.) (July 13, 2007).

105 73 *Federal Register* 7816 (Feb. 11, 2008).

106 *Defenders of Wildlife v. Norton*, No. 97-CV-2330 W(LSP) (S.D. Cal. June 14, 1999).

107 58 *Federal Register* 62,624 (Nov. 29, 1993).

108 *Defenders of Wildlife v. Norton*, No. 97-CV-2330 W(LSP), Opinion at 3-4 (S.D. Cal. June 14, 1999).

109 *Defenders of Wildlife v. Norton*, 258 F.3d 1136, 1140 (9th Cir. 2001). The restrictive measures included monitoring of lizard populations, limitation of habitat disturbance, including off-highway vehicle (OHV) use, and acquisition of private inholdings. 62 *Federal Register* 37,852, 37,854 (July 15, 1997).

110 Id. at 37,852.

111 *Defenders of Wildlife v. Babbitt*, No. 97-CV-2330 W(LSP), Opinion at 7-11 (S.D. Cal. February 25, 2000) (Order Denying Plaintiffs' Motion for Relief from Judgment).

112 Id. at 8.

113 Id. at 8-10.

114 Id. at 11.

115 *Defenders of Wildlife v. Babbitt*, No. 97-CV-2330 W(LSP), Opinion at 12 (S.D. Cal. February 25, 2000) (Order Denying Plaintiffs' Motion for Relief from Judgment).

116 Id.
117 Id. at 13.
118 After the entry of judgment by the district court, plaintiffs filed an application for relief from judgment claiming that publication of a USFWS Final Policy on Candidate Conservation Agreements after the decision of the court constituted extraordinary circumstances warranting further review. The district court denied plaintiffs' motion, finding that the Final Policy was not unexpected and, in any event, was not controlling. *Defenders of Wildlife v. Babbitt*, No. 97-CV-2330 (S.D. Cal. February 25, 2000), order denying plaintiffs' motion for relief from judgment.
119 *Defenders of Wildlife v. Norton*, 258 F.3d 1136 (9th Cir. 2001). The Court of Appeals ruling did not touch on the issue of the propriety of agency reliance on the conservation agreement itself.
120 16 U.S.C. § 1532(6).
121 Subsequently, litigation regarding the status of the species has continued. Most recently, the Ninth Circuit issued a decision in *Tucson Herpetological Society v. Salazar*, 566 F.3d 870 (9th Cir. 2009), and, in response, on March 2, 2010, USFWS issued a Federal Register notice reinstating the 1993 proposed rule to list the flat-tailed horned lizard as threatened. 75 *Federal Register* 9377 (March 2, 2010).
122 *Federation of Fly Fishers v. Daley*, 131 F. Supp. 2d 1158 (N.D. Cal. 2000).
123 *See* 131 F. Supp. 2d at 1160.
124 See Id. at 1167.
125 Id. at 1168. The court noted that the Steelhead Supplement operates as a voluntary-based incentive program rather than like a regulatory scheme.
126 See Id. at 1159.
127 Id. at 1170.
128 131 F. Supp. 2d 1158, 1167.
129 Id. at 1169.
130 Id.
131 *See also Biodiversity Legal Foundation v. Babbitt*, 943 F. Supp. 23 (D.D.C. 1996) (determination of USFWS not to list the Queen Charlotte goshawk could not be based on speculative future regulations of the Forest Service, but only on current Forest Service plan). But see *Friends of the Wild Swan, Inc. v. FWS*, 945 F. Supp. 1388 (D. Ore. 1996); it was arbitrary and capricious for USFWS to conclude that listing the bull trout was warranted but precluded, when, at least in part, it based its decision on "its own speculation" regarding the future effects of federal agency management plans. In fact, the court found that USFWS could not take the actions of other federal agencies into account at all during the listing process.
132 131 F. Supp. 2d at 1170.
133 *Southwest Center for Biological Diversity v. Babbitt*, No. CV 9506471, slip op. (C.D. Cal. July 30, 1996).
134 58 *Federal Register* 51,302 (Oct. 1, 1993).
135 Id.
136 Id. at 51,308.
137 *Southwest Center for Biological Diversity v. Babbitt*, No. CV 9506471, slip op.
138 61 *Federal Register* 52,402 (Oct. 7, 1996).

139 *Southwest Center for Biological Diversity v. Norton*, No. 97-0593 (RWR), slip op. at 8 (D.D.C. December 12, 2001). In fact, the MSCP was finalized before the final decision was issued in this case.
140 *Southwest Center for Biological Diversity v. Norton*, No. 97-0593 (RWR), slip op. at 9 (D.D.C. December 12, 2001).
141 Id. at 11, citing 61 *Federal Register* 52,402 (Oct. 7, 1996).
142 Id. at 12.
143 65 *Federal Register* 37,102 (June 13, 2000).
144 68 *Federal Register* 15,100 (March 28, 2003).
145 Id. at 15,100; 15,101; 15,106.
146 The policy defines agreements and plans to include "conservation agreements, conservation plans, management plans or similar documents approved by Federal agencies, State and local governments, Tribal governments, businesses, organizations, or individuals" (Id. at 15,112).
147 68 *Federal Register* 15,100, 15,114-15 (March 28, 2003).
148 68 *Federal Register* 15,100, 15114-15 (March 28, 2003).
149 Id. at 15,115.
150 64 *Federal Register* 32,706, 32,726 (June 17, 1999).
151 Id. at 32,727.
152 *Western Watersheds Project v. Foss*, 2005 U.S. Dist. LEXIS 45753 (August 19, 2005) (D. Idaho).
153 68 *Federal Register* 42,666 (July 18, 2003).
154 69 *Federal Register* 3,094 (Jan. 22, 2004).
155 See Statement of Julie MacDonald, Deputy Assistant Secretary for Fish and Wildlife and Parks, Department of the Interior, Field Oversight Hearing on the Endangered Species Act, Senate Committee on Environment and Public Works (Aug. 20, 2004), as found at: http://www.epw.senate.gov/hearing_statements.cfm?id=225434 (accessed May 11, 2011).
156 Western Watersheds also challenged two aspects of the manner in which USFWS evaluated the definition of "threatened," i.e., in its definition of "foreseeable future" and its requirement of "high risk of extinction" in determining whether a species is "threatened."
157 2005 U.S. Dist. LEXIS 45753, at 49-50.
158 Id. at 51-52.
159 On remand, USFWS decided not to list the species, in spite of the findings of all members of an expert panel appointed by the agency that that species would likely go extinct in the Snake River Plan portion of its range in 80 years or less and the recommendations of two groups of USFWS managers to list the species. USFWS made its decision based on information not available to the expert panel or managers' groups, but it did not explain how that new information led the agency to set aside the experts' findings and managers' recommendations. Therefore, the court again invalidated the USFWS determination not to list. *Western Watersheds Project v. Kempthorne*, 2008 U.S. Dist. LEXIS 44298 (D. Idaho 2008). Then, in October, 2009, USFWS issued a final decision listing the species as threatened throughout its range. 74 *Federal Register* 52,013 (Oct. 8, 2009).
160 *Western Watersheds Project v. U.S. Fish and Wildlife Service*, 535 F. Supp. 2d 1173 (D. Idaho 2007).

161 Id. at 1178.

162 Id. at 1187.

163 68 *Federal Register* 15,100, 15,103 (Mar. 28, 2003).

164 Landscape-scale multiple species plans, in particular, provide an opportunity to address risks to both listed and unlisted species. For example, the Western Riverside Multiple Species Habitat Conservation Plan covers 146 species, but only 25 were federally listed as of the time USFWS approved the Plan. See http://www.wrc-rca.org/Permit_Docs/Federal%20Fish%20and%20Wildlife%20Permit%20-%20TE-088609-1.pdf (accessed January 19, 2011).

165 336 F.3d 955 (9th Cir. 2003).

166 96 F.3d 434 (9th Cir.1996).

167 168 F.3d 498 (9th Cir. 1999) (table).

168 143 F.3d 515, 522 fn.5 (9th Cir. 1998).

169 6 F. Supp. 2d at 1152.

170 143 F.3d 515, 522 fn.5 (9th Cir. 1998).

171 6 F. Supp. 2d at 1152.

CHAPTER 5

The Evolution of Federalism under Section 6 of the Endangered Species Act

Robert P. Davison

The statutory basis for the relative roles of the federal government and the states in implementation of the Endangered Species Act (ESA) was established largely with enactment of the 1973 law. Most of the provisions in the current ESA concerning responsibilities and involvement of states and their relationships with federal agencies first appeared in the ESA of 1973 and have remained unchanged over the last 37 years.

Legislative History of Section 6 of the ESA

In July 1973, the Senate Committee on Commerce reported legislation (S. 1983) that provided for cooperation with the states (section 6) in carrying out the ESA. Section 6 of S. 1983 built on the provisions of the same section in H.R. 37, which had been introduced in the House of Representatives six months earlier. The Senate Commerce Committee report accompanying S. 1983 stated, "While the Federal government should protect such species where States have failed to meet minimum Federal standards, it should not preempt efficient programs. Instead, it should encourage these, and aid in the extension or establishment of others, to facilitate management by granting regulatory authority and making available financial assistance to approved schemes." Section 6 was described as establishing "mechanisms through which the Federal government and the governments of the States can work fruitfully together toward the mutually accepted goal of protection of endangered and threatened species" (Commerce Committee 1973).

The section 6 provisions of the Senate Commerce Committee bill a) directed the Secretary (of the Interior or Commerce, as appropriate) to cooperate to the maximum extent practicable with the states; b) authorized agreements between the Secretary and a state to manage areas for protection of listed species; c) authorized financial-aid grants to states for programs that the Secretary determines are adequate and active; d) set guidelines for allocation of funds among eligible states; e) required annual review by the Secretary of determinations concerning state programs; and f) made it clear that the legislation did not limit the power of any state to adopt provisions more restrictive than or consistent with its provisions. Section 6(f) also voided any state law or regulation with respect to the import or export of, or the interstate or foreign commerce in, a listed endangered species to the extent that the state law or regulation permitted what S. 1983 prohibited or barred what the bill authorized.

S. 1983, as reported, also included provisions (section 16) that specifically addressed the applicability of the legislation to the states. These provisions a) authorized a state to establish a plan for threatened and endangered species if it met or exceeded the requirements set forth in section 6(c) and represented "an effective response to the Nation's need to conserve those species"; b) required the Secretary to determine whether a state-submitted plan was in accordance with the bill's provisions; c) required periodic review of state performance by the Secretary; and d) provided that the bill's provisions would become applicable in their entirety—except with respect to species listed in Appendix I of the Convention on International Trade in Endangered Species (CITES)—within a state 15 months after enactment unless, prior to that date, the Secretary had made a determination that the state had established a satisfactory plan (Commerce Committee 1973). In addition, section 16(d) provided that at any time the Secretary determined that a state has an acceptable plan, the bill's provisions regarding the management and taking of any species would cease to be applicable or in effect within that state.

Senator John Tunney of California, who managed Senate consideration of S. 1983 for the majority, described the effect of sections 6 and 16 with respect to state programs as follows: "Subject to the provisions of this Act which provide maximum protection for species on the brink of extinction, States with active endangered species programs are given full discretion to manage threatened species which reside within their boundaries" (*Congressional Record* 1973a). Senator Ted Stevens of Alaska, the minority floor manager, described sections 6 and 16 as "the major backbone of the Act." Senator Stevens continued, "This bill will assist those States not yet involved to implement [endangered species] programs and will, if the States do not, provide for Federal preemption" (*Congressional Record* 1973a). Senator Stevens also successfully offered an amendment to section 16(d) that granted the Secretary discretion in implementing the prohibitions against taking in situations in which a particular state does not prevent the taking of species listed as endangered and does not provide adequate protection for the species.

1973 House ESA Bill

The section 6 provisions concerning cooperation with states in the bill (H.R. 37), reported by the Committee on Merchant Marine and Fisheries and approved by the House of Representatives in September 1973, differed very little from those in the Senate-passed bill. As introduced, section 6(c) of H.R. 37 authorized the Secretary "to delegate to a State the authority to regulate the taking by any person of endangered species or threatened species of resident fish and wildlife when he determines that such State maintains an adequate and active program" (CRS 1982). However, the version approved by the Merchant Marine and Fisheries Committee and the full House deleted language that allowed delegation of authority to regulate take. Instead, section 6(c) of H.R. 37 granted authority and issued direction to the Secretary to enter into cooperative agreements to provide financial assistance to qualifying states. More significantly, H.R. 37 contained none of the provisions in section 16 of the Senate-passed bill, or any comparable language, which spelled out the applicability of the new law to the states. In addition, like the Senate-passed bill, H.R. 37 contained the federal preemption of state authority in section 6(f) with respect to the import or export of, or the interstate or foreign commerce in, a listed endangered species (MMF Committee 1973).

Representative John Dingell of Michigan, who introduced H.R. 37, said that he and his colleagues had "given this troublesome question of State and Federal relationships a great deal of thought and consideration." He described the bill as one that provided "a mechanism whereby the State and Federal governments may develop cooperative agreements governing the research, management and/or enforcement of regulations involving endangered or threatened species." It makes far more sense, Representative Dingell said, "to provide an existing framework of rules and then to encourage the States to work cooperatively with the Federal Government in developing regulatory procedures which are consistent with that framework" (*Congressional Record* 1973b).

Confusing Reconciliation of 1973 House and Senate Bills

The Conference Report (1973) approved by Congress in December 1973 largely reconciled minor differences between the House and Senate section 6 provisions, with one exception. The essence and some of the language of section 16(d) of the Senate-passed bill was incorporated into the bill as a new section 6(g), titled "Transition." The resulting provision stated that the federal take prohibitions in sections 9(a)(1)(B) and 4(d) of the ESA do not apply with respect to any resident endangered or threatened species, within any state which is a party with the Secretary to a section 6 cooperative agreement (other than those species listed in Appendix I to CITES or otherwise specifically covered by any other treaty or federal law). In addition, to the extent that taking is contrary to state law, it remains subject to the federal taking prohibition. In addressing only take, section 6(g) appears somewhat narrower in scope than section 16(d) of the Senate-passed bill, which addressed the Act's provisions regarding both "the management and taking" of resident species.

In the Joint Explanatory Statement of the Conference Committee (*Conference Report* 1973), the conferees explained how they had reconciled the House and Senate provisions regarding the states as follows:

> *The House placed the fundamental responsibility for establishing and overseeing an endangered species program in the Federal government, but provided for the development of cooperative programs upon agreement with the agencies concerned. The Senate provided for cooperative state and Federal programs but gave the initial responsibility to the states in Section 16 of its bill. As finally approved, the Act will have the effect of giving the states fundamental roles with regard to resident species for a given period of time (15 months, or 120 days after adjournment of the first legislative session of any state which commences after passage of the Act). The conferees hope that this device will impel the states to develop strong programs to avoid the alternative of Federal preemption. Following the establishment period, during which it is expected that the states will adjust to the new Federal program, the law will apply as it would have under the House bill. Where cooperative agreements are in force, these will of course direct and control the enforcement of endangered and threatened species programs. Where none are then in effect, it will be the responsibility of the states to develop workable programs and to secure cooperative agreements with the Secretary.*

During Senate consideration of the Conference Report, Senator Stevens also described the House bill as one that "placed basic responsibility for establishing and administering the endangered species program in the Federal Government, and provided for the development of cooperative programs with concerned agencies." The Senate, he said, "accepted this and added a new Section 16 onto the bill to shift basic responsibility back to the States." The compromise forged among the conferees, according to Senator Stevens, was approval of "a section giving the States the fundamental roles regarding resident species for up to 15 months, or 120 days after the relevant State legislature has adjourned. This, it is hoped, will encourage the States to develop their own strong programs" (*Congressional Record* 1973c). Representative George Goodling, during House consideration of the Conference Report, described the product similarly and added, "Where cooperative agreements have been entered into, they will control. We are confident that the States will take advantage of this opportunity to avoid Federal preemption" (*Congressional Record* 1973d).

In general, the compromise adopted by Congress in 1973 seemed designed to reconcile two realities they described. The first was the broadly held view that a lack of coherent national and international policies would result in a series of unconnected and disorganized policies and programs by various states detrimental to protection of endangered species. The second generally held tenet was that a program designed to protect endangered species must recognize that there are many times more enforcement agents in the state system than there are in the federal government and, therefore, that the states are far better equipped to handle the problems of day-to-day enforcement than the federal government. The compromise, however, was an imperfect reconciliation because subsections (f) and (g) of section 6 seemed irreconcilable at best and antagonistic at worst. Section 6(g)(2) and section 4(d) of the new law applied federal prohibitions on take of species not listed under Appendix I of CITES, within a state that is party to a cooperative agreement, only to the extent that they also have been adopted by the state. The language of section 6(f), on the other hand, preempted state laws or regulations concerning import, export, or interstate or foreign commerce of federally listed species to the extent they are less restrictive than the ESA.

Regulatory Interpretation of 1973 Section 6 Provisions

Regulations to implement section 6 of the ESA were proposed by the United States Fish and Wildlife Service (USFWS) in April 1975, which coincided with the end of the section 6(g) establishment period (U.S. FWS 1975a). Almost two years after President Richard Nixon signed the 1973 ESA into law, USFWS published a final rule (50 CFR 81.1-81.14) that generally tracked the statutory language of section 6 (U.S. FWS 1975b). Most of the content of the regulations specified the requirements that states would have to meet to receive federal financial assistance for projects. No mention, however, was made of the section 6(c) statutory requirement for the Secretary to make a determination whether a state-submitted program is in accordance with the ESA within 120 days of receipt of such program from the state. More significantly, the regulations promulgated to carry out sections 6 and 9 did not address the application of ESA taking prohibitions during the transition or "establishment" period, which would have expired in any case. Nor did the regulations address the effect of cooperative agreements on federal taking prohibitions under sections 4(d) and 9(a)(1)(B), which appears to have been contemplated by section 6(g) of the ESA.

Narrow Interpretation of Section 6(g)

The effect of section 6(c) cooperative agreements on take of threatened species was addressed in amendments to the section 9 regulations (50 CFR 17.31) in 1975, but only to the extent that it allowed take for scientific or conservation purposes by employees or agents of state conservation agencies operating under such agreements (U.S. FWS 1975c). A similar change was made in 1976 to the section 9 regulations concerning take of endangered species (50 CFR 17.21) to allow, except in four specific situations, authorized state personnel operating under a cooperative agreement to take endangered species for conservation purposes without an individual permit (U.S. FWS 1976). USFWS stated that it believed states operating under approved cooperative agreements "should be accorded a greater degree of flexibility" in the conservation of endangered species, and that "the degree of continual supervision" otherwise required "is unnecessarily burdensome for a responsible state program" operating under these agreements (U.S. FWS 1976a). Of greatest interest was what was left unstated. To the extent that section 6(g)(2), which had been the subject of much debate and interest in Congress two years earlier, was addressed at all, it was given a very narrow interpretation of exempting state employees/agents from section 9 and 4(d) requirements for scientific and conservation purposes.

Creation of Project-based Section 6 Cooperative Agreements

While the regulations issued subsequent to enactment of the ESA of 1973 did not address the provisions of section 6(g), they did construct "project agreements," which were not provided for explicitly by statute, but might reasonably be inferred from it. Project agreements, as defined in the 1975 regulations and subsequently, are statements of a substantial undertaking to conserve various listed species, the benefits derived, cost of actions, and costs to be borne by the federal government and the states (50 CFR 81.1). In this way, the contents of project agreements are defined in the same manner as they are for cooperative agreements under section 6(d)(2) of the ESA and, until 1976, under 50 CFR 81.3. In April 1976, this duplicative regulatory language was resolved by dropping the language that spelled out the contents of cooperative agreements, notwithstanding the explicit language of the ESA's section 6(d)(2) (U.S. FWS 1976b). As a result, part of the meaning given to cooperative agreements by the 1973 ESA was given instead to project agreements. Project agreements under the rules adopted in 1975 and 1976 became the basis on which states with approved cooperative agreements could qualify for section 6 financial assistance.

The ESA of 1973 appeared to contemplate cooperative agreements as a mechanism to assist in the development and implementation of state programs to conserve species. Funding to "assist in the development" of those programs was contingent only on the existence of a cooperative agreement. In contrast, the 1975 section 6 regulations made funding contingent both on the existence of a cooperative agreement and specific project agreements. The creation of two tiers of approval appears to have shifted the emphasis of section 6 from funding the implementation and development of a state program for conservation of threatened and endangered species to funding specific state activities related to the conservation of these species. In the process, the new requirements for project agreements established greater federal oversight of state program funding than may have been intended by Congress.

The 1975 and 1976 regulatory provisions on project and cooperative agreements combined with the silence of these provisions with regard to section 6(g)(2), which concerns the applicability of ESA taking prohibitions within states that are parties to cooperative agreements, either reflected or resulted in a narrow interpretation of section 6. Under this interpretation, which persists today, the primary function of section 6 was to provide project-specific funding to states with qualified programs. There was some disagreement with the 1975 section 6 regulations, although it is not clear that it had to do with the interpretation of section 6 presented in these regulations. Thirty-one states and four other agencies commented on the rulemaking proposal. USFWS reported that 17 of these states and one agency had no objection to the rule. All but one of the remaining relevant comments "were for changes that would depart from the clear meaning of the Act" (U.S. FWS 1975b). The changes sought were not described in the preamble to the final rule. However, the only change made to the proposal was a single insertion of the word "of."

1976 Amendments

If there was substantial disagreement concerning the regulations implementing cooperation with states in the ESA of 1973, it was not reflected in the discussion or language of the amendments enacted in 1976. Witnesses at Congressional hearings indicated the chief limits to greater progress in meeting the obligations imposed by the ESA were "inadequate staff and inadequate budgets in both the Commerce and Interior Departments' endangered species offices"(CRS 1982). In addition to extending and increasing the authorization to fund Interior and Commerce department ESA activities, amendments adopted in 1976 allowed the Secretary to react more quickly to emergency needs of resident endangered species and issue permits for limited disposal of pre-1973, legally obtained endangered species parts and products. Concerns were expressed by witnesses and members of Congress about the rate of listing and critical habitat designation and fulfillment of CITES obligations, but no concerns were raised about the role of states. In transmitting draft legislation to the Congress, Assistant Secretary of the Interior Royston Hughes wrote that major steps taken in implementing the ESA included "guidelines on submission of cooperative agreements and model legislation for use by the states to develop endangered species laws in accordance with the provisions of the 1973 Act have been prepared and distributed" (*Congressional Record* 1976).

State Role in the ESA Less Than Envisioned in 1973

Whatever role was envisioned for the states by the Congress in the 1973 ESA, it appears to have been of greater magnitude than the one that resulted. The relative and absolute amounts of funding for states and the Interior Department authorized by the Congress in the ESA of 1973 for fiscal years 1974, 1975, and 1976 provide some idea of the magnitude Congress envisioned for the federal and state programs. Funding to states under section 6 was authorized at $10 million for the three years, while funding for the Interior Department was allowed to increase from $4 million per year to $10 million per year over that period. In the 37 years subsequent to this action by Congress, appropriations for the Interior Department's USFWS endangered species programs have grown to nearly 18 times the amount of these initial authorization levels, yet funding for states under section 6 in fiscal year 2010 had increased to only $11 million.

1977 and 1978 Amendments Increased Cooperative Agreements

Many states cited section 6 cooperative agreement impediments. The first cooperative agreement was not entered into until 1976, but by the time Congress took up legislation to extend authorization for section 6 appropriations in 1977, 17 states had reached such accords with the Department of the Interior.[1] The Department of Commerce had not yet entered into a cooperative agreement and would not for another eight years. In hearings on this legislation before Congress, the International Association of Fish and Wildlife Agencies (IAFWA) and others testified that a number of states failed to qualify for section 6 funds because they were unwilling to grant their state agency the broad authority to protect any species listed as endangered or threatened that was required by paragraphs 1) and 2) of section 6(c). Many of the 33 states without cooperative agreements, instead, had granted their agencies the authority to protect only certain categories of species such as vertebrates. Other factors cited as responsible for slow state response included lack of experience in endangered species conservation and insufficient state matching funds.

1977 Amendments Removed Obstacles to Cooperative Agreements

The House amended section 6(c) to address these impediments to development of cooperative agreements. The amendment allowed states to qualify for cooperative agreements if they satisfied some of the original requirements and had plans to devote immediate attention to those species most urgently in need of conservation programs. In conference with the Senate, the House language was modified "to assure that adoption of the House language does not have the effect of withdrawing protection for species listed by the Secretary." The conference committee "agreed to a modification which guarantees that applicable prohibitions set forth in or authorized pursuant to Sections 4(d) and 9(a)(1) of the Endangered Species Act with respect to taking of resident endangered or threatened species remain in effect when a State enters into a cooperative agreement under the alternative language of the House amendment" (*Conference Report* 1977). The concern expressed by Congress that protection of some species might be inadvertently lost when a state program became effective under the House-passed 1977 ESA amendments to section 6 appears to underscore its understanding that the takings prohibitions of sections 4(d) and 9(a)(1) cease to apply in a state when that state has entered into a cooperative agreement with the Interior or Commerce departments. As noted earlier, this interpretation had not been reflected in the section 6 regulations adopted by USFWS. Yet, while Congress eased the requirements to enter into a cooperative agreement, it made no changes to section 6(f), which effectively preempts state laws less restrictive than the ESA. Bean and Rowland (1997) note that it seems unlikely Congress would have made no changes in the language of sections 4(d) and 6(g)(2) if it had intended these provisions to take precedence over section 6(f). Although the funding authorization for section 6 remained relatively flat at $4 million annually, the amendments enacted into law in 1977 generally were described as ones that "enlarged and strengthened the role of States in the Nation's endangered species program" (*Congressional Record* 1977).

1978 Amendments Expanded Section 6(c) to Include Plants

Regulations to implement the 1977 changes to section 6 were proposed in August 1978 but were not finalized until six months after additional amendments to the ESA were enacted in November 1978. The 1978 ESA amendments expanded section 6(c) to "permit the States which develop an adequate and active program for the conservation of listed plants to receive Federal grant-in-aid in the same manner as States now receive funds for resident species of endangered and threatened fish and wildlife" (*Conference Report* 1978). In the legislative history concerning the major revisions proposed in 1978, there is no discussion concerning the role of states with cooperative agreements, although Senator Malcolm Wallop of Wyoming cited a number of "instances of less than model cooperation by the Fish and Wildlife Service in working with State game and fish agencies" (*Congressional Record* 1978).

Regulations Promulgated for 1977 and 1978 Section 6 Amendments

The combined effect of the 1977 and 1978 amendments was addressed in the final section 6 regulation published in May 1979. The rule amended existing procedures, under section 6(c) of the ESA, to give states two alternative methods of qualifying for a cooperative agreement for fish and wildlife or plants. A state and the Secretary could enter into a cooperative agreement if the state was able to demonstrate that it had authority and acceptable programs to conserve all species of fish and wildlife or plants that are listed under the ESA. By the time of the final rulemaking, 22 states had signed such cooperative agreements with USFWS on listed species of fish and wildlife. Many other states did not possess such broad authorities. The regulations implementing the 1977 and 1978 amendments made these states eligible to enter into cooperative agreements that covered only those listed species for which they had conservation authority, such as fish and wildlife vertebrates. These so-called "limited authorities" agreements were structured similarly to comprehensive cooperative agreements; they required a state to have authority to conduct investigations and establish programs concerning listed fish and wildlife or plants and provide for public participation. They also mandated that the Secretary and a state conservation agency reach agreement on the species of fish and wildlife or plants "most urgently in need of conservation programs." The final rule required the director of USFWS to consider four criteria in determining which listed species are most urgently in need of protection: 1) degree of threat; 2) recovery potential; 3) taxonomic status; and 4) other appropriate and relevant biological factors. The regulation's preamble makes clear that these criteria are set forth only as "the factors to be relied upon by the Federal Government in negotiations with the States" (U.S. FWS 1979). The rules allowed states to utilize any combination of available cooperative agreement alternatives.

The final 1979 regulations incorporated essentially verbatim the language of the 1977 amendments that continued to apply sections 4(d) and 9(a)(1) take prohibitions in cases in which a state entered into a limited authorities cooperative agreement with the Secretary. Conforming changes were made to the section 9 regulations to limit take by state employees or agents of endangered species (50 CFR 17.21) or threatened species (50 CFR 17.31) to those species covered by a cooperative agreement. Thus, issuance of the 1979 section 6 regulations solidified the Interior Department's interpretation of the ESA—that the effect of cooperative agreements on the applicability of sections 4(d) and 9(a)(1) was limited to largely exempting those who work for states from individual permit requirements in taking listed species for conservation purposes. This extension of the Interior Department's 1975 section 6 interpretation went unchallenged. According to the final rulemaking preamble, all but two of 34 comments received from state or regional organizations expressed unequivocal support for the rules. Concerns of the other two organizations were addressed with amendments to the original proposal.

1980 Amendments Increased Funding Authorization

By 1980 the House Committee on Merchant Marine and Fisheries reported that the 1977 amendment and ensuing regulations had been "dramatically successful" (MMF Committee 1980). Thirty-three states had established their own endangered species programs and signed cooperative agreements with the Interior Department. Agreements were expected in another six states and territories.[2] Acting Assistant Interior Secretary David Hales reported significant accomplishments achieved in cooperation with states. These advances included state acquisition of critical areas, location and protection of peregrine nests, work on egg shell thinning in Colorado, annual survey of brown pelican colonies in Florida, and cooperative efforts to rebuild populations of pelicans, peregrines, and bald eagles in five states (MMF Committee 1980). To expand on this initial success, the IAFWA testified to the critical need of state fish and wildlife agencies for greater endangered species funding through section 6 and suggested that the existing two-to-one federal/state matching ratio be changed to three-to-one. Congress accepted the first of these recommendations, and it again amended section 6 of the ESA in 1980 to extend an authorization of $12 million for fiscal years 1981 and 1982. In doing so, members of Congress reiterated the importance of the section 6 program to fulfillment of ESA goals. The House floor manager of the 1980 bill, Representative John Breaux, noted for example, "it is absolutely vital that the State fish and wildlife agencies, which have the primary authority in this country for the management of resident species, become more involved in the endangered species program" (*Congressional Record* 1980).

Although Congress continued to describe section 6 as allowing "the return of the management of endangered species to the individual State, along with Federal financial assistance, once the State has adopted an endangered species program which is consistent with, and not weaker than, the Federal program," it seemed satisfied with the respective federal and state roles that had been adopted since 1973 (MMF Committee 1980). Indeed, since the ESA floor debate in 1973, no evidence of concerns appears in the legislative history concerning the Interior Department interpretation of the role of states in ESA implementation. The concerns expressed by Congress with implementation of the section 6 program in the seven years following passage of the 1973 ESA instead addressed primarily the extent of state involvement (the number of states with cooperative agreements) and secondarily the level of funding made available for states to conduct conservation activities.

1982 Amendments Focused on Federal Funding Share

In 1982, the ESA was amended substantially, and funding was again reauthorized. Section 6, however, saw only a single modification: the federal/state matching ratio was changed from two-to-one to three-to-one. According to the Senate Committee on Environment and Public Works, the purpose of changing federal matching share is to make the endangered species grant program consistent with the federal aid programs for wildlife and sport fisheries. "It is appropriate," the Committee wrote, "that the Federal Government provide the same support and encouragement to States for endangered species programs" (EPW Committee 1982). It seems that by 1982, if not long before then, that Congress viewed section 6 as a program comparable in scope and effect to the Pittman-Robertson and Dingell-Johnson federal aid programs. Of greatest concern that year were the budget requests of President Reagan's Interior Department, which sought to zero-out funding for section 6. The House Committee on Merchant Marine and Fisheries wrote, "the Committee notes with disapproval efforts to abolish this aspect of the endangered species program. Legislative history of the Act unmistakably shows that Congress intended that a cooperative, Federal–State relationship be initiated and sustained. Efforts by the States to restore endangered species have met with marked success" (MMF Committee 1982). Changes to the section 6 regulations to reflect the 1982 amendments were made in 1984, along with provisions to require state audits at least every two years (U.S. FWS 1984).

1988 Amendments Broadened Scope and Sought Stable Funding

The last legislative changes to section 6 came with the 1988 reauthorization and amendment of the ESA. These changes sought to involve states as partners in newly instituted requirements to monitor the status of candidate and recovered species by allowing funds to be allocated under section 6(d) for these purposes. However, no modifications to the section 6 regulations (50 CFR 81.4) have ever been proposed or adopted by the USFWS to incorporate importance of monitoring the status of candidate and recovered species within a state as criteria to be considered in the Interior Department's allocation of section 6 funds.

Forty-six states and three territories had cooperative agreements by 1987, but the Senate Environment and Public Works Committee expressed the concern that "current Federal/State cooperative efforts to protect endangered species also are inadequate and are in danger of disintegrating altogether" (EPW Committee 1987). The Committee noted that the amount of money appropriated in fiscal year 1988 for matching grants to states under section 6 was roughly the same as it was in 1977, yet there were more than twice as many listed species and four times as many cooperative agreements in 1987 as there were a decade earlier. In 1977 section 6 funding provided about $200,000 for each of the 21 cooperative agreements in effect. Ten years later the amount appropriated provided only about $57,000 for each of 76 agreements. Moreover, under President Reagan, the Interior Department sought to eliminate section 6 funding in five of its seven budget requests, and the legislative proposal it transmitted to Congress to reauthorize the ESA would not have continued the authorization for section 6 grants.

The uncertainty surrounding availability of funds for states in a given year, coupled with the small sums that were made available during the 1980s, resulted in states reducing their requests for grant funds, curtailing their con-

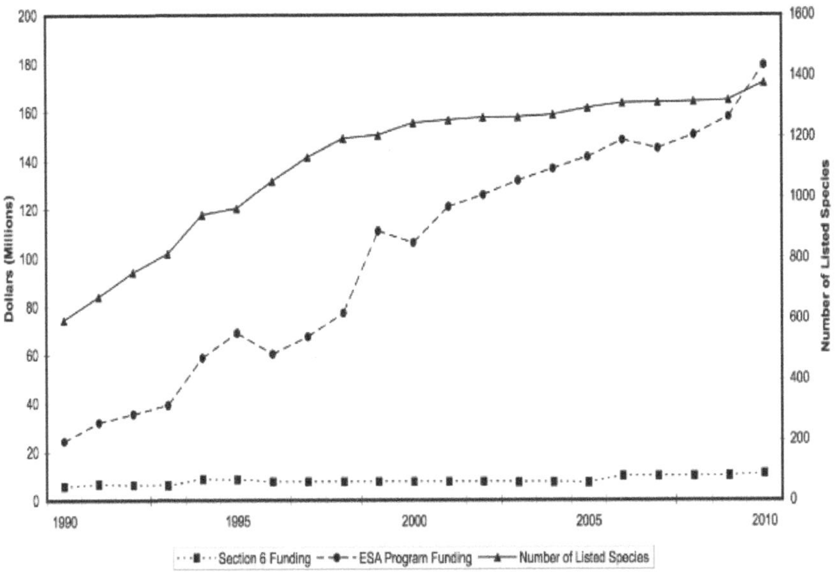

FIGURE 5-1. USFWS ESA Program and Section 6 Funding in Relation to the Number of Listed U.S. Species.

Note: For calendar years 1990 through 2009, the total number of listed species in a given year is the number listed on December 31 of that year. For calendar year 2010, the total number of listed species is the number listed on July 20, 2010. *Sources:* Calendar years 1990 through 2004: Threatened and Endangered Species System (TESS): *http://www.fws.gov/endangered/stats/cy_count_2005.pdf* (accessed February 1, 2006). Calendar years 2005 through 2009: Personal communication between David Harrelson (USFWS) and the author, July 16, 2010. Calendar year 2010: http://ecos.fws.gov/tess_public/pub/boxScore.jsp (accessed July 20, 2010).

servation activities and, in some cases, such as the State of Ohio, eliminating their requests (EPW Committee 1987). Congress also sought to address the uncertainty in funding by creating the Cooperative Endangered Species Conservation Fund, which authorized annual appropriations at an amount equal to 5 percent of the combined yearly federal aid accounts to restore sport fish and wildlife populations. Nevertheless, since 1988 the disparity between section 6 needs and funding has continued to grow. In 1990 there were 596 U.S. species listed under the ESA, and $5.7 million was appropriated for state grants under section 6. At the end of 2009, there were 1,320 listed U.S. species—more than twice the number in 1990, yet the $10.0 million provided that year had only slightly more buying power than the funds provided 20 years earlier. In comparison, the $158 million appropriated for the USFWS endangered species program (listing, candidate conservation, consultation and recovery) for fiscal year 2009 had four times the buying power of the $24.3 million spent in 1990 (Figure 5-1).

Implementation of Section 6 Cooperative Agreements

States and territories have entered into 91 section 6(c) cooperative agreements with the Interior Department (Morgan 2010). Nineteen states and three territories have entered into cooperative agreements with the Commerce Department's National Marine Fisheries Service (NMFS) for species under that agency's jurisdiction (NOAA Fisheries 2010). Every state and four territories have at least one cooperative agreement for some species of fish and wildlife or plants. Many states have multiple agreements. The State of Oregon, for example, has three cooperative agreements that cover vertebrate fish and wildlife, plants, and invertebrate species (Morgan 2010).

Interestingly, no similar development has taken place during the last 30 years in implementation of the management agreement authority created by section 6(b) of the ESA. This provision of section 6 authorizes the Secretary to "enter into agreements with any State for the administration and management of any area established for the conservation of endangered species or threatened species." Actions implementing management agreements can be funded as programs for the conservation of endangered or threatened species in states that have established a cooperative agreement with the Secretary. The intent appears to have been for the states and USFWS to assist each other in the management and administration of specific tracts of land. However, there are no section 6(b) management agreements in effect (Morgan 2010).

The basic language of section 6(c) cooperative agreements has remained almost unchanged since construction of the first one in 1976. What little variation in wording that exists among agreements is due largely to state organizational differences. Agreements establish that USFWS enters into an agreement "for the purpose of assisting in the implementation" of the state agency's endangered and threatened species program. This assistance is achieved principally by USFWS agreeing to provide financial assistance for implementation of acceptable conservation projects, contingent upon available funds, and cooperation on law enforcement activities. The state agency agrees generally to administer its program "in harmony with the terms and spirit" of the ESA. More significantly, "with respect to the importation or exportation of, or interstate or foreign commerce in," federally listed species covered by the agreement, the state agency "agrees not to permit what is prohibited" by the ESA or its regulations. The state agency further "agrees not to engage in, or issue a permit authorizing, the taking" of federally listed species, "without such taking requiring prior issuance of a permit," as may be called for by the USFWS. Exceptions to permit requirements are provided, pursuant to ESA regulations (50 CFR 17.21(b) and 17.31(b)) for take by state agency employees or designated agents related to conservation and scientific purposes. Thus, as Bean and Rowland (1997) note, in entering into cooperative agreements, the Interior Department routinely secured from states an acknowledgment that their laws are preempted to the extent that they are less restrictive than the ESA.

A diverse array of projects is eligible to be funded through section 6, including any that promote the conservation of endangered and threatened species or monitoring the status of candidate or recovered species. Projects generally are selected by USFWS endangered species staff in conjunction with the states. Funding is allocated by formula to USFWS regions based on the number of species covered by state cooperative agreements within that region. Regional offices then further allocate the funding to states within their region by formula or through a competitive process. By way of example, the $9.8 million appropriated by Congress for fiscal year 2008 was allocated by USFWS to assist states in conducting habitat restoration, species status surveys, public education and outreach, captive propagation and reintroduction, nesting surveys, genetic studies, and development of management plans.

Judicial Interpretation of Section 6

More than 20 years after Congress struck its befuddling compromise on federalism under the ESA, two federal courts addressed the extent to which the ESA preempts state authority to conserve federally listed species. In a Montana case (*Swan View Coalition, Inc. v. Turner*), environmental organization plaintiffs alleged that Forest Service road construction resulted in taking of grizzly bears and wolves in violation of the ESA (824 F. Supp. 923, D. Mont. 1992). A forest industry intervenor-defendant argued that Montana law was controlling in this case under the ESA's sections 4(d) and 6(g)(2), because the State of Montana was a party to a full authority cooperative agreement under section 6(c) of the ESA. Given that Montana law did not define take to include "harm" or "significant habitat modification," as defined under the ESA, the forest industry association maintained that the takings claim must fail. The court recognized the existence in sections 4(d) and 6(g)(2) of "express language in the ESA which seems to contemplate less restrictive state law on takings." However, the court concluded that the less restrictive takings provisions under Montana law are preempted by the ESA, "based on the clear language of §6(f) of the ESA combined with the overwhelming priority Congress has given to the preservation of threatened and endangered species." That same year, in a case involving alleged take of salmon due to water pumping, a federal court in California cited section 6(f) and concluded, "to the extent that California law on taking is less protective than the Endangered Species Act, it is preempted" (*United States v. Glenn-Colusa Irrigation District*, 788 F.Supp. 1126, E.D.Ca., 1992).

The holdings in these cases suggest that section 6(f) of the ESA nullifies any exemption from the Act's takings prohibitions provided under section 6(g)(2)(A) or section 4(d). The two courts' interpretations of section 6(f) require that a state's definition of take mirror the definition under the ESA and associated regulations. If section 6(f) requires a state's taking law to be the same as ESA takings prohibitions, neither section 6(g)(2) nor section 4(d) can make the federal prohibitions regarding taking endangered or threatened species through "significant habitat modification" inapplicable in a cooperative agreement state (Davison 1995).

Broadened Use of Section 6 Funding Authorization

In fiscal year 1997, Congress broadened, at Interior Secretary Bruce Babbitt's request, the types of projects funded through the Cooperative Endangered Species Conservation Fund. The extension of section 6, which took place through annual appropriations law rather than amendment of the ESA, established Habitat Conservation Plan (HCP) Land Acquisition Grants. The HCP Land Acquisition Grant program was intended to reduce conflicts between the conservation of listed species and land uses on specific parcels of land. Under this program, USFWS provides matching grants to states for land acquisitions that are associated with approved HCPs. According to USFWS, the HCP Land Acquisition Grant program funds land acquisitions that 1) complement, but do not replace, private mitigation responsibilities contained in HCPs; 2) have important benefits for listed, proposed, and candidate species; and 3) have important benefits for ecosystems that support listed, proposed, and candidate species (U.S. FWS 2004). Although grants are supposed to support conservation actions by the state or local governments that complement mitigation, rather than fund the mitigation required of a HCP permittee, in at least one case (Balcones Canyonlands HCP) the federal grants have, for all practical purposes, constituted much of the mitigation required of developers (Bean 2005). Congress appropriated just $6 million for the program in each of its first three years (fiscal years 1997–1999), but average annual funding rose to $46 million from fiscal year 2002 through fiscal year 2010.

In fiscal year 2001, Congress provided $6.635 million for HCP land planning assistance grants, under the Cooperative Endangered Species Conservation Fund, to assist states in development of HCPs by supporting planning activities such as document preparation, outreach, and baseline surveys and inventories. These grants are intended to encourage development of regional HCPs that allow local governments to incorporate species conservation into local land-use planning, streamline the project approval process, and facilitate economic development (U.S. FWS 2008). The Habitat Conservation Planning Assistance Grants program has allocated to states and territories an average of $7.85 million per year from fiscal year 2002 through fiscal year 2010 to support the development of HCPs.

Congress also funded, in fiscal year 2001, matching grants to states, and through them to other nonfederal entities, for acquisition of habitats in support of approved species recovery plans. These Species Recovery Land Acquisition Grants make it possible for USFWS and states to combine their resources to buy lands from willing sellers before development or other land-use changes impair or destroy habitat values essential to the recovery of listed species. To prevent duplication of the HCP Land Acquisition Grant Program—which provides substantial funding for land acquisitions associated with HCPs—the recovery land acquisition grant funds are limited to acquiring lands that are not associated with permitted HCPs. In its first year of funding, the Recovery Land Acquisition Grant program received $10.427 million. From fiscal year 2002 through fiscal year 2010, the amount allocated annually for these grants averaged $14.3 million.

Participation in the three new programs established within the Cooperative Endangered Species Conservation Fund by appropriations laws is limited, as it is for traditional section 6 grants, to state agencies that have a current cooperative agreement with the Secretary of the Interior. Often, state agencies with such agreements employ sub-grantees, such as land trusts, community organizations, or conservation organizations to implement section 6 land acquisition projects.

Evolution of State Role under Other ESA Provisions

The role of states under the ESA, of course, is not limited to that provided under section 6. Five of the other 17 sections of the ESA address state involvement or authority. The broad policy statement embracing the section 6 cooperative federal–state program is expressed in section 2(a)(5) of the ESA enacted in 1973, which declares that "encouraging the States and other interested parties, through Federal financial assistance and a system of incentives, to develop and maintain conservation programs which meet national and international standards is a key to meeting the Nation's international commitments and to better safeguarding, for the benefit of all citizens, the Nation's heritage in fish, wildlife, and plants." Section 2 of the ESA also stated as policy, "Federal agencies shall cooperate with State and local agencies to resolve water resource issues in concert with conservation of endangered species."

The ESA's section 4 requires consideration of efforts by states in making listing or critical habitat determinations. Actual notice of proposals must be given to conservation agencies in affected states. If a final regulation is issued that conflicts with state agency comments, or a regulation is not adopted in response to a state-petitioned action, a written justification must be provided due to "failure to adopt regulations consistent with the agency's comments or petition." These requirements generally were incorporated into regulations in 1984 (50 CFR 424). Section 4(g), which was added to the ESA in 1988, requires the Secretary to "implement a system in cooperation with the States to monitor effectively for not less than five years the status of all species which have recovered" and been delisted. USFWS has addressed this requirement through adoption of species-specific monitoring plans developed in cooperation with states, recovery teams, and public input.

Section 7(a)(2) of the ESA reiterates the provisions of section 4 by emphasizing that critical habitat may be designated by the Secretary only "after consultation as appropriate with affected States," but otherwise makes no reference to cooperation with states in the interagency consultation process. The section 7 regulations similarly are silent on cooperation with states (50 CFR 402.01-402.48). To the "extent feasible and appropriate," efforts to carry out the Convention on Nature Protection and Wildlife Preservation in the Western Hemisphere under section 8A(e)(2) are to be taken "with the participation of State agencies."

Section 10 of the ESA makes no mention of cooperation with states. In particular, both the provisions of the law and the associated regulations provide no role for states in habitat conservation planning. There is a certain irony to this silence given the increasingly large role played by states and local governments in carrying out HCPs and other agreements for listed and candidate species, which is described below. Notwithstanding the absence of language concerning cooperation with states in establishing experimental populations pursuant to section 10(j) of the Act, however, the regulations implementing this provision mandate consultation with "appropriate State fish and wildlife agencies" in developing and implementing experimental population rules. Moreover, they provide, "Any regulation promulgated pursuant to this section shall, to the maximum extent practicable, represent an agreement between the Fish and Wildlife Service, the affected State and Federal agencies and persons holding any interest in land which may be affected by the establishment of an experimental population" (50 CFR 17.81).

Development of Federal Agency Policy on State ESA Involvement

In 1994, the sparse language in the ESA and its regulations concerning involvement of states outside of section 6 was bolstered with issuance by USFWS and NOAA-Fisheries of an *Interagency Cooperative Policy Regarding the Role of State Agencies in Endangered Species Act Activities* (U.S. FWS 1994). This policy sought "to clarify the role of State agencies in activities undertaken by the Services" under the ESA and its associated regulations. The policy recognizes that "States possess broad trustee and police powers over fish, wildlife and plants and their habitats within their borders" and that, "unless preempted by Federal authority, States possess primary authority and responsibility for protection and management of fish, wildlife and plants and their habitats." The policy also recognizes that "[s]tate agencies often possess scientific data and valuable expertise on the status and distribution of endangered, threatened and candidate species of wildlife and plants" and that, "because of their authorities and their close working relationships with local governments and landowners, [State agencies] are in a unique position to assist the Services in implementing all aspects of the Act."

The role for state agencies described by the interagency cooperative policy is one in which the Services commit to utilizing the expertise and information of those agencies in conducting pre-listing conservation, listing, reclassification, delisting, critical habitat designation, consultation, habitat conservation planning, and recovery. Additionally, with respect to consultation, the Services commit to inform state agencies of any federal agency action that is likely to adversely affect proposed or listed species and proposed or designated critical habitat. In habitat conservation planning and recovery activities, the policy also commits the Services to solicit participation of state agencies in those activities. Overall, the policy appears directed more at making sure the Services utilize the resources of state agencies to improve administration of a federal program than it is directed at building partnerships with states to achieve common ends.

Increased State Involvement under Expanded Use of ESA Section 10

At the time the Services issued their policy on state involvement in ESA administration in 1994, a revolution already was underway in how USFWS used the Act's section 10 authority, a revolution that would have far greater implications for the role of state agencies and local governments in conservation of imperiled species. During the 1990s, administration of the ESA extended its reach at a great rate into incidental take activities that previously had received relatively little attention. From 1982—when section 10 of the ESA was amended to authorize HCPs as a means of permitting, minimizing, and mitigating incidental take of listed species caused by nonfederal activities—until 1992, only 14 HCPs had been approved (Nelson 1999). By July 12, 2010, however, USFWS had approved 667 HCPs (U.S. FWS 2010a). As the number of HCPs grew, the complexity of the plans increased, as did the number of covered species and the size of the areas. In the evolution of habitat conservation planning from a process adopted primarily to address single projects to broad-based, landscape-level planning, the role of states and local governments became far more prominent. For example, the HCP approved in 2003 for western Riverside County, California, resulted from an application by the County of Riverside, Riverside County Flood Control and Water Conservation District, Riverside County Transportation Commission, Riverside County Parks and Open Space District, Riverside County Waste Department, California Department of Transportation, California Department of Parks and Recreation, and 14 western Riverside County cities. It was developed as both an HCP and a subregional plan under California's Natural Community Conservation Planning Act (NCCP), administered by the California Department of Fish and Game (CDFG). It covers 146 species and more than 1.2 million acres. In 2002, an HCP Land Acquisition Grant provided $9 million from the section 6 Cooperative Endangered Species Conservation Fund to acquire key core habitats for the threatened coastal California gnatcatcher and endangered least Bell's vireo and Stephens' kangaroo rat.

As the USFWS greatly expanded the use of HCPs in the last decade, state and local governments increasingly drove their development. At least 17 percent of all HCPs and 53 percent of all HCPs greater than 1,000 acres in size have resulted from applications by local jurisdictions or state agencies or both (U.S. FWS 2010a). The statewide HCP developed for the Karner blue butterfly by the Wisconsin Department of Natural Resources (WDNR) provides an interesting model for states to regain primary management responsibility for listed species. The WDNR is responsible for compliance with the conditions of the federal incidental take permit and HCP implementation (Christenson and Lentz 2001). The state HCP and permit serve as an umbrella that provides incidental take authority to 26 other partner entities that have developed individual conservation agreements with the WDNR. Under this arrangement, the resident ESA-listed species remain under the jurisdiction of the state fish and wildlife agency, which is most knowledgeable about the species, its status, and its existence in the state (Christenson and Lentz 2001). The Wisconsin outcome seems close to what was envisioned by Congressional members and staff who participated in crafting the 1973 ESA, but so far, other states have not replicated this approach.

(For more on the Karner blue butterfly HCP, see Chapter 7.)

State Involvement in Safe Harbor and Candidate Conservation Agreements

State agencies and local governments have played a larger role in the innovative use of authority under section 10 of the ESA to enhance the survival of endangered, threatened, and candidate species through use of safe harbor and candidate conservation agreements. These agreements seek to minimize or remove regulatory disincentives that previously had discouraged landowners from implementing habitat management actions beneficial to imperiled or listed species. Safe harbor permits provide landowners with authority to incidentally take individuals of federally listed species that currently do not occupy their property but may in the future, because of the landowners' beneficial land management practices (U.S. FWS 1999). Therefore, the program encourages voluntary management by private landowners to provide a net benefit for listed species for some period of time, and thus promote recovery, on their lands by giving assurances to the landowners that no additional regulatory restrictions will be imposed, in the future, if the lands are returned to a predetermined baseline condition. State wildlife agencies in Georgia, Alabama, Texas, Louisiana, and South Carolina all have received federal funding, through section 6, to develop statewide programmatic safe harbor agreements (SHAs) for red-cockaded woodpeckers. The states of Texas, South Carolina, and

Georgia entered into early SHAs and received incidental take permits under the HCP authority of section 10(a)(1)(B) to promote recovery of red-cockaded woodpeckers. More recently, states such as Louisiana have entered into these agreements under section 10(a)(1)(A) and received enhancement of survival permits. In either case, USFWS issues an umbrella permit to the state wildlife agency, as opposed to individual permits to each participating landowner (Costa et al. 2001). Once overall baseline responsibilities are identified in the umbrella permits, private landowners, with assistance from state agencies if necessary, are able to fill out a relatively simple evaluation form that documents background information, the baseline, habitat maintenance and enhancement activities, expected net conservation benefits, and implementation schedule. Costa et al. (2001) report that having the state wildlife agency as the delivery system for statewide red-cockaded woodpecker safe harbor programs "has worked very efficiently and has been well received by landowners." They conclude that the lead state role under safe harbor umbrella permit agreements is a successful model, because it minimizes regulatory agency transaction costs and bureaucratic burdens to landowners. (For more on the red-cockaded woodpecker safe harbor program, see Chapter 8.) At least 27 percent of all approved SHAs (20 of 73) involve state agency applicants (U.S. FWS 2010a).

In 1999, in conjunction with establishment of SHAs, USFWS also enlarged the role of section 10(a)(1)(A) of the ESA to further the survival of candidate species through use of Candidate Conservation Agreements with Assurances (CCAAs) (U.S. FWS 1999). Candidate species are plants and animals for which there is sufficient information to propose them as endangered or threatened under the ESA, but for which development of a proposal is precluded by other higher priority listing activities (U.S. FWS 2010a). These agreements offer regulatory assurances as an incentive for private and other nonfederal property owners to implement conservation measures. Nonfederal landowners who voluntarily agree to manage their lands or waters to remove threats are assured that their conservation efforts will not result in future regulatory obligations in excess of those they agree to at the time they enter into the CCAA. Incidental take permits are issued under section 10(a)(1)(A) that allow incidental take by CCAA participants in the course of activities that would return population levels and habitat conditions to those agreed upon.

States have more actively participated in section 10 Candidate Conservation Agreements (CCAs) and CCAAs than they have in SHAs or HCPs, which is not surprising given the primacy of state jurisdiction over most of these unlisted species. Of the 22 CCAAs approved by USFWS, at least 13 are agreements with state wildlife or natural resource agency partners (U.S. FWS 2010a). A similarly large state involvement exists in the 103 CCAs approved by USFWS without any regulatory assurances under section 10 of the ESA. Nearly 60 percent of these agreements are with state agencies (U.S. FWS 2010a). A recent example demonstrating the likely future role of states in candidate species conservation is the CCAA obtained by the Montana Department of Fish,

Wildlife, and Parks in May 2004. The CCAA is an umbrella-style agreement under which voluntary conservation activities will be implemented to benefit Westslope cutthroat trout. In conjunction with the CCAA, a permit has been issued for the future take of the species in conjunction with its recovery in Montana.

Different Direction for State ESA Role under New Agreements

In addition to rapidly expanding state leadership under section 10 of the ESA, a 2002 agreement between the Arizona Game and Fish Department (AZGFD) and USFWS Region 2, which was revised in 2008, represented a different direction for states in ESA administration (U.S. FWS 2010b). In 2007, USFWS Region 6 entered into a similar agreement with Wyoming Game and Fish Department (WGFD), although implementation of most of its provisions was made subject to the availability of sufficient state resources.

The Arizona agreement regarding state wildlife agency participation in implementing the ESA is intended "to facilitate joint participation, communication, coordination, and collaboration." Roles and responsibilities of AZGFD and USFWS are described for candidate species assessments, pre-listing recovery activities, petition management, listing (including reclassification), critical habitat designation, special rules for candidate and listed wildlife, five-year status reviews, recovery plan development and implementation, monitoring of delisted wildlife species, land and water acquisition and management, section 7 consultation, law enforcement, habitat conservation planning, and experimental populations. The agreement between the AZGFD and USFWS brings much greater specificity and sense of partnership to relative federal and state roles and responsibilities than previously forged section 6(c) cooperative agreements or the 1994 USFWS national policy on the subject.

Potentially more far-reaching is a section 6(c) agreement under development between the State of Idaho and the federal ESA agencies (USFWS and NMFS) as part of a settlement agreement resulting from the Snake River Basin Adjudication, which was approved by the United States in the Snake River Water Rights Act of 2004 (Public Law 108-447). Under the proposed section 6 cooperative agreement (IDL 2010), a voluntary Idaho Forestry Program (IFP) would be implemented to supplement existing Idaho forest practice rules by encouraging state and private landowners in the Salmon and Clearwater River basins to improve riparian and stream bank habitat on their lands for aquatic endangered and threatened species (Bogert 2006). Participating landowners would not be liable if listed species are incidentally affected due to their forest management or timber harvest activities because they would receive incidental take authority for those activities by means of consultation on the IFP section 6 cooperative agreement under section 7 of the ESA. Approval of the agreement awaits completion of an Environmental Impact Statement under the National Environmental Policy Act of 1969 and ESA section 7 consultation (Seymour 2010).

Conclusion

Whatever role for states originally was intended by those responsible for the wording of the 1973 ESA, the one that evolved in administration of the law in the 30 years since its enactment is a role that, while expanding greatly in the last decade, is secondary to the primacy of the federal role. To the extent that they are less restrictive than the ESA, state laws or regulations concerning take of federally listed species are effectively preempted by the federal law. This fact is incorporated in the terms of all section 6(c) cooperative agreements. State role in ESA administration was diminished further by the narrow regulatory interpretation of section 6, which characterized the section as a tool to fund specific, approved state projects under cooperative agreements rather than development and implementation of state programs for conservation of threatened and endangered species. Moreover, federal spending in support of state programs over the history of the ESA has been extremely low and in decline relative to spending for federal efforts. Apart from continued debate about section 6 funding, there is scant, if any, evidence in three decades of ESA legislative history that members of Congress or administration officials were sufficiently unhappy with the relative federal and state roles to raise the issue on the six occasions in which ESA amendments were discussed and adopted between 1976 and 1988.

Of greater interest than whether the role of states in ESA administration evolved as Congress intended is the manner in which federal, state, and local government roles have developed independently in the last 15 years, particularly with respect to regulation of nonfederal activities and encouragement of conservation practices on nonfederal lands. The increasing number of instances in which states have sought and obtained umbrella-style ESA section 10 permits—to effectively assume responsibility for minimizing and mitigating nonfederal incidental take activities and promoting nonfederal habitat conservation—are outcomes that Congress, at least as late as 1988, never appeared to have intended or anticipated. But they are ones, which through innovation at the state and federal level, are perhaps close to what was envisioned by many of those who participated in crafting the 1973 ESA. In similar fashion, the innovation of USFWS agreements with AZGFD and WGFD regarding state participation in administration of the ESA serve as a possible model for a new type of agreement. Revised agreements could either augment or replace existing section 6(c) cooperative agreements with ones in which state and federal roles are more clearly delineated and directed toward building a more effective partnership on behalf of imperiled species. Finally, the IFP agreement in Idaho offers a potential example of how states may be able to work with USFWS and

NMFS to develop their own programs under section 6(c) that enlist greater participation in endangered and threatened species conservation by state agencies and private landowners.

Comprehensive state wildlife conservation planning and the state wildlife grants program to support development and implementation of state plans promise a new era of state involvement in conservation of at-risk species that will continue to shape and enlarge the role of states in administration of the ESA. At a minimum, the resulting increased ability of states to collect, synthesize, and easily retrieve data on species and their habitats will increasingly make them key sources of information and expertise in virtually every ESA decision. Increasingly common efforts by states to work together to address rangewide conservation of at-risk species, such as the efforts on behalf of black-tailed prairie dogs and greater sage-grouse, also portend far greater state roles in this arena in the future.

References

Bean, Michael J. and Melanie J. Rowland. 1997. *The Evolution of National Wildlife Law.* Westport, CT: Praeger Publishers.

Bean, Michael J. 2005. Personal communication between Michael J. Bean, Environmental Defense Fund, and the author. January 13.

Bogert, Laurence Michael. 2006. "Nez Perce Water Rights Settlement Article: The Future is No Place to Place Your Better Days: Sovereignty, Certainty, Opportunity, and Governor Kempthorne's Shaping of the Nez Perce Agreement." 42 *Idaho Law Review* 673.

Christenson, Jimmy S. and David R. Lentz. 2001. "Steps for Success in Rare Species Conservation: The Wisconsin Statewide Karner Blue Butterfly HCP." Transactions of North American Wildlife and Natural Resources Conference 1: 443–455.

Commerce Committee (U.S. Senate Committee on Commerce). 1973. 93rd Congress, 1st sess., S. Rpt. 93-307.

Conference Report. 1973. 93rd Congress., 1st sess., Rpt. No. 93-740.

———. 1977. 95th Congress, 1st sess., S. Rpt. 95-607.

———. 1978. 95th Congress, 2nd sess., H. Rpt. 95-1804.

Congressional Record. 1973a. 93rd Congress, 1st sess., vol. 119, July 24.

———. 1973b. 93rd Congress, 1st sess., vol. 119, September 18.

———. 1973c. 93rd Congress, 1st sess., vol. 119, December 19.

———. 1973d. 93rd Congress, 1st sess., vol. 119, December 20.

———. 1976. 94th Congress, 2nd sess., vol. 122, March 10.

———. 1977. 95th Congress, 1st sess., vol. 123, November 30.

———. 1978. 95th Congress, 2nd sess., vol. 123, July 17.

———. 1980. 96th Congress, 2nd sess., vol. 126, May 5.

Costa, Ralph, Susan Ladd Miller, and Steven M. Lohr. 2001. "A Common Sense Approach to Conserving Rare Species and Their Habitats on Private Lands: Safe Harbor and Red-cockaded Woodpeckers—A Case Study." Transactions of North American Wildlife and Natural Resources Conference 1: 498–510.

CRS (Congressional Research Service). 1982. *A Legislative History of the Endangered Species Act of 1973, as Amended in 1976, 1977, 1978, 1979, and 1980*. Washington, D.C.: Government Printing Office.

Davison, Steven G. 1995. "Alteration of Wildlife Habitat as a Prohibited Taking under the Endangered Species Act." *Journal of Land Use and Environmental Law* 10: 155–238. www.law.fsu.edu/journals/landuse/vol102/2davison.html#FNT* (accessed Sept 10, 2010).

EPW Committee (U.S. Senate Committee on Environment and Public Works). 1982. 97th Congress, 2nd sess., S. Rpt. 97-418.

———. 1987. 100th Congress, 1st sess., S. Rpt. 100-240.

IDL (Idaho Department of Lands). 2010. Idaho Forestry Program EIS. www.idl.idaho.gov/eis/eis_index.html (accessed July 20, 2010).

MMF Committee (U.S. House of Representatives Committee on Merchant Marine and Fisheries). 1973. 93rd Congress, 1st sess., H. Rpt. 93-412.

———. 1980. 96th Congress, 2nd sess., H. Rpt. 96-896.

———. 1982. 97th Congress, 2nd sess., H. Rpt. 97-567.

Morgan, Don R. 2010. Personal communication between Don R. Morgan, Chief Branch of State Grants, Endangered Species Program, United States Fish and Wildlife Service, and the author. July 15.

Nelson, Marj. 1999. "Habitat Conservation Planning." *Endangered Species Bulletin* 24: 12–13.

NOAA-Fisheries (National Oceanic and Atmospheric Administration). 2010. Protected Species Cooperative Conservation: ESA Section 6 Grants. www.nmfs.noaa.gov/pr/conservation/states/grant.htm (accessed July 8, 2010).

Seymour, P. 2010. Personal communication between Patrick Seymour, Endangered Species Program Manager, Idaho Department of Lands, and the author. July 21.

U.S. FWS (U.S. Fish and Wildlife Service). 1975a. "Proposed Rulemaking." *Federal Register* 40:18447, April 28.

———. 1975b. "Final Rule." *Federal Register* 40:47509, October 9.

———. 1975c. "Final Rule." *Federal Register* 40:44412, September 26.

———. 1976a. "Final Rule." *Federal Register* 41:19224, May 11.

———. 1979. "Final Rule." *Federal Register* 44:31578, May 31.

———. 1984. "Final Rule." *Federal Register* 49:30074, July 26.

———. 1994. "Notice of Interagency Cooperative Policy Regarding the Role of State Agencies in Endangered Species Act Activities." *Federal Register* 59:34275, July 1.

———. 1999. "Final Rule." *Federal Register* 64:32706, June 17.

———. 2008. "Fiscal Year 2004 Cooperative Endangered Species Conservation Fund Grant Program Notice of Availability of Federal Assistance." www.fws.gov/endangered/grants/section6/ (accessed January 31, 2008).

———. 2010a. "Conservation Plans and Agreements Database." http://ecos.fws.gov/conserv_plans/servlet/gov.doi.hcp.servlets.PlanSelect (accessed July 12, 2010).

———. 2010b. Memorandum of Agreement (MOA) between the U.S. Fish and Wildlife Service and the Arizona Game and Fish Department. "Roles and Responsibilities for Implementing the Endangered Species Act in Arizona." www.fws.gov/southwest/es/arizona/Documents/ESAGuidance/AGFD_MOA.pdf (accessed Sept. 9 2010).

Endnotes

1 The 17 states were Arkansas, California, Colorado, Delaware, Florida, Maine, Maryland, Michigan, Missouri, New Jersey, New Mexico, New York, South Carolina, South Dakota, Virginia, Washington, and Wisconsin.

2 The 14 states without cooperative agreements in 1980 were Alabama, Arizona, Connecticut, Indiana, Kentucky, Louisiana, North Dakota, Oklahoma, Oregon, Rhode Island, Texas, Vermont, West Virginia, and Wyoming.

CHAPTER 6

California's Natural Community Conservation Planning Program: Saving Species Habitat amid Rising Development

GAIL L. PRESLEY

Historically, California has been in the forefront of state initiatives on biodiversity preservation and species conservation. In the same vein, California's Natural Community Conservation Planning (NCCP) program represents an unprecedented state effort, in cooperation with private and public partners, to implement a broad-based ecosystem approach to land-use planning for the protection and perpetuation of biological diversity. The NCCP program identifies and provides for a comprehensive regional approach to protecting plants, animals, and their habitats, while allowing appropriate and compatible economic activity. The California Natural Community Conservation Planning Act (NCCP Act) (California Fish and Game Code Section (§)2800 et. seq.)[1] was enacted into law in 1991, and the responsibility to implement it rests with the California Department of Fish and Game (CDFG).

This chapter describes the history of the NCCP Act and analyzes the conservation planning efforts under NCCP, with particular attention to its first project, the coastal sage scrub pilot program. The chapter further evaluates NCCP's performance during its 19-year span, identifies its continuing challenges, and offers broad recommendations on how other states can execute similar efforts.

NCCP: Origin and Motivations

In California, the conflict between species protection and booming human population growth and urban development began to increase significantly in the 1980s. Some regions of the state were experiencing land-use and environmental conflicts on project after project. During that time, the habitat conservation planning process under the federal Endangered Species Act (ESA) was not well known and was costly and time-consuming for individual project applicants. California was suffering from ESA "gridlock" (Wheeler 2006).

The first habitat conservation plan (HCP) under the ESA was the San Bruno Mountain HCP in the San Francisco Bay area. It offered a single, comprehensive conservation solution for multiple projects. Other developing areas in the state quickly jumped on the bandwagon to develop regional HCPs. These early plans still only dealt with a few listed species, largely ignored sensitive unlisted species that might become listed in the future, and did not assure recovery of listed species.

CDFG and the California Resources Agency—the Governor's cabinet-level agency that oversees natural resource departments (now known as the California Natural Resources Agency)—formed the concept for NCCP in the late 1980s, emphasizing planning for whole ecosystem conservation instead of a species-by-species approach. The idea was to focus on conserving plants and animals on a broad landscape scale, contribute to the recovery of listed species, and prevent declines in species populations that might cause them to be listed. To accomplish this, the new approach needed broad support, which meant accommodating and planning for appropriate and compatible development through a proactive, locally driven, collaborative approach.

By the late 1980s, rapid urban expansion in the south coastal area of California caused several species to decline to the point at which CDFG and the United States Fish and Wildlife Service (USFWS) were considering adding them to the state and federal threatened and endangered species lists. One small songbird, the coastal California gnatcatcher (*Polioptila californica californica*), had become the focal species of an increasingly contentious land-use dispute between land developers and environmentalists for the future use of coastal scrub habitat in Southern California. In the words of then U.S. Secretary of the Interior Bruce Babbitt, urban development and the federal and California Endangered Species Acts were headed for "environmental and economic train wrecks" (Pollak 2001b). The time was ripe for a new approach.

What Is Conservation Planning under the NCCP Act?

At its core, an NCCP is equivalent to creating a land-use plan that goes well beyond the scope of biological conservation. An NCCP focuses on protection of sensitive species, biological diversity, and ecosystem functions while at the same time defining what land uses will be allowed in the plan area and where they can be located. In many cases, the NCCP process defines the full, spatial extent of development that will be allowed in the plan area.

Conservation planning focuses on the future of the privately owned portion of the landscape. Although existing public conservation lands are typically part of the conservation strategy, the main task of the planning process is to define where future development and other consumptive uses of private land will be allowed. In California, private land comprises approximately 51 percent (52 million acres). Of that private land, 67 percent (35 million acres) is covered by natural vegetation communities. Many listed or sensitive species are dependent on habitat on private lands for their survival. In an NCCP, private lands with the most important habitat values are identified through the planning process; if conserved, they "become part of a scientifically validated system of reserves, including corridors and linkages with other natural lands, that will be managed for long-term protection of multiple species and other ecological values" (Murphy 1999).

A conservation plan that limits private property rights through the legal exercise of land-use regulations by necessity becomes a public process. Consequently, the creation of a conservation plan invariably involves a diverse array of stakeholder participation in a negotiated process.

Table 6-1 provides an overview of the NCCP process.

The desired result of the NCCP process is a comprehensive plan that provides for species conservation and management and allows wildlife agencies to issue permits to authorize the take of species under the ESA and NCCP Act. Species whose conservation and management are provided for in the plan are called covered species. The NCCP Act provides CDFG the authority to permit take of any covered species (NCCPA § 2805[e]), whether or not it is listed as threatened or endangered under the California Endangered Species Act (CESA) (Fish and Game Code §§ 2050 et seq.). This authority provides an incentive to local applicants to cover certain species not currently listed so additional take permits would not be necessary should those species become listed in the future. Covering unlisted species requires that they be treated as if they were listed; it can mean the protection of additional habitats, ecological processes, and improved reserve configuration. A plan that covers unlisted species may provide conservation sufficient to prevent species from declining to the point at which they would qualify for protection under CESA or ESA.

Environmental Protection in California

The NCCP Act is administered by CDFG, the state agency charged with protecting wildlife. Additionally, in two primary laws CDFG plays a large role related to protection of the environment and threatened and endangered species—the California Environmental Quality Act (CEQA) and CESA.

Apart from CDFG's specific responsibilities, California has adopted a complex set of statutes and regulations to protect the environment. Cities and counties regulate private land use by way of planning, zoning, and subdivision controls. There are 58 counties and over 470 incorporated cities in California. State agencies regulate the private use of state land, natural resources, and certain activities of statewide significance. There are at least 21 state agencies that are or could be directly involved in the approval of development projects.

TABLE 6-1. Natural Community Conservation Planning Process

Scope—Identify initial geographic planning area and natural communities

Planning Agreement—Convene parties to develop a Planning Agreement which outlines:
Geographic planning area
Species and natural communities to be addressed
Processes for scientific and public input
Interim process for project review to maintain conservation options

Committees—Establish committees as necessary:
Plan stakeholder committee
Biological technical group (agency biologists, consultants)
Specific issues (funding, mitigation strategies, data management, outreach, etc.)

Data Collection—Collect biological and other necessary planning data in GIS format:
Species field surveys
Vegetation mapping and habitat modeling
Soils, hydrology, land ownership, etc.

Conservation Principles—Convene independent science advisors to identify conservation principles for:
Landscape conservation and reserve design
Conservation of specific target species and habitats
Data gaps (may include research recommendations)
Adaptive management and monitoring

Identify Reserves—Use conservation principles, data, and expert opinion to identify reserves.

Plan Development—Develop draft plan with input from committees and interested public:
Avoidance, minimization, and mitigation measures
Conservation obligations of partners
Mechanisms for development approval
Economic analysis and funding strategies
Draft implementing agreement

Plan Review and Approval
Conduct public review of draft plan and CEQA/NEPA* documents
Develop final plan
Jurisdiction(s) approves final plan (local government), signs implementing agreement
Issue necessary permits (CDFG, USFWS), sign implementing agreement

Implementation
Assemble the reserve system
Biological monitoring
Adaptive management program
Compliance monitoring

Note: *CEQA = the California Environmental Quality Act; NEPA = the National Environmental Policy Act.

California's environmental review of proposed land-use projects employs rigorous standards. In most cases the requirements extend beyond federal statutes established under the National Environmental Policy Act (NEPA). CEQA was enacted in 1970 as a system of checks and balances for decisions in California affecting land and water uses. Every project that is not exempt from CEQA must be analyzed to determine its potential environmental effects. Local governments (cities and counties) typically act as the lead agencies for conducting CEQA reviews on land-use projects. The environmental review process records the scope of the applicant's proposal and analyzes all its known environmental effects. Project information is used by state and local permitting agencies and the public in their evaluation of the proposed project and for public disclosure of potential environmental impacts.

CDFG reviews proposed projects for their potential impacts and provides information about resources that might be adversely affected. CDFG often also provides recommendations for avoiding, minimizing, and mitigating project impacts. CDFG's input in the CEQA process is advisory, and thus suggested measures to reduce impacts do not become mandatory unless adopted by the lead agency.

Drawbacks of Project-by-Project Review and Permitting

Although CEQA is an important tool for minimizing and mitigating environmental effects of proposed projects, the project-by-project application of CEQA falls short of the comprehensive conservation needed for sensitive and listed species. Table 6-2 compares project-by-project permitting and large landscape-scale NCCP. Project-by-project mitigation has two typical results: resources on-site are avoided, or off-site resources are protected, often in small, disjunct habitat patches. While helping to forestall species decline, both of these outcomes can result in protected areas that are isolated from larger patches of important habitat and, consequently, are unsustainable over time. Such approaches fall short of meeting the recovery needs of sensitive and listed species.

According to the CDFG website "CESA allows for take [of threatened and endangered species] incidental to otherwise lawful development projects. CESA emphasizes early consultation to avoid potential impacts to endangered and threatened species and to develop appropriate mitigation planning to offset project caused losses of listed species populations and their essential habitats."[2] (CDFG 2011a).

CESA incidental take permits are most common for permitting the take of listed species for individual projects. The incidental take permit process involves identifying, minimizing, and mitigating the impacts of the take. Similar to the CEQA process, this project-by-project approach does not result in a comprehensive mitigation and conservation strategy for sensitive and listed

TABLE 6-2. Comparison of Project-by-Project Permitting and Large Landscape Conservation Planning (NCCP)

Issue	*Project-by-Project (CESA, FGC §2081)*	*Natural Community Conservation Planning (FGC §2835)*
Planning scope	Typically individual projects, some larger scale	Biological regions, large, multi-jurisdictional
Biological scope	Single species or groups of listed species	Ecosystems and natural communities, multi-species
Focus of conservation	Mitigation for impacts to highly imperiled species	Recovery of species, habitat conservation, ecological function
Scientific input	Agencies and consultants/ applicants	Agencies, consultants, and independent scientists
Parties involved	Agencies and applicants	Local, state, federal, public, private landowners
Public participation	Only through CEQA process	Workgroups, workshops, hearings, public comment
Use of agency resources	Processing of hundreds of individual permits	Servicing and enforcing several large-scale plans
Reserve design	Can lead to fragmented set-asides	Large, interconnected habitat blocks
Duration of agreements	Typically short-term, project construction, < 5 years	Long-term, > 20 years
Land management	Endowed by permittee	Collaborative, adaptive, biological monitoring
Monitoring oversight	Mostly passive, relies on permittee	Active by agencies, oversight by public
Landowner assurances	Predictable, project-specific	Predictable, broad, long-term, statutory
Permit issuance standard	Full mitigation and avoid jeopardy	Provide for the conservation and management of the species, recovery within the plan area
Cumulative impacts	Component of species jeopardy determination	Comprehensive, all covered species and habitats

Source: Adapted from O'Connell and Johnson, 1997.

species. Individual mitigation projects often result in "postage-stamp" preserves that might never fit into a well-managed and monitored conservation landscape.

As development spreads across the landscape, the cumulative habitat loss results in declines of species and can lead to their protection under CESA. As more development is proposed for the remaining listed species habitat, more projects need incidental take permits under CESA. The project-by-project incidental take permit process becomes an overwhelming workload problem, but more importantly, falls short of comprehensive conservation for the species, as Table 6-2 illustrates.

Moving from CESA to NCCP

CESA was first enacted in 1970 to protect individual species that have declined to either threatened or endangered status. The 1984 amendments to ESA authorized an incidental take permitting process, which allowed for permits to be issued for "otherwise lawful activities" such as a legally approved development project. CESA was amended in 1984 to add key provisions modeled after ESA; however, explicit statutory language for issuing permits to private parties that would allow incidental take of state-listed species was not included. The only take that could be authorized under CESA was for "scientific or management purposes."

Many projects in California had the potential to impact state-listed species. CDFG, in the absence of official incidental take permitting authority, used CESA §2081 to issue "management authorizations" for take. Management authorizations included measures that mitigated for the take, compensated the habitat loss by permanently protecting habitat in another location, and assured management of the protected habitat. A court case in the mid-1990s[3] stopped CDFG from using management authorizations as take permits. Amendments to CESA in 1998 authorized CDFG to issue permits for incidental take of state-listed species under specified conditions, including the following: impacts of the take must be minimized and fully mitigated; measures must be roughly proportional to the impact of the take and capable of successful implementation; funding must be assured for monitoring; and permit issuance is subject to CEQA compliance (CDFG 2011b). CDFG also adopted regulations in 1998 addressing incidental take permits.

Management authorizations and incidental take permits issued under CESA were used to offset impacts to state-listed species by individual projects and regional HCPs. Most of the early HCPs only addressed a few listed species and their habitats. For example, the San Bruno Mountain HCP permit provided incidental take for only three species, and the Coachella Valley Fringe-toed Lizard HCP only addressed that single species. There was little emphasis and guidance on addressing ecosystems, ecological functions, and maintaining biological diversity. In the meantime, the NCCP Act was passed in 1991, providing a mechanism to authorize take of both listed and unlisted species whose conservation is assured in an ecosystem-based NCCP.

CESA / NCCP

CESA—Allows take of listed species, full mitigation standard, best applied to single projects, small areas, or few species.
NCCP—Allows take of both listed and unlisted species, recovery standard for plan area, designed for complex regional multi-species plans.

At this point, CDFG had two mechanisms, CESA and NCCP, each with a different standard, for authorizing take of listed species. CESA's standard is to minimize and fully mitigate for the impacts of the proposed take, as opposed to the more demanding recovery standard under NCCP. A CESA permit can be used to complement an HCP at a regional scale, but the long-term conservation benefits are not required to be as high. In practice, CDFG recommends that CESA incidental take permits only be used for individual projects or for larger-scale projects that clearly cannot or do not intend to meet the NCCP standard. CDFG also has determined that take of unlisted species could no longer be authorized through CESA permits. Such take is explicitly authorized through the NCCP Act.

The NCCP approach favors large-scale, regional, ecosystem-focused planning efforts. The rewards of taking on a large landscape planning effort under the NCCP Act include receiving take authority for unlisted species, assistance with implementation of conservation actions from the state, and assurance that no additional mitigation measures or funding will be requested, so long as the plan is implemented as agreed. Starting in the late 1990s, CDFG began to emphasize its preference for NCCP as the permitting mechanism for regional conservation plans.

Relationship of NCCP to ESA

The NCCP Act was modeled after the HCP process of section 10 of the ESA. However, it was intended to take earlier and more comprehensive actions in order to overcome perceived shortcomings of the ESA and the HCP process. A common criticism of the ESA is that by the time a species is considered and eventually listed, it could have declined so severely as to be impossible to recover (Wilkinson 1999). Similarly, a common criticism of the HCP process is that HCPs focus on mitigating impacts to listed species, rather than on conserving habitat at the ecosystem or landscape scale, which would lead to recovery of listed species and stability for unlisted species (Hood 1998).

The focus of NCCP is to conserve the ecosystems upon which sensitive species depend, rather than just their occupied habitat. The NCCP Act also goes beyond CESA and ESA requirements for mitigation of impacts. The standard for approval of an NCCP is to "provide for the conservation and management" of covered species. Conservation is defined in the NCCP Act as all measures necessary to return a listed species to the point at which listing as threatened or endangered is no longer needed, or, for unlisted species, to maintain or enhance the condition of a species so that listing will not become necessary (NCCPA §2805[d]). In contrast, the approval standard for HCPs is to minimize and mitigate impacts to the maximum extent practicable (ESA §10[a][2][B]).

There are several other requirements in the NCCP Act that go beyond those in the ESA. An NCCP must provide for the conservation of natural communities, incorporate independent scientific input, provide for public participation, establish a permanently protected reserve system (or equivalent conservation measures), and include monitoring and adaptive management programs. The reserve system and conservation strategy must address conservation of biological diversity, ecological functions, and ecosystem integrity. To further illustrate the similarities and differences, Table 6-3 provides a comparison of HCPs and NCCPs programs.

Notwithstanding the different standards, the two acts can still be blended together almost seamlessly. They complement each other so well that every NCCP permitted to date has also been an HCP. There are features of each law that help fill gaps in the other law. For instance, the definition of "take" in ESA is more protective than in California code because it includes the terms "harm" and "harass." These terms give USFWS broader authority to protect habitat for listed species than does the California prohibition of take. Under NCCP, the plan must provide conservation above and beyond mitigation. By combining the best features of the ESA and NCCP, integrated plans have effectively "raised the bar" on standards for regional HCPs. Thus, California NCCP/HCPs have the highest standards for regional conservation plans in the nation.

Supporters and Opponents

The NCCP Act, like virtually every environmental law, has both supporters and opponents. In the days before NCCP, although environmental advocates and development interests rarely agreed, both were dissatisfied with ESA and CESA implementation. The environmental groups believed that these laws were failing to protect species, because populations were still declining, and often listings did not occur until the species were on the brink of extinction. Development interests believed that the process of individual permitting was too costly and time-consuming, which threatened the economic viability of proposed projects. Most developers in NCCP areas have come to appreciate a program that provides them with clear conditions and streamlined permitting, with assurances that mitigation requirements will not change in the future (Whalen 2006).

TABLE 6-3. Comparison of HCPs and NCCP

Issue	*Habitat Conservation Plan (ESA §10(a))*	*Natural Community Conservation Planning (FGC §2835)*
Permit issuance standard	Minimize and mitigate impacts to the maximum extent practicable	Provide for the conservation and management
Take authority	Listed species only	Listed and unlisted species
Biological scope	Single species or groups of listed species	Ecosystems and natural communities, multi-species
Focus of conservation	Mitigation for impacts to highly imperiled species, contribute to recovery	Recovery of species within the plan area, natural community conservation, ecological function
Planning agreement	Not required	Required component
Scientific input	Agencies and consultants/ applicants	Agencies, consultants, and independent scientists
Parties involved	Agencies and applicants	Local, state, federal, public, private landowners
Public participation	Recommended in five-point policy	Required component
Interim project review	No process required	Process must be established in planning agreement
Adaptive management	Recommended in five-point policy	Required component
Monitoring	Recommended in five-point policy	Required component
Obligations of permittees	Mitigation for impacts	Affirmative commitment to provide conservation beyond mitigation
Funding	Must be assured for mitigation	Must be assured for conservation
Plan assurances	"No surprises" regulation, full assurances provided for all HCPs for unforeseen circumstances	Statutory authority, commensurate with level of conservation

According to Pollak (2001a, 14), the legislation authorizing the NCCP Act "received formal support from local governments, including the City of San Diego, the San Diego Association of Governments, and the Riverside County Habitat Conservation Agency. In addition, it was supported by the California Building Industry Association and the Lusk Company (commercial builders). It also received support from a conservation organization, the Mountain Lion Foundation. Although there was no organized opposition, some environmental groups expressed reservations about the proposal."

Legal Challenges to NCCP

There have been four legal challenges to NCCPs.

The City of Poway NCCP was sued by a property owners group for inadequate CEQA documentation (*Poway Property Owners Group v. The City of Poway*, February 1999). The city used a mitigated negative declaration instead of a full environmental impact report (EIR) to document potential environmental effects. Although this case was successfully defended, CDFG now always recommends that the local governments prepare an EIR for any NCCP, since it must also serve as CDFG's CEQA document for issuing the NCCP permit. An EIR provides a more rigorous assessment of the potential environmental effects of the proposed plan.

The Lake Matthews NCCP (Metropolitan Water District [MWD]) was challenged on inadequate CEQA documentation and an inappropriate basis for authorizing incidental take of listed species under CESA (*San Bernardino Valley Audubon Society v. Metropolitan Water District of Southern California*, April 1999). The CEQA challenge was affirmed upon appeal, but in subsequent settlement talks, an EIR was not required because the scope of projects that would be covered was reduced to just MWD's projects. The challenge was eventually dismissed because CESA was amended to provide an incidental take permitting process.

In two separate lawsuits, the San Diego Multiple Species Conservation Program (MSCP) was sued for inadequate conservation of vernal pool species (*Southwest Center for Biological Diversity et. al. v. Berg et. al.*) and one rare plant (*California Native Plant Society et. al. v. Babbitt and Clark*). Both cases were settled out of court through agreements for additional protective mechanisms or habitat protection.

Now, however, as NCCPs have been brought to governing boards for approval, hearing records indicate that most stakeholders have supported approval of individual NCCPs.

The Southern California Coastal Sage Scrub Pilot Program

Initially, the California Legislature agreed to a pilot program for protecting the coastal sage scrub ecosystem, which occurs across the coastal portions of five Southern California coastal counties (Los Angeles, Orange, Riverside, San Bernardino, and San Diego). This area is identified as a national and arguably international "biodiversity hotspot," due to the total number of species and the number of endemic species (Stein et al. 2000). The ecological significance, number of listed and sensitive species, and the increasing rate of urban development made the area an ideal candidate for NCCP.

CDFG, USFWS, and city and county governments started working together on a series of NCCP programs for the coastal sage scrub ecosystem. This ecosystem consists of significantly fragmented habitat (less than 340,000 acres; 531 square miles), scattered over more than 6,000 square miles, and is home to approximately 100 potentially threatened and endangered species.

One of the first steps in the planning process was to provide a sound scientific foundation for conservation. A scientific review panel consisting of independent, widely respected conservation biologists was convened to provide scientific guidance.[4] The panel developed a set of coastal sage scrub conservation guidelines (CDFG 1993a) that identified biologically based planning subregions. Knowing that a plan of this magnitude would take several years to complete, the panel recommended interim measures be applied to new development during the planning phase that would protect sufficient habitat to assure that a robust reserve network could be designed. The panel also identified conservation goals, providing a scientific foundation that could guide the participants through the rest of the planning process.

To formally launch the planning efforts, CDFG and USFWS entered into enrollment agreements with individual local jurisdictions and landowners. Approximately 25 NCCP programs were initiated to cover the activities of all the local jurisdictions, water districts, and utility providers in the coastal sage scrub region.

CDFG, the California Resources Agency, and USFWS collaborated to develop the Southern California Coastal Sage Scrub NCCP Process Guidelines (CDFG 1993b). The process guidelines explain the roles of local, state, and federal governments and describe how the planning process should proceed, including enrollment, scientific input, habitat loss during the planning phase, steps for planning and approval, and monitoring and evaluation. In 1998, a set of NCCP guidelines that could be applied statewide was distilled from the original coastal sage scrub guidelines.

The listing process under ESA provides an option to establish special rules to adjust the general protective measures available for threatened species. These special rules, established by USFWS under section 4(d), can define conditions under which take may be authorized. USFWS designated the coastal California gnatcatcher as a threatened species in 1993, meaning that it was likely to become endangered in the foreseeable future. USFWS also adopted a special rule under section 4(d) that allowed the NCCP program to provide the approach for conservation and recovery of the gnatcatcher. In the section 4(d) rule for the coastal California gnatcatcher, USFWS defined the conditions associated with certain land-use activities under which take of the gnatcatcher would not be a violation of the ESA. Under this special rule, USFWS permits take of gnatcatcher associated with land-use activities during the preparation of a plan covered under the NCCP Act if the take results from activities conducted in accordance with the conservation guidelines developed by the Coastal Sage Scrub Scientific Review Panel (U.S. FWS 1993).

The most important aspect of the use of the 4(d) special rule in San Diego was that the USFWS essentially endorsed the NCCP approach to conservation planning for federally listed and sensitive species. It sent a clear message to local stakeholders that planning for conservation under the state NCCP program would also satisfy the federal standards. This position provided reassurance that the NCCP approach was a worthwhile investment for the state, and subsequently the legislature authorized new staff positions and funding for CDFG to implement the program.

Fifty-nine local government jurisdictions, scores of landowners from across five counties, federal wildlife authorities, business and community groups, and environmental advocates actively participate in the NCCP program. Ten plans in Southern California have been approved, with permits issued; several others are nearing completion. Figure 6-1 shows the progress toward assembling the reserve system for the San Diego Multiple Species Conservation Program after 12 years of implementation.

The partnerships that formed during planning have proved crucial for plans to succeed in the implementation phase. CDFG, USFWS, and local plan participants are committed to an ongoing partnership that will last at least the life of the permits—from 20 to 80 years in the existing approved plans.

NCCP Program Performance

The NCCP program has expanded from the pilot Southern California project and is now available statewide. Plans are being prepared that cover urban development in central and northern coastal areas, inland deserts, and the Sierra Nevada foothills, agricultural areas of the Central Valley, and the operations for and conveyance of water through the San Francisco Bay–Delta region. Figure 6-2 shows regional conservation plans underway in California as of 2010. Following are some of the notable NCCP program accomplishments:

1996–1997: The first two NCCP programs were approved—the Central/Coastal Orange County Natural Community Conservation Plan and the San Diego Multiple Species Conservation Program (covering southwestern San Diego County).

Late 1990s: Twenty-five NCCP programs were underway in San Diego, Orange, Riverside, Los Angeles, and San Bernardino counties.

2000: A programmatic NCCP was approved for the massive CALFED Bay–Delta Program, covering water infrastructure and habitat restoration projects throughout the Sacramento–San Joaquin Delta, San Francisco Bay, and Central Valley.

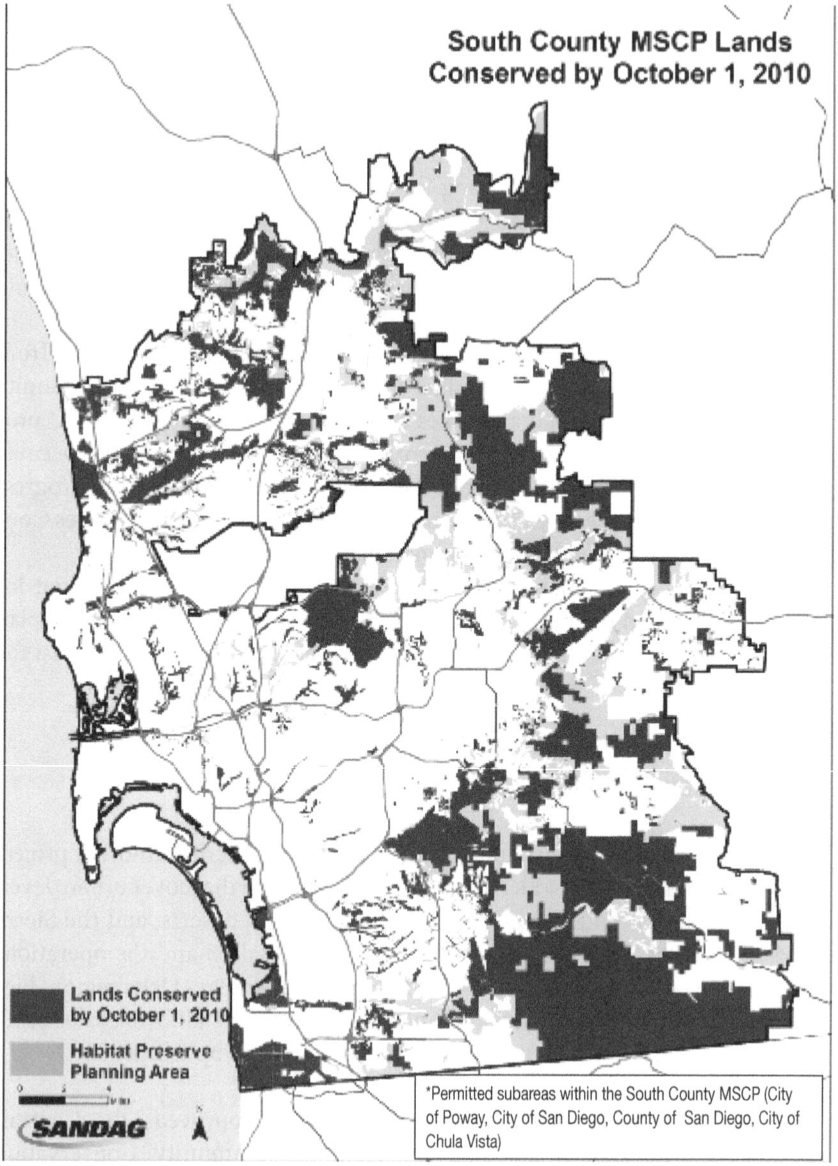

FIGURE 6-1. As of October 2010, more than 108,400 acres of the total reserve system (172,000 acres) have been protected, providing coverage for 85 listed and sensitive species.

Source: San Diego Association of Governments.

2002: The NCCP Act was significantly revised to incorporate new standards.
2003: The first "working landscape" NCCP was started by the Mendocino Redwood Company to address timber harvest.
2004: The Western Riverside County Multiple Species Habitat Conservation Plan, covering 1.2 million acres and 146 species, was approved. The NCCP approach was initiated to resolve Colorado River water transfer issues for the Salton Sea Ecosystem Restoration Project.
2006: By mid-year, the count of Northern California regional conservation planning efforts to sign NCCP Planning Agreements reached seven. A planning agreement was finalized in December to launch the Bay Delta Conservation Program, an effort between water users in the delta and the state and federal water and wildlife agencies.
2007: Permits were issued for the first Northern California NCCP, the East Contra Costa County NCCP.
2008: The Coachella Valley MSHCP was approved, covering 1.1 million acres and 27 species. Planning agreements were signed for the 1.6 million acre San Diego East County MSCP and the 312,000-acre San Diego North County MSCP.
2009: A Planning Agreement was signed for the Orange County Transportation Authority.

Table 6-4 and the NCCP Scorecard (in the following bulleted list) contain some illustrative statistics about California's NCCP plans. Table 6-5 lists several approved NCCP plans.

NCCP "Scorecard"

- As of 2010, there are 35 active NCCP programs of varying scope and complexity, covering more than nine million acres.
- Eleven counties are participating in NCCP programs.
- The number of species covered by NCCP programs ranges from 13 (Palos Verdes Peninsula) to 146 (Western Riverside MSHCP).
- NCCP programs range in size from 8,861 acres (Palos Verdes Peninsula) to 1.2 million acres (Western Riverside MSHCP).
- Twelve NCCP programs have been approved, with permits issued. Table 6-5 lists the approved plans.

Grant Funding for Planning

Creating an NCCP is an expensive process that typically lasts several years. From the beginning, USFWS supported the NCCP effort with generous funding. In the five Southern California counties, CDFG received $3.5 million in grant funds from the National Fish and Wildlife Foundation, and USFWS provided more than $25 million in grants to local governments for planning.

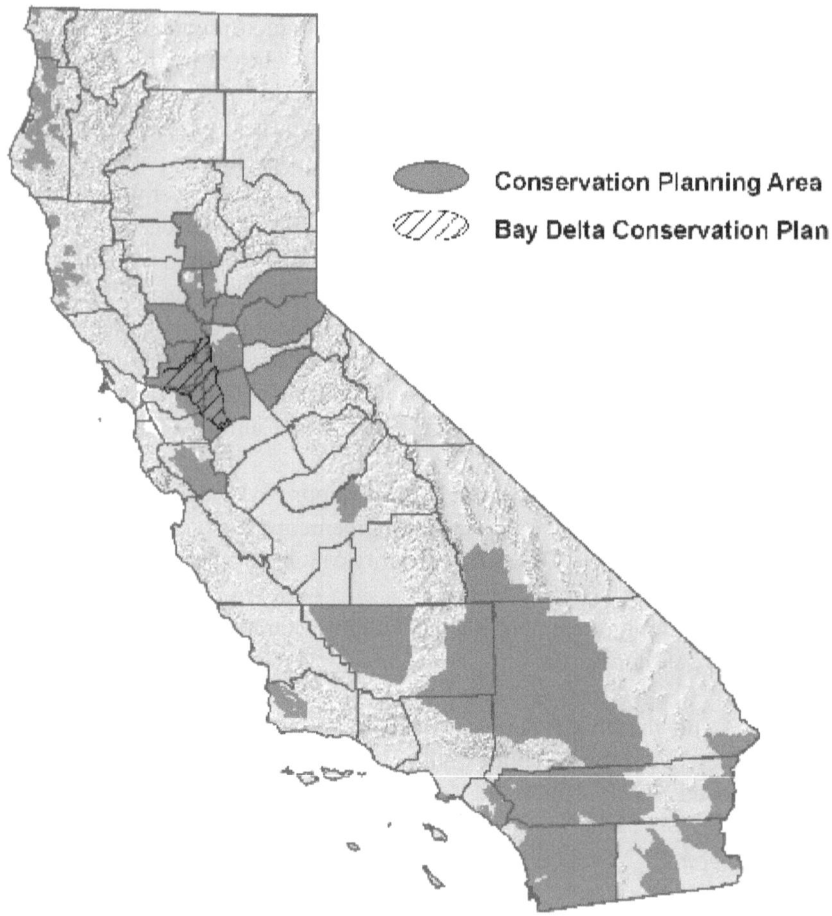

FIGURE 6-2. Regional Conservation Plans in California

Source: California Department of Fish and Game.

Through 2005, a coalition of local government, business, and environmental community representatives from the five counties in Southern California traveled annually to Washington, D.C., to solicit support for federal NCCP funds. The group met with individuals who could affect the Department of Interior budget: their local senators and congressional representatives, members of the Appropriations Committee, the Council on Environmental Quality, Office of Management and Budget, and the Department of Interior (Whalen 2006). Their coalition demonstrated that diverse stakeholders were united behind the NCCP approach, and they were consistently successful in securing funds for Southern California planning. As planning activities wound down in this region, the 2005 appropriation for Southern California was reduced to $1 million, and it was eliminated in 2006.

TABLE 6-4. NCCP Planning Area Statistics

County	*Population**	*Plan Name*	*Planning Area Size (acres)*	*Reserve Acreage*	*Covered Species*
Contra Costa	948,816	East Contra Costa NCCP	173,680	74,950	28
Orange	2,846,289	Central Coastal Orange County NCCP	208,000	37,380	39
Riverside	1,545,387	Coachella Valley MSHCP	1,136,400	747,600	27
		Western Riverside MSHCP	1,200,000	500,000	146
San Diego	2,813,833	Multiple Habitats Conservation Program (northwest county)	111,908	20,593	77
		Multiple Species Conservation Program (southwest county)	582,000	172,000	85

Source: Population information from U.S. Census Bureau, 2009

TABLE 6-5. Approved NCCP Plans

Multiple Habitats Conservation Program (northwestern San Diego County)	Multiple Species Conservation Program (southwestern San Diego County)
• City of Carlsbad	• City of Chula Vista
CALFED–Environmental Water Account	• City of La Mesa
East Contra Costa County	• City of Poway
Orange County Central–Coastal	• City of San Diego
San Diego Gas and Electric Company	• County of San Diego
Western Riverside County	Coachella Valley

In 2001, USFWS initiated a nationwide grant program for habitat conservation planning. The ESA section 6 HCP Assistance Grants program allows states to apply for planning funds on behalf of HCP applicants. From 2001 to 2010, CDFG successfully competed for $12.6 million for NCCP planning (Table 6-6). Section 6 funds constitute the bulk of funding available for large, complex regional conservation plans. California has been able to provide critical ESA section 6 grant funds to 23 different planning efforts. The Southern California plans were not included in the national grant program until 2005, when their line-item appropriation was reduced and then eliminated in 2006.

The section 6 grant program requires matching funds from nonfederal sources (25 to 55 percent of the project costs). Typically, local applicants must provide these funds, which can be in the form of cash that they contribute to the effort (including other nonfederal grants), or in-kind services. Local sources of funds for NCCP have largely been in-kind services of existing

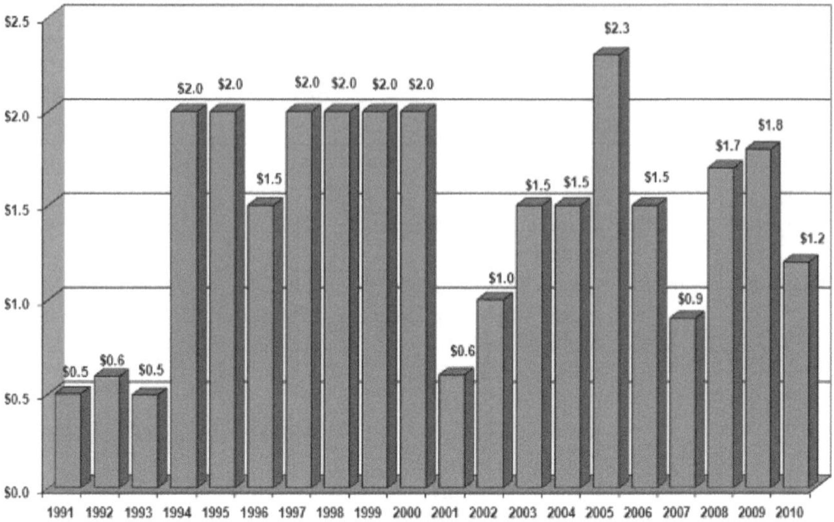

FIGURE 6-3. Federal ESA section 6 HCP Assistance funds have been critical for encouraging local governments to initiate NCCP planning, and for maintaining momentum in the process.

planning staffs, funds from the local government's general fund, landowner and co-applicant contributions, nonfederal grants, and contributions from stakeholder groups. In 2006, Proposition 84 provided up to $20 million in state funds for development of the Bay Delta Conservation Plan. Figure 6-3 shows federal funding provided for NCCP plans from 1991 to 2010.

Grant Funding for Implementation

Since 1998, the California Legislature has appropriated $10.3 million for NCCP local assistance grants. These grants are available to public and nonprofit entities for urgent tasks associated with implementation of NCCP programs. Grants have been awarded for land acquisition, Geographic Information System (GIS) development, biological monitoring, habitat restoration, focused research to fill important information gaps, habitat management, and public education.

Significant state and federal funding has been invested in land acquisition to support NCCPs. Proposition 12, a state bond act passed in 1999, provided $100 million of state funds for acquiring habitat to complement approved NCCP programs. All of these funds were spent in San Diego and Orange counties (the only counties with approved plans at the time). Proposition 84, passed in 2006, included an additional $90 million in state funds for habitat acquisition in NCCP programs. In 1997, USFWS initiated a grant program, available nationally, for habitat conservation planning land acquisition

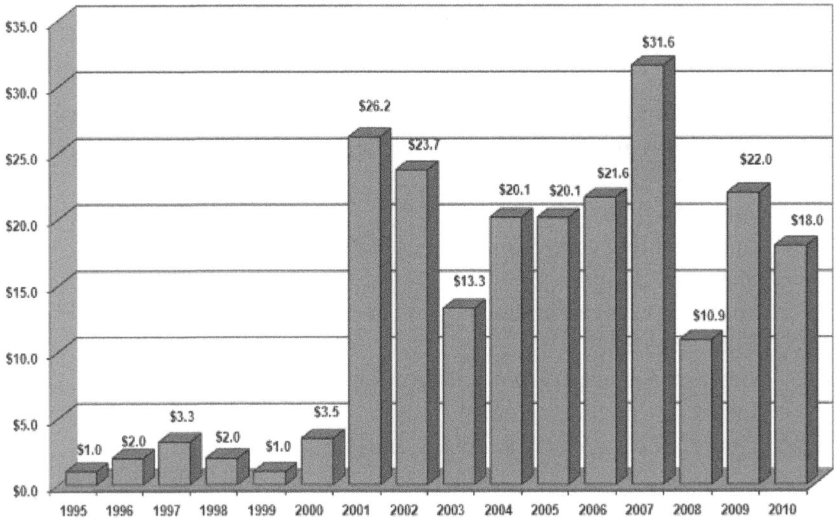

FIGURE. 6-4. Federal ESA section 6 HCP Land Acquisition funds have been essential to support the high conservation standard for NCCPs.

grants. From 1995 through 2010, California received more than $220 million for NCCP land acquisitions (Figure 6-4). Many more millions of dollars of state funds were invested in additional land acquisitions in NCCP areas and to match federal grant funds.

Evolution of the NCCP Act

The evolution of the NCCP program over the last 19 years has been marked by significant changes in CDFG's ability to accomplish effective conservation that will persist in perpetuity.

The original NCCP Act of 1991 was broad in scope, but lacking in details. The Act authorized CDFG to develop nonregulatory guidelines; thus, guidance for the early years of the program came from the NCCP Process and Conservation Guidelines. Some observers believed that the standards in the law insufficiently assured that the plans would meet the intended recovery of covered species within the planning area. They also did not believe that the nonregulatory guidelines provided enough of a mandate on CDFG to meet the standards. Proposed revisions to the NCCP Act were championed by the environmental community (that is, the Defenders of Wildlife, Audubon Society, and California Native Plant Society) and gained momentum as NCCP began to expand to areas in Northern California.

The first substantive changes to the NCCP Act were enacted in 2000. This revision added introductory legislative intent language about NCCP, new

NCCP Act 2002 Changes

- New definitions (including adaptive management and monitoring)
- Additional guidance for independent scientific input
- Interim project review process
- Approval findings
- Implementing agreement content requirements
- Statutory assurances
- Species coverage standards
- Injunctive relief
- Grounds for suspension or revocation of permits
- Future optional regulations
- Grandfathering of plans already underway

emphasis on and standards to address the inclusion of independent scientific input and analysis in the development of plans, and a requirement for CDFG to establish a process for public participation throughout the development and review of plans.

The most sweeping changes to the NCCP Act were passed in 2002 and actually resulted in repealing the existing NCCP Act and replacing it in its entirety. The new Act mirrors the historical program practices and codified the previous nonregulatory guidelines. Many of the sections remained unchanged or were reorganized; however, new sections were added that more clearly specified the standards for new NCCP programs. Two additional pieces of legislation made changes to the NCCP Act in 2002 and 2003; both were clean-up bills.

Collaboration with USFWS and National Marine Fisheries Service

In Southern California, blending the ESA and NCCP has resulted in a very close working relationship between USFWS and CDFG. The agencies have a solid working knowledge of each other's endangered species acts and environmental laws, and they are able to combine these laws to make a very effective case for protecting sensitive wildlife resources. It is the goal of the two agencies to come to consensus on issues, thus presenting a united perspective to applicants and stakeholders when reviewing projects and designing conservation plans.

The working relationship between CDFG and USFWS in Southern California is more collaborative than in other areas in the state, largely because the programs were cemented together during the coastal California gnatcatcher listing and 4(d) process. At the time, Interior Secretary Bruce Babbitt, the USFWS Regional Director, and the California Resources Agency were committed to integrating the ESA with NCCP. This commitment was translated to all staff levels within both agencies. The partnership led to new ways of interacting and sharing workload and costs. It is common practice to send joint comment letters that display both agencies' letterheads and signatures. This has multiple benefits, such as allowing staff to divide up review and writing tasks, and demonstrates that both agencies concur on the content. The agencies partner in many ways, such as providing work space in each other's offices for visiting staff; housing the primary NCCP GIS system for both agencies at USFWS; jointly contributing staff and funding to focused research or data collection; establishing interagency agreements to address staffing shortfalls; and agreeing to represent each other at meetings when limited staffing prevents both agencies from attending.

Although biologists at both agencies interact on a daily basis, the agencies also have regularly scheduled coordination meetings that are essential to the continuing partnership and success of the plans. These meetings cover a diverse range of topics of varying sensitivity including: state and federal funding for planning and implementation; land acquisition strategies; problem-solving on individual projects and plans; coordination with other agencies, such as the Bureau of Land Management and U.S. Geological Survey Biological Resources Division; joint policy development as it relates to NCCP; implementation oversight of approved plans; coordination of monitoring and land management; and political contacts with local governments. These management meetings are especially valuable because issues of local importance can be discussed in detail and problems solved from a broader source of resources.

By the early 2000s, several regional efforts were underway in Northern California to comply with the ESA and CESA by developing HCPs. Most of these plans began to explore NCCP as the state permitting mechanism. Seven NCCP agreements have been executed in northern California, and of those plans one has been approved (East Contra Costa County). Two additional NCCP agreements are in preparation. Three other counties are assessing local interest. With the expansion of NCCP to the northern part of the state, CDFG and the Sacramento Field Office of USFWS are holding similar management coordination meetings.

The National Marine Fisheries Service (NMFS) has primarily been involved with CDFG in NCCP in CALFED and the Bay Delta Conservation Plan. Several of the plans in Northern California are looking to cover species under NMFS jurisdiction, and NMFS is an active participant.

Effect on State and Federal Wildlife Agencies

The NCCP program dramatically affected the conservation arena in California. It facilitated new conservation partnerships across the state and among diverse interests. It brought a higher conservation standard to regional planning that addresses conservation at the landscape and ecosystem scale, unparalleled by any other state laws.

Commonly, new legislation directing CDFG's activities is enacted without the necessary new personnel. With the passage of the NCCP Act in 1991, the California Legislature agreed to fund new staff dedicated solely to NCCP. The new personnel were assigned to the statewide NCCP program, administered from the headquarters office in Sacramento, but a new satellite office was opened in San Diego to immerse them in the day-to-day activities of the program. This new staffing was a strong message from the state about the importance of the new program. All CDFG staffing and program support for NCCP is funded with general funds, not hunting and fishing license revenues, which make up the bulk of CDFG's budget. General funding reinforces the benefits of NCCP to the general public, but it also makes it vulnerable during budget reductions.

As the NCCP program expanded statewide, CDFG recognized the need for additional staff to keep up with the increasing workload. State agencies must propose changes to their budgets through an annual process. CDFG wrote and submitted several proposals for budget increases in the late 1990s that were never approved. Feedback from the Senate Natural Resources Committee was that funding for new NCCP positions would not be approved until the NCCP Act was modified to add new standards. Unfortunately, passage of the revised NCCP Act in 2002 coincided with the energy crisis and huge state budget deficits, so funding for new NCCP positions was not available. With continued growth in the number of plans statewide, and a recovering state economy, CDFG's NCCP staffing levels were increased slightly for the first time in 2006. However, significant statewide budget problems in the following years resulted in cuts to NCCP staffing.

The NCCP program also affected implementation of federal ESA habitat conservation planning in California. The marriage of NCCPs and HCPs raised the standard for HCPs, because each combined plan has to also meet the NCCP standard of providing for the conservation of all covered species within the plan area. This standard has led to plans that call for larger reserve acreages, better connectivity between core habitat areas, enhanced public input, improved monitoring and adaptive management programs, and more state and federal contributions to implementation. Although the USFWS's five-point policy (see Table 6-3) recommends some of these features for HCPs, they did not become de facto requirements until HCPs were combined with NCCPs. The NCCP requirements address many of the concerns about scientific sufficiency and public participation raised by the environmental community, as evidenced by its support of legislation that revised the NCCP Act in 2002. Addition of the NCCP requirements may sufficiently enhance the quality of regional HCPs to reduce the number of lawsuits brought against the USFWS by environmental groups.

The higher conservation standard of NCCP includes the concept that the public shares in the responsibility to pay for a portion of habitat conservation. The 2002 revision to the NCCP Act added specific language (NCCPA §2802) that it is the intent of the legislature to acquire lands consistent with NCCP programs and to provide assistance with implementation of the plans. Mitigation must still be funded by the sectors benefiting from the take authorization (for example, developers), but conservation is the responsibility of all citizens. As a consequence, the state and federal governments have agreed to acquire, monitor, and manage a portion of the needed reserve system in most of the regional NCCP programs approved to date. In San Diego County, USFWS established a new national wildlife refuge to meet this commitment. In other areas, the wildlife agencies' contributions have come through grants for habitat acquisition, monitoring, adaptive management, and staffing commitments for these activities.

Continuing Challenges

The NCCP program has grown and evolved since 1991 with the goal of improving each new plan. Improving the plans does not necessarily mean increasing mitigation requirements. The improvements desired by CDFG include more secure implementation mechanisms, more meaningful and realistic monitoring and adaptive management programs, and better documentation of how the plans meet the intent of the NCCP Act. Yet there remain many challenges to improving the process.

Funding for Planning and Implementation

The availability of money dedicated to the planning effort will dictate the pace and outcome of the process. Local governments do not typically have the significant amounts of money necessary to complete the NCCP process, estimated from \$3 to \$6 million depending on complexity and the length of time it takes. This problem is exacerbated during difficult state budget times, when state funds to local governments are often reduced. Grants for conservation planning are essential to maintaining momentum in NCCP programs. The USFWS's section 6 grant program has enabled many jurisdictions to initiate and make significant progress on their plans. As previously discussed, CDFG has been very successful in securing grants for plan proponents. However, many of the plans could use more funds than they receive, which would facilitate quicker completion.

The cost to implement a plan that provides for conservation of biological diversity at the ecosystem scale will be much greater than a plan that simply mitigates impacts. A comprehensive approach offers environmental, social, and economic benefits, but local jurisdictions continue to struggle to provide a funding strategy that can be assured over the life of the plan. The state and federal governments can assist with the costs for conservation beyond mitigation, but the local partners must be able to cover the bulk of the costs for implementation. Some plans in Southern California are still struggling to meet their funding commitments five to ten years after the permits were issued.

Leadership

For some time now, the NCCP program has recognized the importance of "champions" in the success of the planning process. A champion is a local leader who acts as a vocal advocate for the plan and moves it forward by keeping everyone at the negotiating table. The champion works behind the scenes to make sure key products and decisions occur on schedule so milestones are met. In some plans, a natural leader emerges as the planning process progresses. In others, a champion must be cultivated to assure the long-term success of the process. Most often, a champion is a local elected official or influential staff central to the management and coordination of local land-use planning. Plans with no clear champion have often faltered, resulting in intermittent delays, loss of momentum and progress, or outright failure.

Time Frame

Completing a complex regional conservation plan is typically a multi-year exercise. The development of an NCCP program averages about six years. This represents a significant improvement over the pre-NCCP, species-by-species, project-by-project approach, but additional streamlining would be very beneficial. Many things affect schedules and cause delays in planning: available funding; timing of biological surveys; insufficient wildlife agency staffing; and inadequate outreach. Lack of political support, especially from locally elected officials, has significantly delayed plan progress, and in fact it has stopped planning altogether in some instances, as in San Bernardino County.

Perhaps the most significant problem with delayed permitting is the ongoing loss of habitat while projects continue to be approved and developed. All projects approved by the local jurisdictions before the plan is approved and permitted are not subject to the conditions of the plan. These interim projects do not necessarily contribute the same share of conservation as they would if approved pursuant to the plan. Some projects may be on lands with very high conservation value, and their loss could impede meeting the NCCP standard. The value of these lost conservation opportunities can be in the millions of dollars.

It is also important to be mindful of local political cycles and the timing of elections. It is typical to have strong political support from at least a portion of the board or council. If those supporters leave office before the plan is finished, the plan could lose the votes necessary for approval.

Institutional Capacity

The wildlife agencies have embraced the benefits of regional conservation planning as one of the most effective conservation tools they currently possess. The wildlife agencies therefore encourage local jurisdictions to consider creating comprehensive plans to address all their sensitive wildlife issues, even when the agencies know their staffing levels are inadequate. Sufficient agency staffing is necessary to maintain the plan's progress and momentum. After permits are issued, the wildlife agencies need to stay engaged with plan partners to facilitate successful implementation.

Similarly, the local jurisdiction must build capacity for implementation. Staff that interact with project proponents need to be trained to understand how to apply the requirements of the plan to new development projects. The jurisdiction must enact new ordinances and pursue new funds for implementation. New preserve lands need to be acquired, monitored, and managed. The habitat losses and new preserves must be tracked in a GIS, which is used to generate annual compliance and progress reports. Partnerships for acquisition, monitoring, and management must be actively sought and nurtured.

Implementation Commitments

As stated earlier, the higher conservation standard of NCCP includes the legislative intent that the public shares in the responsibility to pay for a portion of the conservation. In most of the NCCP programs approved to date, the state and federal governments have agreed to contribute acres to the reserve system and assist with management and monitoring. For example, the state and federal agencies agreed to contribute significant acreages to reserve systems: 13,500 acres to the San Diego MSCP; 50,000 acres to the Western Riverside MSHCP; 8,000 acres to the East Contra Costa County NCCP; and 21,000 acres to the Coachella Valley MSHCP. In addition, the agencies agreed to conduct biological monitoring and adaptive management to the plan standards on lands they acquire and hold. Inherent in these commitments are funding for habitat acquisition and the wildlife agency staff positions that will be needed to acquire, manage, and monitor lands.

Similarly, the cities and counties with approved plans struggle to stay committed to the plans as staffs and elected officials change. Pressure is relieved after permits are issued, and the attention of elected officials shifts to other urgent matters. Institutionalizing implementation processes and funding at the local level are critical to long-term commitment and success.

Other Initiatives

CDFG strives to assess the NCCP program's successes and failures and apply the new knowledge to all NCCP programs. This approach keeps the planning efforts functioning at the "state-of-the-art" level, while at the same time it enhances consistency among plans. In addition to management-level coordination meetings among the wildlife agencies in the north and south parts of the state, two main initiatives at CDFG are in use: an in-house handbook and intern and fellowship projects. In 2004, NCCP program staff began creating an in-house handbook, *NCCP in Practice—Tools and Tactics*. The handbook is a collection of the lessons learned in the NCCP program since its inception. It is designed as informal guidance for CDFG employees who are working on NCCP programs, to help them learn from the experiences of others and translate that information and guidance into an effective planning process in their local areas. NCCP staff continue to add to an intranet version of the handbook available only to CDFG employees, with an eye towards a possible web-based version available to the public, as many local planners seek similar guidance. Finally, CDFG has initiated several research projects to address key questions that have emerged from regional conservation planning. A partnership with the Environmental Careers Organization from 2000–2004 allowed CDFG to sponsor four graduate fellows to conduct NCCP program evaluations. The reports are available on the NCCP web site[5]. In the 2009–2010 academic year, CDFG hosted two interns from the Conservation Management Training Program, a new doctoral training program in the University of California Davis Graduate Group in Ecology, a program described on the group's website.

Conclusion

California is endowed with highly varied ecosystems, landscapes, and rich biological diversity. Because of the intrinsic value of these assets, California has a long history of enacting laws to protect its natural resources. It had strong wildlife protection and state and local land-use planning laws in place long before the passage of the ESA and CESA. The evolution from CESA to NCCP reflects both the state's commitment to resource protection and the tremendous pressures placed upon the resources from conversion of natural lands to human uses.

Regional conservation planning is a time-consuming and resource-intensive program for all involved. California is an economically wealthy state with booming urban growth and productive working landscapes. NCCP programs flourish in the areas of the state where business and local interests have sufficient capital to invest in the process.

The combination of a strong history of environmental protection, significant pressures on natural resources, and adequate business and local community wealth to engage in planning might not be unique to California. It certainly was the motivation for the successful passage and implementation of the NCCP Act. It is also the motivation for the state to work in partnership with USFWS and NMFS on ESA implementation. Having a strong set of state laws to complement ESA has allowed the state and federal agencies to work together on common goals. Although the relationships have not been without their challenges, the efforts to work collaboratively on issues of mutual concern and interest have proven to be invaluable to conserving California's rich natural heritage.

Other states might consider developing strong environmental legislation that complements the ESA. Such legislation could clearly indicate the state's intentions, effectively giving the state equal footing with USFWS and NMFS on conservation issues. The legislation should provide incentives for conservation at the ecosystem level thereby giving communities the opportunity to address political and social issues because of the long-term, broad-landscape application that affects so many sectors. Such legislation would also be complemented by legislation similar to CEQA. Significant conservation gains in California at the local level were never fully realized until local land-use planning through CEQA was blended with the NCCP/HCP approach to conservation.

From the California experience, a few key features, if included in legislation, would provide equivalency to NCCP and help assure a strong blended process with the ESA:

- Conservation of biological diversity should be a stated goal if we are to achieve true conservation of the country's diverse natural resources. It must be planned at an ecosystem scale that takes into consideration natural communities, native species, and ecological functions and processes in the plan area.
- Built-in incentives should include coverage for unlisted species, long permit duration, statutory assurances, grants for planning and implementation, state funding for habitat acquisition, and management and monitoring of state-owned reserve lands.
- Enforcement powers are a necessary component that encourages compliance, motivates people to participate in planning, and addresses concerns of the environmental community. Illegal take of state-listed species must be prohibited, and the state wildlife agency must be given the power to enforce criminal takings of species and their habitat.
- Because conservation of biological diversity and resolution of local land-use conflicts are in the interests of all citizens, state general funds are an appropriate funding source.
- Plans should have a strong scientific foundation incorporating independent scientific input that addresses conservation strategies for species and natural communities, reserve design principles, management principles and conservation goals, identification of data gaps and uncertainties, and recommendations for targeted studies.

- Since public support is essential to success, participation should be encouraged through early and ongoing outreach and education, stakeholder working groups, and public-review timelines of sufficient length for the complexity of the plan.
- Conserved habitat must be protected in perpetuity (through conservation fee title ownership or easements), and monitoring and adaptive management programs should extend (at some level) indefinitely beyond the term of the permits. Plans should be required to demonstrate adequate funding for these features.

In addition to these program features, CDFG has compiled a set of the lessons learned throughout the years of evolution of the NCCP program:

- **Provide leadership**—Leadership from the wildlife agencies is essential in these endeavors. The people assigned to regional conservation planning should be given the skills and authority to make decisions. Staffs need the skills to be able to think "outside the box" and "close the deal," so that planning decisions can be made and plans completed. Wildlife agency staffs should focus on the "big picture" gains and not the small losses. Managers should encourage and support creativity.
- **Build trust**—Strong relationships with the local jurisdictions (cities and counties) are crucial to program and plan success. The wildlife agencies should remember that the regulatory way is not always the best way. Each stakeholder has an interest to represent and a valid role in the process. Time invested in building relationships with elected officials and local management will foster open communication.
- **"Mortar" the scientific foundation**—Front-load the planning process by consulting with independent science advisors early. A solid scientific basis to the plan builds credibility with constituents and provides the foundation for the ultimate plan decision. To assure the plan follows scientific principles, plan participants should "check in" with the science advice at decision points. Use a robust GIS to store and analyze data. Almost all the data, conservation problems, and solutions relevant to regional conservation planning have a spatial component. A map-based conservation strategy is unsurpassed at conveying information—especially to laypersons—and is very important for generating broad support and stakeholder buy-in.
- **Seek partnerships for implementation**—The interactions with wildlife agencies should not end when the plan is approved and permitted. To be successful, the planning partners must stay committed to implementation. The energy and motivation necessary to make the plan successful comes from continuing the partnerships established during planning and seeking out new partners. Implementation activities such as habitat acquisitions, monitoring and management, data collection and management, and regional funding can be accomplished most effectively if they are shared

among a diverse group that can bring its own expertise (and funding contributions) to the effort. Strong, ongoing, collaborative partnerships, coupled with a science-based plan, can qualify for significant federal funding.

A large part of the success of the NCCP approach in California is having strong federal counterparts interested in similar conservation outcomes. The ESA and NCCP Act complement each other such that blending them together into a single conservation planning process makes an excellent case for robust conservation. The equal partnership between CDFG and USFWS/NMFS is critical to their mutual efforts, and these agencies would acknowledge that they have accomplished more conservation together through these acts than would ever have been accomplished independently.

References

CDFG (California Department of Fish and Game). 1993a. *Southern California Coastal Sage Scrub Natural Community Conservation Planning Conservation Guidelines.* Sacramento, California: CDFG. July.

———. 1993b. *Southern California Coastal Sage Scrub Natural Community Conservation Planning Process Guidelines.* Sacramento, California: CDFG. November.

———. 2011a. "State Laws, Regulations and Policy for the Incidental Take of State Listed Species under the California Endangered Species Act." www.dfg.ca.gov/habcon/cesa/incidental/cesa_policy_law.html

———. 2011b. "California Endangered Species Act (CESA): Sections 2081 (b) and (c)—Incidental Take Permit Process." www.dfg.ca.gov/habcon/cesa/incidental/incid_perm_proced.html

CEQA (California Environmental Quality Act). California Public Resources Code. Div. 13, §§ 21000 et seq.,1970, as amended.

CESA (California Endangered Species Act). 1970. California Fish and Game Code. §§ 2050 et seq., 1973, as amended.

Hood, Laura C. 1998. *Frayed Safety Nets: Conservation Planning under the Endangered Species Act.* Washington, D.C.: Defenders of Wildlife.

Murphy, Dennis D. 1999. "Southern California Natural Community Conservation Planning: Case Study." *Bioregional Assessments.* Washington, D.C.: Island Press.

NCCPA (Natural Community Conservation Planning Act). 1991. California Fish and Game Code. Section 2800 et seq., as amended 2003.

O'Connell, Michael. A., and Stephen P. Johnson. 1997. "Improving Habitat Conservation Planning: The California Natural Community Conservation Model." *Endangered Species Update*, 14(1&20). School of Natural Resources and Environment, The University of Michigan.

Pollak, Daniel. 2001a. "Natural Community Conservation Planning (NCCP): The Origins of an Ambitious Experiment to Protect Ecosystems." *California Research Bureau Report* No. CRB-01-002. California State Library. March.

———. 2001b. "The Future of Habitat Conservation? The NCCP Experience in Southern California." *California Research Bureau Report* No. CRB-01-009. California State Library. June.

Stein, Bruce A., Lynn S. Kutner, and Jonathan S. Adams, eds. 2000. *Precious Heritage: The Status of Biodiversity in the United States*. Oxford University Press.

U.S. FWS (U.S. Fish and Wildlife Service). 1993. "Coastal California Gnatcatcher Section 4(d) Special Rule, Question/Answer Fact Sheet." Carlsbad, California: U.S. FWS.

Whalen, James. 2006. Personal communication between James Whalen, Principal, J. Whalen and Associates, and the author. February 7.

Wheeler, Douglas P. 2006. Personal communication between Douglas Wheeler, partner, Hogan & Hartson L.L.P., and the author. January 25.

Wilkinson, J. B. 1999. "The State Role in Biodiversity Conservation." *Issues in Science and Technology*, XV (3).

Endnotes

1 The full text of the NCCP Act can be found at www.dfg.ca.gov/habcon/nccp (accessed September 11, 2010).

2 Information on CESA can be found at www.dfg.ca.gov/habcon/cesa/ (accessed September 11, 2010).

3 The appellate court opinion, *Planning and Conservation League et al. v. Department of Fish and Game et al.* (1st Cir. 1997), which was previously published at 67 Cal. Rptr.2d 650, was superseded when the California Supreme Court granted review, 65 Cal.Rptr.2d 346. When enactment of Senate Bill 879 rendered the case moot, review was dismissed and the case was remanded (71 Cal.Rptr.2d 212) to end the litigation.

4 The Coastal Sage Scrub Scientific Review Panel members were Dr. Peter Brussard (University of Nevada, Reno), Dr. Michael Gilpin (University of California, San Diego), Dr. Dennis Murphy (Stanford University), Dr. Reed Noss (Stanford University), and Dr. John O'Leary (San Diego State University).

5 www.dfg.ca.gov/habcon/nccp

6 http://ecology.ucdavis.edu/programs/Conservation_Management

PART IV

Case Studies

CHAPTER 7

The Karner Blue Butterfly: Wisconsin's Statewide Habitat Conservation Plan

David R. Lentz and Jimmy S. Christenson

The case of the Karner blue butterfly in Wisconsin represents one of the first statewide habitat conservation plans (HCPs) to conserve an endangered species. The story of Karner blue butterfly conservation in Wisconsin holds several important lessons for state–federal collaboration in conserving an endangered invertebrate species. This chapter analyzes the role of the Wisconsin Department of Natural Resources (WDNR) in recovery of the species and offers suggestions to better facilitate similarly effective species conservation efforts in the future.

Karner Blue Butterfly Habitat and Range

The distribution of the Karner blue butterfly (*Lycaeides melissa samuelis*) is discontinuous and broadly follows the northern limits of the wild lupine, a forb of the pea family (Figure 7-1).

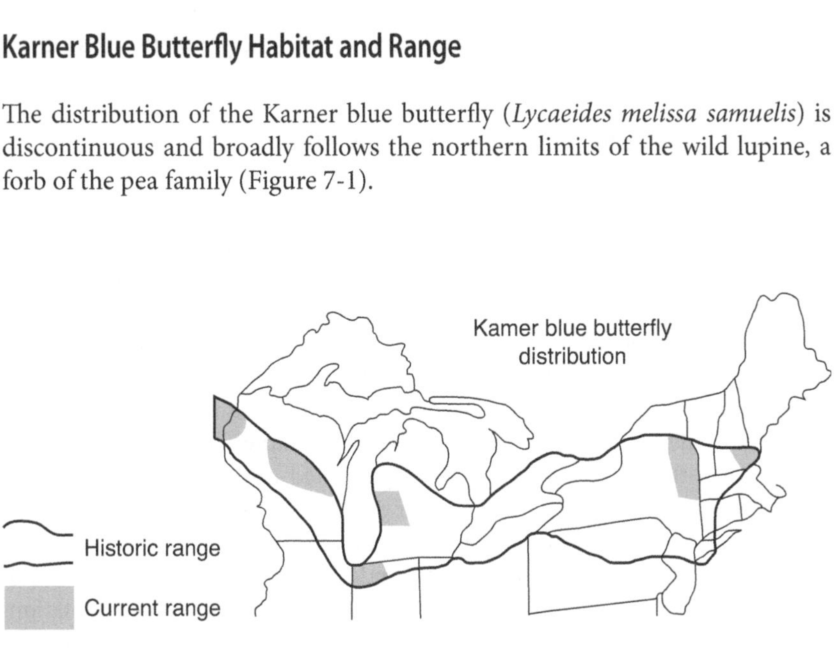

FIGURE 7-1. Karner Blue Butterfly Historic and Current Ranges

Wild lupine is the only known larval food plant for the Karner blue butterfly and is, therefore, closely tied to the butterfly's ecology and distribution. Recorded information indicates that the Karner blue butterfly originally occurred in shifting clusters of populations, or metapopulations, across a vast fire-swept landscape covering thousands of acres. While the fires resulted in localized extirpation, post-fire vegetational succession promoted colonization and rapid population buildups (Schweitzer 1989). Periodic disturbance is necessary to maintain openings in the forest canopy for wild lupine to thrive. A variety of other understory plants associated with the fire-induced forest openings serve as nectar sources for the adult butterflies.

Over the past century, as fire patterns and forestry practices have changed in the upper Midwest, it is estimated that Karner blue butterfly numbers, across the species' historic range, have declined by 99 percent or more. Most of the decline is deemed to have occurred in the last two decades of the twentieth century. Historically, eight major population clusters of Karner blue butterfly were known to exist in parts of Wisconsin, Michigan, Minnesota, Indiana, Illinois, Ohio, Massachusetts, New Hampshire, Pennsylvania, New York, and Ontario. The butterfly is now extirpated from Massachusetts, Pennsylvania, Illinois, and Ontario. Unconfirmed reports indicate that individual Karner blues might have been sighted at a historic Ontario site in 1990 or 1991 and at Illinois Beach in 2005 or 2006. Minnesota, Indiana, New Hampshire, New York, and Ohio support small remnant populations. Wisconsin boasts the largest viable population of the butterfly in the nation, and, according to the state authorities, the butterfly was never in danger of extinction within the state. Michigan supports a number of significant, sustainable populations as well.

Karner Blue Butterfly Is Deemed Endangered

On January 21, 1992, the United States Fish and Wildlife Service (USFWS) published a proposed rule listing the Karner blue butterfly as an endangered species (57 F.R. 2241). The USFWS, in listing the Karner blue butterfly as endangered, indicated as the primary causes leading to the species' imperiled status the following: habitat modification and destruction due to development; forest succession in the absence of natural disturbances; silviculture; and fragmentation of remaining habitat. WDNR recommended that the Karner blue butterfly be designated as a threatened species, rather than endangered. In calling for a threatened status, WDNR presented results from state surveys conducted from 1990 to 1992 which showed that Wisconsin supported a relatively large number of populations of Karner blue butterflies with seemingly good potential for long-term viability under favorable management. Populations occurred on federal, state, and private lands comprising populations from 100 to 1,000 individuals. The sites exceeded well over 100 or 1,000 acres, and one property had the potential to exceed 50,000 acres of habitat. Further,

WDNR argued that given the wide distribution of the butterfly in Wisconsin, a flexible conservation regime allowed under threatened status, pursuant to section 4(d) of the Endangered Species Act (ESA), would be most effective in meeting recovery goals.

Les Ferge, a professional lepidopterist in Wisconsin, shared the recommendation that the species be listed as threatened. Nonetheless, USFWS affirmed the endangered status for the Karner blue butterfly in its final ruling in December 1992 (57 F.R. 59236). According to USFWS, endangered status was warranted in situations where the species is in danger of extinction throughout all or a significant portion of its range. USFWS did note that a few sizable populations with potential for long-term viability were still extant in Wisconsin. However, the Service argued that immediate federal protection and habitat restrictions were crucial to the short- and long-term viability of the Karner blue populations, even at these larger sites. USFWS emphasized the fragility of remaining populations, based on the loss of a population in Ontario due to adverse weather conditions.

Shortly after the Karner blue butterfly was listed as an endangered species, USFWS approached WDNR about Karner blue conservation and recovery strategies. The Service harbored a few concerns as to how Wisconsin would respond to the listing, considering the widespread distribution of the species across the state and the large numbers of landowners affected. To this day, the Karner blue butterfly is an unlisted species under the Wisconsin Endangered and Threatened Species Act. Because of the relative abundance of the species in Wisconsin, the threshold for listing either as endangered or threatened under the Wisconsin Act has not been met. USFWS was eager to avoid the controversies with the listing that it was experiencing in the Pacific Northwest with the northern spotted owl.

Wisconsin's HCP for the Karner Blue Butterfly

In 1993 and 1994, WDNR, in response to contacts by landowners and USFWS, began studying the possibility of obtaining an incidental take permit on behalf of the Wisconsin landowners, and developing an HCP. The permit and associated HCP would preserve Karner blue habitat in Wisconsin, while allowing the continuation of land management practices like forestry, recreational and wildlife land management, and right-of-way maintenance.

The relative abundance of the Karner blue in Wisconsin and its association with actively managed habitats made it essential for Wisconsin landowners and managers to have regulatory relief from the take provisions of ESA. Wisconsin is the nation's largest manufacturer of paper. Paper mills in the state have historically relied on timber from county, state, and private forestlands. Private forests comprise about 57 percent of forestlands in the state, and Karner blue are often found in forest openings following timber harvest.

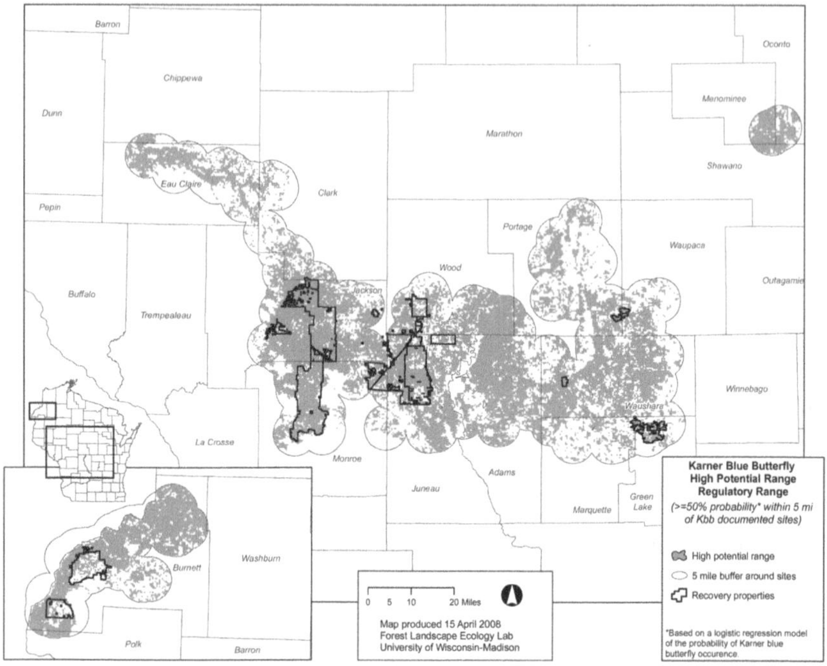

FIGURE 7-2. Karner Blue High Potential Range

Similarly, utility rights-of-way provide excellent butterfly habitat, due to recurring maintenance, for example, mowing, etc. WDNR, as a large public landowner and forest manager in the state, required incidental take authority for the purpose of managing its lands, especially forests and wildlife management properties. Consequently, there was a strong interest and demand within the state to procure an Incidental Take Permit for the Karner blue butterfly.

An HCP must accompany any application for a permit under the ESA. The HCP describes the proposed strategies and conservation measures to be applied if a permit is issued. In developing the HCP, WDNR consciously decided to focus its conservation efforts on particular habitat types associated with the Karner blue butterfly. The vision, early on, was to offer incidental take protection to all citizens of the state who might be affected by the presence of the species. The geographic focus, due to habitat requirements, was on the central and northwestern parts of the state, which offer the highest potential for Karner blue habitat. The area is referred to as the "Karner Blue High Potential Range" (Figure 7-2).

The Wisconsin initiative focused on use of an HCP process, rather than a safe harbor agreement (SHA) or other ESA tools, because the USFWS representatives involved were more familiar with HCPs and were of the opinion that use of other alternatives was impractical. The biology of the species, given its dependence on habitat disturbance, called for a conservation strategy involving active management. In the absence of historic natural disturbance, Karner blue survival relied on lands actively managed to mimic historic disturbance.

Various state landowner groups supporting a Karner blue conservation initiative expressed greater comfort and trust in working with WDNR than USFWS to protect a federal endangered species that most locals viewed as neither rare nor declining. Private forest owners, such as International Paper, Georgia Pacific, Consolidated Papers, and others, initially contacted WDNR to develop an HCP and file an application for an incidental take permit. Subsequently, counties with county forests, utilities, other state agencies, and The Nature Conservancy (TNC) also joined in the initiative. Later, organizations such as the Wisconsin Woodland Owners Association (WWOA), the Audubon Society, and the Sierra Club also joined the conservation effort. Following a public participation process, a coalition of partners for the Wisconsin HCP for Karner blue came together, including 26 landowners who committed more than 250,000 acres to the conservation effort.

The HCP concept developed by the WDNR was the first of its kind to apply to the entire state. It sought to obtain incidental take protection for qualified landowners and land managers; by so doing, it would eliminate the need for all affected to individually develop HCPs and apply for incidental take permits. The WDNR approach emphasized conservation commitments from landowners of large land holdings, ensuring that small landowners need not be drawn into the permit application process. The focus on large landholdings was important, for it was ecologically unsound and physically impractical to address every small land parcel in the Karner blue butterfly's potential range.

WDNR and USFWS agreed there were simply not enough resources to address Karner blue conservation in the manner that incidental take permits, issued under section 10 of the ESA, had done, even though few permits had been issued prior to 1994. Because of the size of the lands involved, a two-tiered conservation approach was agreed upon. The first tier focused on larger land holdings within the range and required specific management practices designed to preserve habitat value for the butterfly. In the second tier, small landowners were targeted through outreach and education, encouraging them to conserve the species along with their neighbors, local garden clubs, and so on. USFWS was receptive to this concept and encouraged WDNR to take the lead in developing the HCP.

In 1994, WDNR and USFWS distributed a letter to about 3,000 people in and around Wisconsin's Karner blue range, explaining the listing and the proposed HCP concept. Several public meetings followed. WDNR, with the assistance of state and local groups, then identified major landowners in the area as potential HCP partners. In these early deliberations, it was again reasserted that most small, private landowners would receive permit coverage automatically, without needing a certificate of inclusion.

WDNR and USFWS had jointly determined that extending coverage to small landowners, without a formal certificate of inclusion, would not violate the ESA or related regulations. Lawyers for both agencies became comfortable with the approach and opined that existing guidelines and past implementation of the ESA did not, and should not, prevent use of this strategy. Satisfied with the legal opinion, interagency partners focused on Karner blue biology and developing a sound risk analysis of how the proposed two-tier strategy would affect species viability. The analysis determined that the conservation plan detailed in the HCP—including coverage for small endeavors without a certificate of inclusion—was reasonable and posed no significant risk to the species. The final test was notification to the public, through the HCP/EIS (environmental impact statement) process, to determine if there were further concerns. No concerns surfaced, probably due to the thorough explanation of the participation strategy and associated risk analysis described in the HCP/EIS.

When the statewide HCP strategy was agreed upon, the next step was to identify partners with large land holdings, agree on sets of commitments, and establish a process to enter into agreements. All commitments and responsibilities of each partner were captured in a contract between the partner and WDNR, referred to as the Species and Habitat Conservation Agreement (SHCA). WDNR then submitted all the SHCAs as part of its HCP, coupled with an application for an incidental take permit. Partners included representatives from the forest industry, counties, utilities, TNC, and state agencies. All pledged to apply conservation measures to benefit the species on more than 250,000 acres. They also committed to engage in public education and outreach efforts to encourage other landowners and managers to partake in Karner blue conservation. Several partners' historic land management practices had already benefited the Karner blue, through the incidental creation or maintenance of disturbance habitat preferred by the butterfly. The newly developed strategies in the HCP authorized landowners to manage the land for their desired purposes while minimizing incidental take. Importantly, the desired results in the HCP included both the landowners' economic objectives and the conservation objectives for the Karner blue butterfly. Survey and monitoring protocols were established to determine whether Karner blue conservation objectives were being met.

From 1995 to 1998 the partners—including WDNR, with technical assistance from USFWS—met to develop an acceptable plan, including the participation plan affording incidental take coverage to all landowners and land users in the state. The WDNR also committed to conduct a review following three years of implementation of the HCP to determine if changes were needed in the participation plan. That review was conducted, and the USFWS determined that no change was needed.

The development and execution of the Wisconsin HCP for Karner blue was marked by a high degree of transparency and public participation.

Representatives of the Sierra Club, the Wisconsin Audubon Society, the WWOA, and the Wisconsin Paper Council participated regularly throughout the plan development. Other groups—including the National Wildlife Federation and the Defenders of Wildlife—were notified of plan developments and actively participated in select discussions. The application for an incidental take permit was filed with USFWS on April 1, 1999. Although the period from 1994 to 1999 was lengthy, it proved necessary for developing a level of trust between the landowner partners and WDNR sufficient to make the partnership viable. Unfortunately, during the seven years between the listing and the issuance of an incidental take permit, disturbance activities that might have benefited the habitat and the species were prohibited, because those management activities could have resulted in the incidental take of the species. A permit was issued by USFWS on September 27, 1999 following a thorough public review of the application and completion of an HCP and EIS review process.

WDNR and USFWS Roles in the Wisconsin Initiative

WDNR is the lead agency and an implementing co-partner in the initiative, the applicant for the permit, and the entity responsible for implementing the HCP. It is the responsibility of the WDNR HCP Implementation Coordinator, in conjunction with the Implementation and Oversight Committee (IOC), to administer the HCP. The HCP coordinator also has the task to audit partners to assure compliance with the partner-state agreements (SHCAs), the HCP, and the permit. In addition, WDNR receives and compiles data on surveys, monitoring, and land transactions, as required by the permit. Twice a year, WDNR and the USFWS Green Bay Field Office staff meet to assess data, performance, and need for changes to the plan or protocols. WDNR is also a member of the partners' IOC. IOC members represent the various categories of partners and make decisions on implementation of the HCP on behalf of the partnership. Early in development of the initiative, WDNR decided to exercise its responsibilities as a partner and facilitator and not just be a command-and-control regulatory agency. Several key partners commended WDNR on its role in the IOC and asserted that WDNR's decision not to exercise a traditional regulatory command-and-control role was crucial to the success of the endeavor.

USFWS provided technical support and assistance throughout the development of the HCP. This support was especially helpful in informing WDNR and partners of USFWS guidelines and procedures they needed to follow in applying for a permit. The Service was also helpful in pointing out potential concerns with the proposed conservation strategy and identifying information needed in reviewing HCP applications and issuing appropriate biological opinions.

The funds needed to develop and administer the HCP for the Karner blue were obtained from diverse sources. The estimated in-kind value/cost of developing the HCP exceeded $1.5 million. The major contribution for the Wisconsin HCP came from partners' in-kind services—approximately one third of which came from WDNR staff. These contributions included landowner representatives' and participants' time and expertise, as well as partners' implementation of conservation activities to comply with commitments made in the HCP. Their activities included continuing participation in HCP implementation meetings and processes, pre- and post-management surveys, effectiveness monitoring, planning, and enhancement or maintenance of habitat. In addition, WDNR contributed funds through regular state budget processes and grants. USFWS assisted with funds for activities associated with species recovery, including surveys, research, and land acquisition, and it took care of publishing costs for HCP and EIS. Some partners also donated funds for administration costs.

Accomplishments of the Wisconsin Initiative

This statewide HCP constitutes an innovative approach to species conservation under the ESA. The Wisconsin initiative is designed to move regulated communities beyond compliance and toward proactive conservation measures on the land for the betterment of the Karner blue butterfly. Conservation practices are applied to the lands in the course of routine land management activities. Management programs are designed to ensure that the habitat is not lost because of natural succession in the vegetation communities. Management actions call for necessary and beneficial disturbance activities to the habitat, rather than the more traditional "hands-off" approach given to protecting individual animals and preserving existing vegetative communities. To accomplish a disturbance-driven management regime statewide, this strategy includes all Wisconsin landowners and managers who might affect the species, regardless of land size and use; thereby building a statewide partnership in conservation.

Performance

The primary HCP objective of "no net loss of habitat" continues to be met. Karner blue habitat is always in flux because of the transient nature of disturbance habitat. As such, it is difficult to precisely quantify the extent of suitable Karner blue habitat at a given time. This challenge is further compounded when transient disturbance habitat must be measured across the entire state. However, one way of ensuring that no net loss of habitat occurs is through continuous audits and reviews of land management practices and habitats in large landholdings participating in the HCP. Previously unidentified habitat continues to be found, and large Karner blue habitat areas on public land have not experienced a decline.

Another objective of the Wisconsin initiative was to offer small, private landowners incidental protection, while simultaneously encouraging them to assist in Karner blue conservation through education and outreach efforts. A three-year review of this participation strategy, as required by the incidental take permit, was completed and submitted to USFWS for review on September 10, 2003. Landowners throughout the state—as well as others, through such events as a "Karner Blue Butterfly Festival" celebration—have come together to support the species and conservation effort. WDNR, in the Participation Plan Three-year Review and Assessment Report, concluded:

> *The Karner Blue Butterfly HCP has done more than help the Karner blue butterfly; it has bolstered statewide enthusiasm for rare species conservation and ecosystem restoration. The plan has also rebuilt trust between landowners and natural resources agencies, at least state agencies. This is a significant accomplishment given the storied history of traditional Endangered Species Act implementation. The results of these achievements are beginning to show. The WDNR is confident that the relationships developed with landowners will encourage even greater conservation collaboration in the future.*

USFWS, upon review of the report, concluded that it fulfilled the requirements of the permit and constituted an impressive overview of the good outreach and education work being done by the Wisconsin DNR and partners to the HCP. (Results of that review are available on the WDNR Karner blue butterfly web site at http://dnr.wi.gov/forestry/karner/.)

The Karner blue butterfly is better off in Wisconsin as a result of actions taken under the statewide HCP. Measurable successes of the Wisconsin initiative include: a) increases in known occurrences of the species, from 103 locations in 1991 to over 500 known subpopulations in 2007; b) increase in lupine (habitat) sites for all shifting mosaic landowner categories (such as industrial forestlands) of 238 percent and 300 percent since 1998; and c) consistent increases in voluntary landowner participation in this initiative, documented by WDNR, USFWS, and partners to the HCP.

The conservation success of the Wisconsin HCP is due, in large measure, to efforts of the HCP partners in executing conservation strategies, conducting surveys, monitoring habitat conditions, and carrying out education and outreach for Karner blue conservation among small landowners and managers in the state. A large portion of Wisconsin forest landowners manage the working landscape while accounting for the Karner blue's needs. Forest managers employ management surveys and timber harvests in a manner that maximizes the species' ability to move from one habitat patch to another when forest succession eliminates its disturbance habitat. Karner blue butterflies are most commonly found on early successional forest habitats, as in the period soon after a timber harvest. After eight to twelve years, habitat patches are often lost, as trees grow and the canopy closes. Therefore, the goal of many forest landowners—referred to as a "shifting mosaic" strategy—is to harvest timber close to the declining habitat so that Karner blue butterflies can easily travel to newly created habitat in the harvest area. Historically, Karner blues have occurred in forest stands because of this dynamic; however, now there is increased planning for Karner blue habitat.

Highway corridor partners, and some utility corridor partners, have committed to managing their rights-of-way with frequent surveys and mowing and clearing operations designed to maintain beneficial disturbance of habitat, consistent with HCP guidelines. Other partners have committed to enhancing and maintaining habitat by conducting necessary habitat maintenance practices such as burning—especially on wildlife areas where resulting habitat offers benefits to an array of species, including Karner blues. New occurrences of the Karner blue, and its preferred habitats across large landscapes in Wisconsin, continue to be discovered as new collaborators join the effort and lands not previously surveyed are treated with management actions beneficial to Karner blues.

The participatory approach used by WDNR in the development and execution of the HCP contributed to alleviating—in part—landowner fears of being prosecuted under the ESA's take prohibition. The approach seems to have liberated landowners to practice good land stewardship, whereas before they might have been unwilling to admit the presence of listed species on their land. The Karner blue butterfly is abundant in Wisconsin because of historic and ongoing landowner management activities. The Wisconsin initiative was designed to educate landowners about the species and how they can help conserve it. Through educational efforts, landowners found they could develop strategies to benefit rare species. During the implementation of the HCP participation plan, WDNR staff were frequently surprised by new voluntary conservation efforts for the Karner blue conducted by partners, of their own volition. The task for WDNR and the HCP partners was making landowners aware of the species, informing them of efforts that could help conserve the species, and educating them on the regulatory relief offered to them through the incidental take permit.

WDNR's Division of Forestry played, and continues to play, an integral role in the success of the Karner blue HCP. The division has been instrumental in building upon its existing relationships with paper mills and private forestland owners to advance Karner blue conservation in county, state, and private forests in the state. The division was also integral in educating the public that forest landowners were practicing ecosystem management and were not focused solely on "boards and cords" management. The participatory approach, in addition, allowed other state agencies—which are not traditionally involved in natural resource management—to take a more proactive stance in species conservation. For instance, the Wisconsin Department of Transportation (WDOT) has modified mowing regimes in Karner blue butterfly habitat to ensure that its maintenance activities result in conserving butterfly habitat.

The collaboration between WDNR and USFWS has been productive throughout the development and execution of the HCP. Field representatives of the two agencies worked very hard to gain appropriate approvals to initiate the HCP process. On occasion, however, the different mandate and regulatory framework under which the two agencies were operating led to disagreements. These are discussed in more detail in subsequent sections.

Present Situation

The Wisconsin HCP for the Karner blue butterfly is in its tenth year of implementation. In 2009 the HCP partners assessed the program for the first 10 years, updated the HCP in line with adaptive management improvements, and subsequently applied to renew the permit for another 10 years. An assessment of the periodic reviews of partner commitments under the HCP, and WDNR audits of the SHCAs, reveal that most, if not all, stated goals of the Wisconsin initiative in conserving Karner blue populations have been met. Most public and private groups in Wisconsin, including WDNR, believe that no further "recovery" of the Karner blue butterfly in the state is called for because of its widespread presence and persistence. However, the USFWS Recovery Plan states otherwise. The recovery goal, according to the Recovery Plan, is to perpetuate viable metapopulations of the Karner blue butterfly in the major ecological regions throughout its historic range—a band between 41 degrees and 46 degrees north latitude extending from Minnesota to Maine (Dirig 1994).

Wisconsin and western Michigan now harbor the largest metapopulations of Karner blues that occur in the greatest amount of area in the geographic range of the species. The goal for these areas is to stabilize and maintain, and in some cases expand, the populations that now occur. Because of the significance of these two states as the centers of Karner blue abundance, more recovery units (RUs) and more metapopulations are established in these areas than in other parts of the range. These multiple RUs should protect the species against wide-scale declines in either state.

Although WDNR and partners continue to disagree with the USFWS decision that further recovery efforts are needed for the species in Wisconsin, they have agreed to assist in achieving the recovery goals, while continuing to suggest their modification.

WDNR continues to maintain that the endangered status of the Karner blue is driven by its decline in the eastern states, even though the species flourishes in Wisconsin. According to WDNR, a recovery strategy should more clearly state recovery targets to be expected in the core and peripheral range of a listed species. The present case of the Karner blue butterfly produces a rather perverse result: Wisconsin, with the largest and most expansive Karner blue populations, is asked to carry a heavier regulatory load than the eastern states, where the Karner blue populations are stagnant or declining. In short, the good work of the Wisconsin landowners and WDNR should be rewarded in Karner blue conservation and not be held hostage to lack of progress elsewhere.

During the development of the Karner blue butterfly Recovery Plan, WDNR strongly argued that recovery criteria should emphasize the amount, quality, and distribution of habitat disturbance over time across the landscape rather than the number, sizes, and spatial configurations of individual Karner blue populations. Since Karner blue is a species associated with disturbed habitat, the abundance and location of individual butterfly populations will vary across space and time. It would be difficult to keep track of precise butterfly populations that fluctuate and move over time. Furthermore, it would not be economically or ecologically sensible to preserve every butterfly population in the state, such as small populations occurring on declining habitat where the population is beyond the tipping point in natural succession and not associated with a larger metapopulation to support its continued existence. The recovery strategy should focus on statewide butterfly populations associated with disturbed habitats, rather than on individual populations that, for very sound ecological reasons, could vanish and then reappear elsewhere. Therefore, according to WDNR, it is ecologically more sound to focus on the distribution of disturbed habitats preferred by Karner blue over time, instead of emphasizing individual Karner blue populations. A habitat-based strategy, WDNR argues, would better account for normal population fluctuations and benefit other rare species associated with disturbed habitats.

Most state groups—public and private—hold that the current population and occurrence of the Karner blue in Wisconsin demonstrates a healthy, viable metapopulation, and that there is no need to maintain or be concerned with every individual population. The future outlook for the butterfly looks good, as it continues to benefit from disturbed habitats caused by historic land-use practices in the state, such as forestry, which are expected to continue. However, the present language of the ESA and its regulations does not offer any regulatory relief for Wisconsin landowners. The Act does not presently offer an opportunity to distinguish geographically separate populations for invertebrate species, as it does for vertebrate species, by designating distinct populations segments. In the absence of such a regulatory tool, and a very low probability of butterfly recovery in the eastern states, the prospect of formally recognizing and rewarding Wisconsin's conservation success with Karner blue butterfly seems slim, although recent cases and USFWS interpretations of the ESA could offer some relief, should the recovery goals be met. The Karner blue's incongruous endangered status within Wisconsin will continue to incur increasing land management costs that add no conservation value but divert scarce local, state, and federal resources from being used to protect more imperiled species.

Another challenge faced by WDNR and other parties to the Wisconsin initiative is to maintain a consistent level of commitment and resources to Karner blue conservation. Since the issuance of the permit in 1999, several key players have retired or moved on, and forestlands under corporate ownership have changed hands. Like other states, Wisconsin needs to develop institutional frameworks that allow for adaptive management in species recovery in the face of new information, new personnel, and new economic realities, among other factors.

Observations

Wisconsin's HCP for the Karner blue butterfly has been an effective state strategy, in cooperation with USFWS, to both conserve butterfly habitats and populations and provide regulatory relief to state landowners and managers. The initiative offers an effective model for species conservation that can be emulated by other states. In the development and execution of the Wisconsin HCP for the Karner blue butterfly, several regulatory, financial, philosophical, and institutional issues facilitated or hindered effective conservation efforts.

Administration of the ESA should recognize and reward successful species conservation efforts. Wisconsin's comprehensive and successful state effort to conserve the Karner blue butterfly has neither been recognized nor rewarded by changing the endangered status of the Karner blue butterfly. The ESA has historically prohibited variable or different treatment of a listed invertebrate species. Distinct population treatment, under the language of the ESA, is restricted to vertebrates. Under past interpretations by USFWS, recovery of invertebrates must be accomplished throughout the entire range for the Service to propose its delisting, although recent USFWS interpretations might offer some hope for relief in the state. Therefore, even with growing healthy Karner blue populations in Wisconsin, the state can expect no reduction in restrictions on state landowners and their land management until the species is recovered throughout the range. This policy is both politically and ecologically unsound. It offers a disincentive to individual states to invest in conservation of a particular species if the likelihood of that species recovery is slim to poor in other states within the species range.

Species ranges are often dynamic. The ebb and flow in natural cycles depends on the availability of suitable food and habitat conditions for that species. All species ranges include core and peripheral areas, which, for wide-ranging species, may lie in different states. The health and abundance of a species population in its peripheral range is usually more vulnerable to the vagaries of natural fluctuations in habitat conditions. It is imprudent to use the same standard to assess species conservation success for the core and peripheral range. There is no cogent rationale for Wisconsin to bear widespread land-use restrictions because it supports the largest and healthiest Karner blue populations, while states with fewer butterfly populations, in practice, enjoy a lesser regulatory burden. This anomaly in the administration of the ESA should be remedied, rewarding landowners and administrative agencies for success in species conservation by authorizing variable regulatory standards for species in different geographic areas. Such a change could constitute a powerful incentive to landowners and state agencies to initiate proactive efforts for imperiled species within their state jurisdictions, if they are assured that their efforts will result in regulatory relief.

The key ingredient in encouraging state efforts directed to species conservation is to recognize and reward conservation success when it is demonstrated. USFWS, in administering the ESA, should direct more focus on successes in species conservation. It must emphasize and act with comparable haste in downlisting and delisting species as it does in listing species that are threatened or endangered. State agencies, landowners, and others should be rewarded with regulatory relief for successfully recovering a species or preventing its further decline. State fish and wildlife agencies, as well as landowners affected by a listed species, should be consistently encouraged to play a greater role on recovery teams and in recovery planning. To an ecologically reasonable extent, a broad set of options should be available to encourage innovative strategies to achieve species recovery. Similar flexibility is also needed in developing nuanced recovery measures, which are better aligned with species distribution, to reward successful conservation programs. Federal agencies should make greater use of distinct population segments and geographical exclusions in both listing and delisting species, including invertebrates, rather than solely relying on blunt and rigid rangewide treatment. States embracing species conservation and enhancement should be treated differently and provided incentives to continue their conservation efforts, even if a neighboring state chooses not to or is unable to produce comparable conservation results.

It is increasingly difficult to support the Wisconsin initiative primarily out of state funds. Thus far, funds for the administration of the HCP and permit compliance have come largely from the WDNR Division of Forestry. In light of ensuing overall state budget cuts, that source of funding is probably set to decline. USFWS has, thus far, been unable to substantially contribute to the Wisconsin Karner blue initiative. Section 6 funding appears unavailable, because Karner blue is no longer in peril in Wisconsin. Wisconsin ethics and conflict of interest laws prevent WDNR from soliciting or accepting funds from entities or individuals it regulates, including most of the private landowners in the state. Over the last five budget cycles, WDNR has sought to lift constraints and allow partners to donate funds to the Karner blue initiative. Due to the unique circumstances of this conservation program, the Wisconsin legislature recently remedied this barrier. In these circumstances, WDNR finds itself in a peculiar situation; USFWS offers no regulatory relief by delisting the Karner blue butterfly in Wisconsin despite its healthy populations there, and the Service does not offer much financial assistance, because the species is not at risk in the state.

Conservation philosophies of state and federal agencies are often different and can effectively complement each other under the right circumstances. WDNR is largely a resource management agency, and USFWS is primarily a resource regulatory agency. Broadly speaking, traditional regulatory agencies such as USFWS often approach species conservation with the premise that the threat of regulatory penalties will protect or further the health and welfare of imperiled species. By contrast, most state wildlife agencies such as WDNR, through experience with law enforcement programs, have learned that the threat of a penalty is often an unreliable tool to obtain active assistance and cooperation in achieving conservation objectives. Different conservation philosophies of USFWS and WDNR affect their respective management strategies. USFWS is partial to relatively rigid application of its rules and regulations, while a state wildlife management agency such as WDNR is more given to exploring a wide range of relatively flexible management options in meeting its resource conservation goals. USFWS, as a regulatory agency, might prefer to advocate a cautious approach, minimizing the threats to a single species, while WDNR, with a broad resource management mandate, might prefer a range of alternative management activities that could benefit a broad range of species, including the ones listed under the ESA, and thus have the potential to create additional habitats. Over time, the latter approach might boast a high likelihood of resulting in greater species abundance and diversity even if, in the short run, it resulted in take of listed species. This is particularly true when a listed species is dependent on habitats that were regularly refurbished by natural events such as wildfires or floods. An effective national species conservation strategy should recognize and foster the complementary philosophies of management and regulatory agencies and not imprudently force the two to be congruent.

In particular, the ESA should be increasingly administered in a manner that encourages willing land stewards to engage in species conservation by removing regulatory barriers. USFWS employees must be encouraged to exercise sound judgment to authorize the use of innovative initiatives and strategies, based on sound ecology, that hold promise for species conservation. The target audience for most, if not all, species conservation efforts is landowners and managers. To obtain landowners' cooperation, it is important that they understand the reasons for species decline and the conservation steps suggested to reverse the course. When wildlife professionals are unable to make that case, they lose credibility and support from landowners. For example, it is increasingly difficult to rationalize the restrictive land-use regimen in Wisconsin, when all can see that Karner blue butterfly populations within the state are doing well. It is always important to bear in mind that federal and state agencies cannot achieve success in species conservation by relying solely on their own limited resources. Therefore, ESA rules and policies that allow for credible partnerships with private landowners, based on reliable, factual information, should be strongly encouraged. In this regard, USFWS policies that empower state agencies, with regulatory authority and financial assistance, to forge credible and meaningful partnership with private landowners can be very effective.

Sorely needed are institutional structures that facilitate collaborative state and federal identification and execution of species conservation priorities within a state on a sustained basis. In a world of competing demands on the public treasury, there will never be enough resources to address all actions possible in conserving threatened and endangered species. Effective species conservation is smart conservation that enables scarce resources to go further through economies of scale, elimination of redundant actions, and reduction of transaction costs. The present structure of state and federal partnerships under the ESA relies too heavily on extraordinary individual efforts that make collaborations succeed despite bureaucratic and institutional hurdles. This structure is unsustainable, as individuals retire or move to new challenges, and collaborative processes lose the energy that sustains them unless replacements are found who have a desire to finish the job.

Conclusion

Federal agencies, under the ESA, have historically provided a valuable safety net for imperiled species, especially in cases where state agencies were either unable or unwilling to engage in efforts to conserve them. Looking at the future, a new conservation strategy is called for—one that encourages a stronger commitment by state agencies to lead species conservation efforts within their states. Only if a state is unable or unwilling to accept this responsibility should the federal agency step in. State fish and wildlife agencies have valuable expertise, knowledge, and data regarding resident species and their historic ranges. They must be given the opportunity to take a lead role, or at least a coequal role, in applying species conservation programs and bringing landowners and others together to forge collaborative conservation efforts.

Compared to federal regulatory agencies, state agencies enjoy a stronger presence within their states and work closely with landowners through state fisheries, wildlife, forestry, or endangered and threatened species programs. It is not impossible for federal regulatory agencies to achieve a similar level of credibility and trust with state private actors, but it is substantially more difficult and costly in time and funds. Effective, proactive species conservation programs require close collaboration with public and private landowners and managers, and state agencies are better positioned to achieve that. Consequently, all efforts should be made to harness species conservation resources available at the state level.

ESA regulations can and should be bolstered to recognize and build upon the strengths of state agencies in species conservation. In doing so, regulatory, financial, philosophical, and institutional hurdles, discussed above, should be addressed with haste. Regulatory relief under the ESA should accompany demonstrated success in species conservation. Frequent and innovative use of distinct population segment and geographical exclusion provisions under the ESA should be made for all listed species. The primary weakness underlying state initiatives for species conservation will continue to be lack of reliable funds. Federal administration of the ESA should strongly encourage a robust state grant system—one that makes federal financial assistance and regulatory relief commensurate with a state's accepted responsibility in species conservation and demonstrable conservation successes. The newly created federal funding programs available to private landowners offer a good start, but much more needs to be done. Further, dedicated attention should be paid, by both state and federal agencies, to developing a permanent institutional structure—one that allows them to forge their complementary philosophies and resources to achieve effective species conservation.

Section 6 of the ESA recognizes the potential for cooperation between federal and state agencies in implementing the ESA. Unfortunately, thus far federal agencies have used section 6 agreements primarily as a funding mechanism. Section 6 cooperative agreements should be used in a much broader context, since they were originally intended to encourage and authorize comprehensive state species conservation initiatives and programs. Robust section 6 agreements could well free more resources for species conservation by eliminating needless duplication of efforts. A robust and broad section 6 agreement between WDNR and USFWS, addressing the conservation of the Karner blue butterfly, among other species, can result in more efficient, effective conservation of all associated species.

The real challenge for state wildlife agencies in species conservation is often political or economic in nature. In an era of tight state budgets, state wildlife agencies are forced to prioritize their conservation efforts. Regulatory restrictions under the ESA, including land-use controls, are of great concern to landowners. To elicit landowner cooperation and political and economic support for species conservation, agencies need to alter widely held negative perceptions regarding conservation of endangered and threatened species. They need to be able to reassure landowners and managers that successful conservation efforts will be rewarded by regulatory relief, and that the federal government will contribute funds for effective conservation strategies. Federal administration of the ESA should pursue policies that enable state agencies to obtain necessary support within, and funding for the states to implement effective conservation strategies. If state fish and wildlife agencies are given the authority to pursue proactive rare species conservation, without unreasonable federal restrictions, and if private landowners can be brought into these efforts in a collaborative manner, therein lies a formula for success. The availability of stable federal funding as an incentive to engage in species conservation programs will greatly help make such conservation a reality.

State wildlife agencies must be recruited and funded to address species decline prior to listing. It is arguable that Wisconsin's initiative for the Karner blue butterfly could have been equally effective if USFWS had given the state an opportunity to enact a conservation strategy that would preclude listing within the state. USFWS, in encouraging state species conservation efforts prior to listing, should also attempt to broaden the single-species focus in the

ESA to emphasize ecosystem habitat and natural community management and enhancement.

The value and vigor of the ESA should not and need not be lost. However, to sustain its value and garner further acceptance, the present approach to species conservation needs to be modified. Sole reliance on a command-and-control system will not protect and conserve imperiled species. Federal and state agencies alone do not have enough resources to meet the challenges of species conservation. The full purpose and vision of the ESA, to conserve threatened and endangered species and their ecosystems, will only be accomplished through collaboration. State agencies can help achieve that collaboration. State wildlife agencies must be actively recruited to take the lead in species conservation efforts. In several instances, doing so could be as easy as loosening some of the reins on ESA administration and reaching out to state agencies and landowners in a meaningful manner to gain their support and collaboration.

References

Dirig, R. 1994. "Historical Notes on Wild Lupine and the Karner Blue Butterfly at the Albany Pine Bush, New York." In *Karner Blue Butterfly: A Symbol of a Vanishing Landscape*, D. A. Andow, R. Baker, and C. Lane eds. Minnesota Agricultural Experiment Station, University of Minnesota–St. Paul. Miscellaneous Publication 84-1994: 23–36.

Endangered Species Act of 1973, as amended (16 U.S.C. 1531 et seq.).

Federal Register. 1992. (Listing of the Karner blue butterfly as endangered.) Vol. 57, No. 42; December 14, 1992. Rules and Regulations, pp. 59236–59244.

Lane, C. P. 1994. "Habitat Preferences of the Karner Blue Butterfly in Minnesota." In *Karner Blue Butterfly: A Symbol of a Vanishing Landscape*, D. A. Andow, R. Baker, and C. Lane eds. Minnesota Agricultural Experiment Station, University of Minnesota–St. Paul. Miscellaneous Publication 84-1994: 63–72.

Schweitzer, D. 1989. "Fact Sheet for the Karner Blue Butterfly with Special Reference to New York." The Nature Conservancy. Cited in *Karner Blue Butterfly: A Symbol of a Vanishing Landscape*, D. A. Andow, R. Baker, and C. Lane eds. Minnesota Agricultural Experiment Station, University of Minnesota–St. Paul. Miscellaneous Publication 84-1994.

U.S. FWS (U.S. Fish and Wildlife Service). 2000. "Handbook for Habitat Conservation Planning and Incidental Take Permitting," as amended. June 1.

CHAPTER 8

The Red-cockaded Woodpecker: Conservation through Statewide Safe Harbor Agreements

Ralph Costa and Richard Gooch

In this chapter, we discuss the evolution, use, and success of statewide safe harbor agreements (SHAs) for enhancing the management and conservation of the federally listed, endangered red-cockaded woodpecker (*Picoides borealis*). We cover our topic in five main sections. First, we provide brief overviews of the history, conceptual framework, and implementation of these statewide programs. Second, we detail the development of the first such program and how other states became involved. In our third section, we focus on the various challenges encountered and how they were, or are being, addressed. Our fourth section provides details of the programs' performance and accomplishments to date (as of March 2010). Finally, in our conclusion, we summarize our thoughts regarding the programs' value, success, and future.

We hope to convey several key points. First, and perhaps foremost, statewide safe harbor programs have proven to be a valuable conservation tool for addressing the potentially controversial interface between the Endangered Species Act (ESA) of 1973, as amended, and protection of the federally listed red-cockaded woodpecker on private property. We develop and expand upon this theme throughout various sections of the chapter. Second, such relatively complex, broad-scope, statewide programs are not without challenges. Real or perceived, these must be addressed and include, but are not limited to, program development time, implementation logistics, and adequate funding. We expound on these and other related issues in our discussion and observations section. A third, critical point is that we believe the relationships developed between private landowners, state wildlife agencies, and the United States Fish and Wildlife Service (USFWS/the Service), via these SHAs, have fundamentally changed the paradigm of how management, conservation, and indeed recovery of red-cockaded woodpeckers on private lands is addressed. This relationship is initially outlined in the implementation overview section and further detailed in our discussion and observations section.

Background

The Service's official safe harbor policy, and associated implementing regulations, were published on June 17, 1999 (64 F.R. 32717), with amended regulations made on May 3, 2004 (69 F.R. 24084). However, the safe harbor concept for conservation of an endangered species was first implemented in April 1995 for red-cockaded woodpeckers in the Sandhills Physiographic Region of North Carolina, encompassing six counties (Costa and Kennedy 1996, Bonnie 1997a, Costa 1999).

Safe harbor's use at the statewide (or regional) scale—that is, within the red-cockaded woodpecker's range in any particular state—draws upon the Service's red-cockaded private lands conservation strategy outlined by Costa (1995).[1] One of the three key components of that strategy included development of statewide habitat conservation plans (HCPs). Although analogous to SHAs, HCPs are not the same and usually involved an attempt to resolve an immediate conflict between the presence of red-cockaded woodpeckers and a landowner's economic objectives.

These HCPs had three objectives: 1) minimize and mitigate the permanent loss of woodpeckers in small isolated populations; 2) provide a source of woodpeckers for other public or private lands committed to woodpecker conservation and recovery; and 3) relieve private landowners of the potential economic hardship of managing minor woodpecker populations on small, isolated properties. Although providing incidental take for isolated woodpecker groups, using statewide HCPs, has generally failed (with one exception, Georgia; see below), the concept of developing statewide red-cockaded woodpecker conservation strategies for private lands can be traced to USFWS private lands conservation strategy.

The transition of the states' focus from HCPs to SHAs was likely due to the evolving controversy surrounding HCPs. More importantly, however, this shift was also attributable to states' acknowledgment that the need for SHAs was much greater than that for HCPs. In other words, more landowners would be interested in SHA regulatory assurances in order to grow and manage the type of forests that are woodpecker-compatible than in mitigating the loss of woodpecker groups using the HCP process. Additionally, state wildlife agencies likely realized the comparative advantage of offering an incentive-based approach, as opposed to the regulatory-based HCP option.

Comparing and Contrasting Safe Harbor Agreements and Habitat Conservation Plans

Key features distinguish the safe harbor approach from an HCP. The objective of the SHA is to facilitate an interested private (nonfederal) property owner to provide for conservation of a threatened or endangered species without incurring additional regulatory burden, beyond what is currently present at the time the SHA is signed. An HCP is a process available within the ESA to allow nonfederal property owners to incidentally take protected species associated with an immediate economic use, usually offsetting the loss of the species by some negotiated mitigation strategy. These mitigation strategies might involve purchase of off-site habitat (occupied or unoccupied by the species of concern), with a financial commitment to manage those lands for some length of time, or establishing replacement individuals, as in the case of red-cockaded woodpeckers, for those lost. HCP applications seek authority to adversely affect listed species through land or water development actions. SHAs typically focus on integrating the needs of a species into the existing management and economic plans of the landowner. SHAs have been used to promote reintroductions and improve the habitat conditions of species. Another important difference is that SHAs are expected to enhance the survival and recovery of the species, whereas the regulatory standard for HCPs is to mitigate or minimize adverse affects on the species.

Conceptual Framework Overview

The attraction for landowners of participating in red-cockaded woodpecker safe harbor programs is that they have a clear and up-front responsibility for addressing the presence (or potential presence) of red-cockadeds on their lands. This "baseline" is identified in the SHA—usually expressed as "active clusters" (one or more), and the required foraging habitat associated with each cluster. An active cluster is the aggregate of cavity trees occupied and defended by one red-cockaded woodpecker group.[2] The baseline is zero for many participating landowners. In most cases, management of the baseline is identified, as are other conservation measures necessary to maintain the presence of the species on the landowner's property. For red-cockadeds, the conservation measures entail some combination of active forest management—for example, prescribed burning to preclude hardwood mid-story encroachment, and sufficient pine basal area (stocking or density) of the size and age of trees necessary to provide adequate foraging substrate and roost/nest cavity excavation.

The regulatory assurance of safe harbor essentially requires no more than baseline habitat from the landowner. Landowners are free to improve the habitat on their properties so as to have more woodpeckers on their land. Then, if their economic or management objectives for the enrolled property change, they are free to return to the baseline habitat conditions. Also, the assurances address the fact that USFWS will not change the intensity or type of conservation measures without the landowner's consent. The incidental take authority is for "above" baseline conditions—for example, new active clusters that are established via habitat enhancement or restoration during the life of the SHA.

These assurances allow the property owner to alter or modify habitat some time in the future, even if such actions result in the take of a listed species. Habitat alteration can be undertaken to the extent that it returns the species or habitat back to the original, agreed-upon baseline conditions. It is important to note that since approving the first use of safe harbor for red-cockaded woodpeckers in April 1995, USFWS has had only two landowners, of 267 enrolled, leave the program. In one case, the single active cavity tree comprising the baseline died of natural causes, allowing reduction of the baseline to zero. With this change, the landowner decided to leave the program and pursue harvesting of the tract. In the other case, the landowner changed his management objectives and harvested the forest habitat. With the second landowner, the baseline was also zero.

For red-cockaded woodpeckers, USFWS does not approve SHAs unless it can make a finding that such approval can reasonably be expected to provide a net conservation benefit. The net conservation benefit assessed by USFWS for red-cockaded woodpeckers was initially addressed by its red-cockaded private lands conservation strategy in the early to mid-1990s. It was later supplemented by technical and policy decisions made on individual applications, for both HCPs and SHAs, during the late 1990s and early 2000s. This net conservation benefit guidance has been recently formalized in the Red-cockaded Woodpecker (*Picoides borealis*) Recovery Plan, Second Revision (U.S. FWS 2003). In other words, landowners were well aware of the Service's internal thought process as to what role private lands had regarding species recovery and the biological and regulatory standards that would affect this role.

For example, in the 1990s, while USFWS recognized that private lands would play a minor role in actual recovery criteria for red-cockaded woodpeckers, it simultaneously acknowledged that those private lands still maintained an indirect recovery role. That role included serving as habitat corridors, donor populations, genetic reserves, and increasing the size of adjacent, extant public land populations. Regarding biological and regulatory standards, USFWS, since the early 1990s, established and provided clear guidance as to the quantity and quality of habitat that should be provided for each group of red-cockaded woodpeckers on private lands (Costa 1992). This habitat was intended to provide each woodpecker group its minimal ecological needs, that is, its nesting and foraging habitat requirements. Simultaneously, landowners providing this habitat were meeting their ESA regulatory responsibility to avoid take. Additionally, beginning with the development of the first red-cockaded woodpecker HCP in 1993, USFWS established guidance for a "no net loss" policy as the mitigation standard. That is, each woodpecker group to be lost would be replaced by at least one other group at another location. All of the above standards, and several others, helped establish a clear understanding among landowners of how USFWS intended to integrate management and conservation of red-cockaded woodpeckers with private landowner ESA responsibilities.

Implementation Overview

Following is a brief overview of how statewide red-cockaded woodpecker safe harbor programs are implemented. The respective state wildlife agency, in coordination with USFWS at three levels (that is, with the state field office, Atlanta regional office, and the Red-cockaded Woodpecker Recovery Coordinator), works with statewide red-cockaded woodpecker safe harbor steering and scientific advisory committees to develop and submit an SHA. The structure and content of SHAs are governed by the safe harbor policy and implementing regulations (64 F.R. 32717; as amended, 69 F.R. 24084). Other intersecting federal and state laws come into play during the development of an SHA, including the National Environmental Policy Act (NEPA). A variety of analysis and decision documents must be prepared (for example, a biological opinion, environmental assessment, and set of findings) to satisfy SHA regulations and NEPA requirements. The SHA and NEPA documents are submitted for public review and comment. If the SHA is approved, USFWS simultaneously issues an enhancement of survival permit.

The permit provides authority to the wildlife agency for "incidental take" of red-cockaded woodpecker groups and their habitat, above baseline conditions. The state wildlife agency is the primary delivery system of the program to landowners. Individual landowners are enrolled in the program by cooperatively drafting safe harbor management agreements (SHMA) with the state wildlife agency. The SHMA is tiered off the requirements of the SHA and documents baseline conditions, specifies conservation measures the landowner agrees to undertake, identifies and quantifies the expected net conservation benefits, and details the landowner's monitoring and reporting requirements. A standard, fill-in-the-blank evaluation form documents that information and is appended to each SHMA. The Red-cockaded Woodpecker Recovery Coordinator and the local state USFWS field office review all evaluation forms in the enrollment process. Once all parties agree that the SHMA is acceptable, landowners are officially enrolled in the program via a "Certificate of Inclusion." The certificate gives the landowner the same regulatory assurances and incidental take authority provided to the state wildlife agency, via the so-called "umbrella" permit issued to the state by USFWS (Lohr 2000).

North Carolina Sandhills Red-cockaded Woodpecker Safe Harbor Program

A brief summary of how the Sandhills safe harbor program came into existence is instructive, as it became the operational guide to how USFWS facilitates interest in statewide efforts eventually led by state wildlife agencies.

In response to ongoing loss of longleaf pine habitat in the Sandhills—and the need to address the recovery relationships and responsibilities among the

TABLE 8-1. Status of Red-cockaded Woodpecker Statewide Safe Harbor Agreements

State	*Initial Contact*	*First Meeting*	*Permit Issued*	*Status*
Alabama	October 22, 1993	February 28, 1994	November 30, 2006	Complete
Florida	June 9, 1995	July 20, 1995	May 19, 2006	Complete
Georgia	June 8, 1993	September 1, 1993	May 24, 2000	Complete
Louisiana	June 12, 1995	September 11–13, 1995	January 25, 2005	Complete
North Carolina	December 11, 1995	January 11, 1996	October 2, 2006	Complete
North Carolina (Sandhills)	February 24, 1993	March 3, 1993	April 17, 1995	Complete
South Carolina	June 8, 1993	July 26, 1993	March 26, 1998	Complete
Texas	July 27, 1993	June 1, 1995	March 20, 1998	Complete
Virginia	March 4, 1997	August 27, 1997	July 5, 2000	Complete

private, state, and federal property owners in the region—a Red-cockaded Woodpecker Interagency Working Group (IWG) was formed in March 1993 to develop and implement a Red-cockaded Woodpecker Sandhills Private Land Initiative (Campbell et al. 2004) (see Table 8-1). It was later discovered that the primary reason for longleaf pine habitat loss in the Sandhills (and in other areas of North Carolina) was ESA-related landowner fear of land-use restrictions, due to existing or potential occupation by red-cockaded woodpeckers (Michael 1999; Drake 2000). The IWG, recovery coordinator, and other USFWS and state agency personnel throughout the Southeast were intuitively aware of this landowner fear and the regulatory disincentive it created. The IWG's desire to respond to the "more-is-worse" (that is, more woodpeckers on your land) disincentive was a primary, driving force that eventually lead the group to formulate a so-called "safe harbor" concept. By April 1995, as a result of this IWG's efforts, the nation's first safe harbor permit was approved.

Unlike other statewide SHAs later approved for the red-cockaded woodpecker, the Sandhills program was unique in the sense that USFWS itself was, and remains, the administrator/permit holder. There were some brief, initial discussions regarding potential issuance of the permit to the North Carolina Wildlife Resources Commission. Because of the many unknowns associated with this first permit and other reasons, the state, IWG, and USFWS chose not to pursue this option. The permit was ultimately issued to the recovery coordinator for several reasons. First, at the time, and primarily due to the experimental nature of the program, no suitable applicant was identified or came forward. Additionally, USFWS wanted to maximize the program's potential success by implementing it internally, while simultaneously testing it as a viable conservation tool. In other words, before USFWS could ask others, such as state wildlife agencies, to support and operate the program, it had to demonstrate that safe harbor was a credible and functional conservation effort. Second, without a participating state agency, USFWS itself represented the best administrative vehicle to delegate incidental take authority to participating landowners.[3] Third, the concept from the start, at least as it evolved in the Sandhills, was viewed as a "recovery" program because the area of concern, that is, the Sandhills, was incorporated within the designated North Carolina Sandhills recovery population, centered on Fort Bragg. For this "recovery-rationale" reason, and others mentioned above, USFWS made a decision to issue the permit to itself, the logical recipient being the recovery coordinator.

Involving States in Safe Harbor

After issuance of the Sandhills permit, the promotion and evolution of statewide red-cockaded woodpecker safe harbor programs was primarily directed and coordinated by the Service's recovery coordinator. The foundation had been laid when, from June 1993 to December 1995, the recovery coordinator made initial contact with key state wildlife agency administrators and wildlife biologists regarding development of statewide HCPs in Alabama, Florida, Georgia, Louisiana, North Carolina, South Carolina, and Texas (see Table 8-1). The recovery coordinator had been making these contacts and promoting statewide HCPs for two primary reasons—first, to address the ongoing loss of demographically isolated woodpecker groups, and second, to respond to the potential economic impacts of these groups on the landowners who harbored them. Typically, landowners harboring isolated woodpecker groups were "mom-and-pop" landowners. That is, they owned relatively small tracts of land that could be significantly impacted by the presence of even one woodpecker group. USFWS had made the decision not to support or pursue large-scale, statewide HCPs for larger "populations" of red-cockaded woodpeckers. Landowners harboring non-isolated, larger populations were free to develop individual HCPs if they chose.

All states contacted in the early to mid-1990s expressed interest in learning about statewide red-cockaded HCPs. Subsequently, from July 1993 to January 1996, the recovery coordinator organized and chaired initial, individual meetings of interested and affected constituents for each of those states (see Table 8-1). In each state, the wildlife agency staff and recovery coordinator identified the agencies, organizations, institutions, and individuals who they believed had—or represented a constituency who had—a potential stake in the issue and therefore should be represented at the first meeting.

Initially, the groups formed as a result of these statewide meetings were called Red-cockaded Woodpecker Conservation Coalitions. These coalitions represented a diverse group of potential partners, all of whom had a role or interest in, or were indirectly or directly affected by, red-cockaded woodpecker management and conservation on private property. Typically, by the close of the initial meeting, the group had decided to move forward with development of statewide HCPs for isolated red-cockaded groups. Another important outcome of some initial statewide meetings was agreement that one or two committees should be formed to move the initiative forward—typically, a steering committee and a scientific advisory committee. Many original members of coalitions subsequently became members of one or both of these committees. Over time, other key individuals were asked to participate with coalitions or on committees.

The primary roles of steering committees were to 1) investigate funding sources; 2) conduct outreach promotion, that is, serve as the primary point of contact for their constituents' interests in and education regarding the initiative; and 3) identify and resolve any administrative, legal, or regulatory issues to ensure efficient program implementation. The primary roles of scientific advisory committees were to 1) determine current status and trend of red-cockadeds on private land; 2) develop outreach and educational materials; and 3) identify suitable mitigation properties and procedures. In some states, coalition members took on the above roles.

The simultaneous establishment of coalitions (1993–1995) to produce statewide HCPs, and the development and launching of the Sandhills safe harbor program (1995), caused a reexamination by USFWS and the coalitions of what statewide red-cockaded woodpecker "conservation" plans should address. Additionally, opposition to HCPs from a few influential environmental groups (Bonnie 1997a, 1997b)—as well as the scientific community's incomplete understanding, at the time, of red-cockaded demographic isolation—created further uncertainty for coalitions on how to proceed. While coalitions recognized the value of relieving some landowners of their red-cockaded woodpecker-related ESA responsibilities via a statewide HCP, they also understood that the pool of potential applicants (those harboring isolated woodpecker groups) was relatively small. Additionally, at the time, HCPs in general were under significant scrutiny from the environmental community. As a result, state wildlife agencies were leery of getting involved in a possibly controversial listed species regulatory program. In contrast, the Sandhills safe harbor program was "in the news" as an emerging success story that had the potential to benefit *all* landowners, not just a limited number harboring isolated groups. Thus, from a coalition's point of view, safe harbor was a known, acceptable, and—importantly—successful program; statewide HCPs were an unknown "risk."

In spite of these differences between the programs and the potential hurdles associated with HCPs, USFWS informed each state that it would continue supporting development of a statewide HCP. However, given the initial success of the Sandhills safe harbor program (Costa 1997), the Service also offered to help each state develop a combination statewide HCP/safe harbor permit or simply a statewide safe harbor permit. A combination HCP/safe harbor permit would provide both incidental take for isolated groups and safe harbor enrollment opportunities for any qualifying landowner. Eventually, all involved states—Alabama, Florida, Louisiana, North Carolina, South Carolina, and Texas (Virginia had no need for an HCP)—except Georgia decided to drop development of statewide HCPs and instead move forward with statewide safe harbor programs. Georgia decided to continue development of an HCP/safe harbor combination program and was issued such a permit in May 2000 (Ozier et al. 2004). Georgia desired such a combination permit to provide 1) incidental take authority for the 23 landowners known, or thought, to harbor isolated woodpecker groups, and 2) safe harbor assurances for the approximately 35 families in the Red Hills region of south central Georgia who collectively harbored approximately 180 woodpecker groups, as well as their neighbors or others in the state who were interested.

In addition to the formal statewide coalitions, one particularly valuable contributor to the regional success of red-cockaded woodpecker SHAs was and continues to be Environmental Defense (Bean et al. 2001). Not only was Environmental Defense a member of the Sandhills IWG, it also was and is an active member of several coalitions (North Carolina, South Carolina, and Virginia). Additionally, over the years it has contacted the recovery coordinator and state wildlife agencies to offer assistance to develop statewide SHAs, for example, in Florida and Louisiana. Environmental Defense has also helped significantly with public outreach for safe harbor programs for red-cockaded woodpeckers and other species (Environmental Defense 1999).

Discussion and Observations

The time it has taken, from initial USFWS contact with the state wildlife agency, or from the initial Coalition meeting, to the issuance of permits has been long—very long in some cases (see Table 8-1). Initial meetings for establishing statewide SHAs in Alabama, Florida, and North Carolina began in 1994–1996 and were completed during calendar year 2006 (encompassing between 128 and 153 months). The other five permits took from 33 to 115 months to issue. Ironically, the nation's first safe harbor permit, the Sandhills, was issued 25 months after the first meeting, and before any policy guidance or regulations were codified.

Various factors have contributed to the inordinately long period of time it has taken to issue states safe harbor permits under section 10(a)(1)(A) of the ESA. Significant delays can occur at the state level for two primary reasons. First, after states agree to draft a statewide SHA, either an existing employee has to be assigned the job or someone new has to be hired. In the first case, drafting an SHA becomes one more "task" among a suite of others which compete for priority status. In the second case, funding a new position specifically to draft, and then implement, an SHA has been challenging for each of

TABLE 8-2. USFWS Funding Provided to State Wildlife Agencies and Private Landowners to Develop and Implement Red-cockaded Woodpecker Statewide Safe Harbor Agreements, 1997–2004.

States Participating									*Total*
Virginia	North Carolina (Sandhills)	Texas	South Carolina	Georgia	Louisiana	Alabama	Florida	North Carolina	**$3,326,200**
$51,000	$230,000	$150,000	$1,810,000	$547,500	$55,000	$33,500	$134,200	$315,000	

the states involved. Fortunately, USFWS funding (see Table 8-2) has helped states to largely overcome this issue, at least for the planning and development stage of SHAs. Additionally, the Service has attempted to standardize, to the extent possible, a working template of both the programmatic SHA and the individual SHMA documents. We believe standard templates will reduce the time required to develop agreements, because the majority of each document contains previously approved language and reflects the best legal and policy interpretations of the statute.

A second reason for state-level delays has been turnover of key personnel. Successful development of statewide red-cockaded woodpecker conservation initiatives requires the involvement of at least two key positions in state wildlife agencies. Those positions are the mid- to upper-level administrator, who oversees the nongame and/or endangered species program, and the field or species-specific biologist (for example, "red-cockaded woodpecker" biologist, avian ecologist, etc.) who is tasked or hired to draft the SHA and ultimately put the program into action. When initiatives begin to drag out over years, it is inevitable that turnover in one or both of these positions will occur, and indeed, that has happened. Logistical delays are exaggerated by staff replacement and the resulting loss of program momentum.

On the USFWS side, timing issues are also related to multiple factors, including staff turnover at field offices. Document preparation and SHA reviews are major time-consuming activities for the Service. Required documents—including an environmental assessment, biological opinion, set of findings, and ultimately a permit are prepared by the respective field office and then reviewed by the recovery coordinator, various regional office staff, and the solicitor's office at the Department of Interior. Similarly, SHAs are reviewed, in multiple drafts, by the appropriate USFWS field office, recovery coordinator, regional office staff, and solicitor's office. These multi-office, multi-draft reviews, along with the associated editing and reformatting occurring among drafts going back and forth to the state, consume significant time. Additionally, SHAs must be published in the Federal Register. This process alone—preparing the Notice of Availability, publication in the Federal Register, and responding to any comments—requires months to complete. Eventually—given the present level of state and USFWS staffing, the document preparation, reviews, and Federal Register requirements—time to permit issuance is measured in years.

Several key lessons have been learned regarding the timing issue. First, USFWS should inform the state agency up front that patience will be required, because it will take several years to complete the administrative procedures necessary to issue a permit.[4] Second, USFWS must identify, early on, key players in each state and work very closely with them throughout the entire process. We have found that without a local state/agency champion for such large-scale programmatic conservation initiatives, it is difficult to get going, sustain, complete, and implement the project. Third, USFWS should strive to have its own project "champion," an individual who will likely remain in place long enough to witness permit issuance.[5]

By having a recovery coordinator, USFWS provided vital consistency and continuity throughout the long process in each state—while maintaining critical momentum, which would otherwise have been lost—as both key state and USFWS field office personnel retired, transferred, resigned, or otherwise moved on. Very few state agency key players or USFWS field office biologists originally assigned to these initiatives are still in place. The recovery coordinator has assumed the role as the source of information on the history, background, progress, current status, and key contacts (private, state, and federal) for these programs. This substantially reduces both the learning curve of new staff and further potential delays.

In summary, key individuals have played important roles in the establishment and management of statewide safe harbor programs. The state SHAs cannot be implemented without the creation of a collaborative work environment to fully engage the stakeholders. The ability of USFWS or state wildlife agency personnel to explain the safe harbor process, with its technical and legal requirements, and articulate the economic responsibilities—managing baseline habitat—and conservation value to a landowner in clear, concise "kitchen table" discussions is a critical component of safe harbor marketing. Without this leadership, large-scale safe harbor programs will not be successful, and their numerous benefits to landowners and listed species will go unrealized.

State Wildlife Agency Participation and Motivation

In the early 1990s, potentially explosive conflicts between red-cockaded woodpeckers and forest landowners (including economic impacts, fear of the ESA, etc.) were common throughout the range of the woodpecker (Vandersteen 1992; Cone 1994). A reasonable person might ask: "What would motivate a state wildlife agency to insert itself into such a potent controversy?" State wildlife agencies realized the value of engaging landowners with a simple, incentive-based conservation program, and the demand for the safe harbor concept became very evident with the initial successes of the Sandhills safe harbor program.

Several factors influenced states' decisions and a particular state's interest in pursuing SHAs. First, the red-cockaded is a species particularly well suited to a safe harbor approach. The threats to, and appropriate conservation measures for, red-cockadeds are well established and accepted by most, if not all, constituencies. Species relocation and reintroduction technologies and methods (which have become a vital tool in consolidating or shifting safe harbor baselines) are well documented and proven as appropriate tools to advance recovery (Hagan and Costa 2001; Costa and DeLotelle 2006). Most importantly, various land management objectives of private landowners (notably the economies of pine straw raking, pole timber production, and hunting of bobwhite quail) are consistent with red-cockaded woodpecker management.

Second, most experienced agency personnel (and for that matter, others seeking a solution) understood that this compatibility between the economics of red-cockaded woodpecker management and existing landowner desires needed only the creation of regulatory incentives to achieve a complete "win-win" program. The survival of red-cockaded woodpeckers on private lands today is a direct result of the land management activities that have been practiced on those properties for decades. Typically, private lands harboring red-cockadeds can be put in one of two primary categories—industrial forests and quail hunting plantations. Woodpeckers on industrial forestlands are usually restricted to "natural" pine tracts that are not yet, or were never, slated for industrial forest management, that is, short rotation plantations. Landowners managing land for quail hunting conduct frequent prescribed burns, and maintain, via selective timber harvest, open, park-like, mature pine stands. Additionally, some landowners rake pine straw for additional income, ensuring a hardwood-free mid-story. These activities and resulting habitat conditions create ideal forests for red-cockaded woodpeckers. Thus, the recreational and economic use of their lands is totally compatible with recommended habitat management practices for red-cockaded woodpeckers. The use of safe harbor provided a strong regulatory assurance, especially for the nonindustrial landowner, to continue their woodpecker-compatible management without fear of additional ESA responsibilities beyond their baseline.

Third, the states' interest in being the "administrator" on a statewide basis eliminated the transaction costs of a landowner wanting or needing to pursue an individual agreement and obtain an individual permit (Kennedy et al. 1996). Therefore, by participating, the state agencies effectively remove the final obstacle of transaction costs to a successful program.[6] By performing this important function, the agency benefits from the resulting goodwill of landowners. We believe USFWS and state wildlife agencies, by assuming the transaction costs up front, substantially facilitate landowner participation in the program.

Adequate State Authority

One significant obstacle that several southeastern states have had to address is the need for clear legislative authority to implement safe harbor programs. One may presume this process would be long, aggravating, and potentially problematic. However, the influence of cooperating coalitions is instructive for its ability to advocate and achieve almost instantaneous solutions in this regard. The coalition that produced the South Carolina red-cockaded woodpecker safe harbor plan (including large forest products companies, conservation groups, and quail hunting plantation owners) scheduled a meeting in early 1998 with the appropriate legislative staff members regarding the need for a legislative solution. The coalition was well aware that the Service's approval was imminent, adding a bit of urgency to the situation. USFWS was invited to offer its support to the coalition and respond to any information needs about the SHA. When the invited legislators and staffers saw the unequivocal support for the SHA across the spectrum of interests represented by the coalition, the necessary legislative fix was introduced, and the state authority to implement an SHA was clarified. This succinctly illustrates the power of partnerships and the attractiveness of a well-executed SHA.

Value of Certainty for Participating Landowners

Essentially, engaging with USFWS or a state wildlife agency regarding federally listed species issues is a form of risk analysis for a landowner. The essential part of that decision process is the landowner asking: What will the government require me to do? Depending upon the clarity of that answer, a reasonable person may then proceed to engage or not engage in the process.

In our collective experience, many landowners would not even ask the question, because they feared the consequences of being in the "spotlight." For the red-cockaded woodpecker, this question was answered early, and without the landowner coming forward, through the creation of a regulatory guidance document for landowners wanting to avoid a section 9 take situation (Costa 1992). Subsequently, USFWS developed a set of consistent biological standards on appropriate management, mitigation, and conservation measures that eventually coalesced into a conservation strategy, as outlined in the recovery plan. These efforts advanced and sustained private landowners' interest in safe harbor by virtue of identifying what the Service expected from them in order to conserve red-cockaded woodpeckers.

The message here is that safe harbor's continued success, and perhaps other listed species' private land conservation strategies, will likely be based upon the Service's (and state wildlife agencies') ability to articulate what essential habitat requirements are necessary to avoid take, and to do so in terms familiar to a landowner. Additionally, this approach would likely have merit and be viewed as beneficial in situations where proactive partnerships have been formed to address a candidate species' conservation requirements, hopefully eliminating the need to formally list the species under the ESA.

Adequate Funding

The success of red-cockaded woodpecker safe harbor programs would probably not have been realized without a commitment by USFWS of financial and technical resources to both the state wildlife agencies and enrolled landowners. The safe harbor enrollment process can be made easy and relatively quick for landowners. The process can have easily understandable biological objectives, and the landowners can receive regulatory assurances. After all these elements are in place, significant numbers of landowners will participate, and many have (see Table 8-3). However, our experience has shown that the presence of financial assistance will motivate and enhance participation in safe harbor programs. Unfortunately, to date, USFWS does not have a dedicated source of funding for state agencies and landowners to implement approved SHAs. Nevertheless, in the Southeast region USFWS has made the conscious decision to discover creative ways to reward the partnerships created by SHAs.

USFWS provided substantial amounts of funding from 1997 to 2004 to state wildlife agencies for development of statewide red-cockaded woodpecker safe harbor programs (see Table 8-2). Additionally, upon permit approval, both state agencies and participating private landowners have received funding to implement SHAs (see Table 8-2). States use USFWS funds to hire wildlife biologists who assist with developing and drafting the SHA; implementing (for example, enrolling landowners in) the statewide programs; conducting outreach; and completing monitoring and reporting requirements. Landowners use USFWS funds to implement beneficial habitat management practices on enrolled properties. These conservation measures include prescribed burning, hardwood mid-story control, longleaf pine planting, and artificial cavity installation. Overall, USFWS funding sources, typically grants, have been varied; they include HCP Funds, Endangered Species Landowner Incentive Program, Section 6, Landowner Incentive Program, Partners for Fish and Wildlife, Flexible Funds, and a Private Stewardship Grant Program. As of September 2004, USFWS had provided a total of $3,326,200 to states and individual landowners to develop and implement statewide red-cockaded woodpecker safe harbor programs (see Table 8-2).

USFWS also has allocated staff resources directly for the Sandhills safe harbor program. The Service representative administering that SHA is in direct contact with constituents enrolled, or interested, in an SHMA. The representative administers the program, provides the appropriate outreach, and performs the most important task in the success of any safe harbor—that is, establishing, maintaining, and continuously improving the one-to-one relationships with participating landowners (Miller et al. 2004).

TABLE 8-3. Accomplishments for Red-cockaded Woodpecker Safe Harbor Agreements as of March 2010

State[1]	*Permit Issued*	*Agreements*	*Acres*	*Baseline Groups*	*Safe Harbor Groups*
North Carolina (Sandhills)	April 17, 1995	108	51,028.5	59	9
Texas	March 20, 1998	23	15,232	31	4
South Carolina	March 26, 1998	102	396,980	278	17
Georgia	May 24, 2000	18	129,906	103	22
Virginia	July 5, 2000	2	2,986	5[2]	0[3]
Louisiana	January 25, 2005	2	10,245	1	0
Alabama	November 30, 2006	2	17,947	10	0
Florida	May 19, 2006	6	15,013	2	0
North Carolina	October 2, 2006	2	2,287	0	0
Regional (Total)	9 Permits Issued 1995–2007	265	641,624.5	489[4]	52

Notes: [1]Data is current as of March 2010. [2]Three original baseline groups and two safe harbor groups sold in a compensatory mitigation exchange. [3] Three original baseline groups and two safe harbor groups sold in a compensatory mitigation exchange. [4]This represents 36 percent of known groups that occur on private lands rangewide (n = 1,346).

Performance and Accomplishments to Date

Since issuing the first safe harbor permit in April 1995, USFWS has processed and issued eight additional red-cockaded woodpecker statewide safe harbor permits: in Texas (1998), South Carolina (1998), Georgia (2000), Virginia (2000), and Louisiana (2005). In 2006, USFWS approved the statewide SHAs in Alabama, Florida, and North Carolina (outside the Sandhills). Exact dates for the nine permits are presented in Table 8-1.

Overall, the success of the red-cockaded woodpecker safe harbor program has been impressive. In 1992, prior to initiation of the Service's red-cockaded woodpecker private lands conservation strategy, USFWS had no formal conservation agreements with any landowners harboring red-cockadeds. As a result of the execution and implementation of the statewide agreements, 265 landowners, 489 woodpecker groups, and 641,625 acres are now enrolled in the program. Totals by state are presented in Table 8-3. The 489 woodpecker groups enrolled in the current SHMAs represent 36 percent of the groups known on private lands throughout the southeast (n = 1,346). In addition to the woodpecker baseline groups enrolled in the program, 52 new safe harbor groups have been established on numerous properties. Without the safe harbor program, these groups would not be in existence. Significantly, dozens of additional recruitment clusters (territories created with artificial cavities for occupancy by red-cockadeds) have been established to create many additional safe harbor groups.

In addition to SHAs, USFWS has approved HCPs or memorandums of agreement (MOAs) with six partners (primarily industrial forest landowners) who are protecting 93 woodpecker groups and managing 43,820 acres. The motivations for these partnerships (most of which pre-date safe harbor programs) include population consolidation, legal and economic predictability, and good stewardship (Costa and Edwards 1997). In total—including all partnerships—as of January 2010, 271 private landowners, harboring 582 red-cockaded woodpecker groups (43 percent of total known on private land) are managing 685,440 acres for the species.

Statewide red-cockaded woodpecker SHAs function on, and provide outputs at, three levels. First, they are a primary conservation component of the Service's comprehensive and dynamic private lands conservation strategy. That is, safe harbor is one part of a larger, broader, and more comprehensive regional (Southeast-wide) strategy, which includes other key elements. The larger strategy includes regulatory components, including private landowner guidelines (Appendix 5 of the Recovery Plan); monitoring components, for example, the USFWS Regional Annual Red-cockaded Woodpecker Property Data Report (Edwards and Miller 2004); other conservation elements, for example, HCPs and MOAs with forest industry (Costa and Edwards 1997); and population growth components such as translocation cooperatives (Saenz et al. 2002; Costa and DeLotelle 2006). Second, safe harbor—developed and implemented at the statewide scale—is a stand-alone conservation partnership among the private, state, and federal parties in each state. As such, it addresses state-specific red-cockaded woodpecker conservation goals and objectives, state-level political, social, and economic concerns and needs regarding red-cockadeds, and it fosters significant, long-term partnerships between the respective parties. Third, safe harbor functions at the individual property and woodpecker group-specific scale in several ways: 1) as a voluntary "contract" between an individual landowner and both the state wildlife agency and USFWS, and 2) to provide protection and management for a) woodpecker "populations" ranging in size (based on current enrollees) from 1 to 75 groups or b) unoccupied habitat with potential for future occupancy by woodpecker groups. At this level, safe harbor's output provides assurances from additional government regulation for individual landowners. Simultaneously, safe harbor provides conservation benefits, via active management, for individual groups of woodpeckers and their associated habitat.

Conclusion

The performance of the red-cockaded woodpecker statewide SHAs has been exemplary. The program's overall success can be linked to, and discussed in the context of, the specific objectives originally articulated for the red-cockaded woodpecker private lands conservation strategy, statewide HCPs, and safe harbor programs. The strategy identified four major objectives for a proposed red-cockaded woodpecker private lands regional conservation program: 1) minimize loss of genetic diversity; 2) increase woodpecker populations on public lands and on appropriate nonfederal lands; 3) foster and develop cooperative relationships among federal, state, and private parties, with the goal of maintaining and establishing large populations beyond those targeted as recovery populations; and 4) minimize the economic impacts of red-cockaded habitat retention to small private property landowners. The strategy further identified several objectives for red-cockaded woodpecker safe harbor programs: 1) encourage conservation and restoration of longleaf pine habitat; 2) increase red-cockaded populations; and 3) maintain no net loss of existing woodpecker groups (Costa and Kennedy 1996).

In the southeastern United States, there exists almost universal interest by state wildlife agencies in pursuing red-cockaded woodpecker SHAs targeting private landowners. An SHA offers a natural partnership between a state wildlife agency and USFWS, where the state assumes the responsibility to implement the agreement. State SHAs reflect a progression of relationships with affected parties that began with conflict or crisis situations and progressed to cooperative conservation programs that have virtually eliminated the probability of a serious conflict. This evolution can be explained by the strength and forbearance of key players, development and use of clear guidance on the species' needs, and consistent application of policy, law, and science.

With state leadership, it is expected that the majority of landowners interested in pursuing positive red-cockadeds-compatible conservation efforts on working forest landscapes will use the simplified procedural mechanism called SHMAs issued under statewide SHAs. This simplification of administrative processes and subsequent meaningful reduction in transaction costs makes it easier for landowners to participate in a safe harbor program. One challenge is that states must make room for this "new" responsibility and discover or otherwise produce the financial and staff resources necessary for the monitoring and implementation of the tasks of a safe harbor program. However, to date, and in support of this endeavor, USFWS has been successful in directing monies to support states' efforts (see Table 8-2). A positive result for USFWS is that state implementation of these programs releases Service staff and resources, which can then refocus on other listed species and habitat recovery programs.

To understand the reasons behind SHA's success in the Southeast, it is important to note that "[s]afe harbor works because it does not punish but rewards, is voluntary not mandatory, ensures flexibility not rigidity, provides assurances not uncertainly, and promotes proactive conservation not reactive habitat destruction" (Costa et al. 2001). Indeed, conservation and management of red-cockaded woodpeckers on private lands has been substantially advanced beyond SHMAs due to the successful partnerships brought about by safe harbor. In other words, interestingly and significantly, numerous landowners enrolled in safe harbor are now involved in other red-cockaded conservation-related initiatives with USFWS, state, or other conservation partners. Through these efforts, they are making important recovery contributions at the local, state, and regional scale. For example, various safe harbor properties are serving as research sites. Several properties serve as recipient or donor populations within their respective woodpecker translocation cooperatives, while other participants have expressed an interest in serving as woodpecker reintroduction sites. Several landowners have subsequently enrolled their safe harbor properties harboring red-cockaded woodpeckers in conservation easements, furthering the protection for woodpeckers. Finally, some safe harbor properties have served, and others are willing to serve, as mitigation banks in habitat and woodpecker "trading" programs related to HCPs (Bonnie and Bean 1996; Kennedy et al. 1996; Bonnie 1997c).

Following the Service's HCP "no net loss" guidance, to replace each woodpecker group authorized for incidental take, some safe harbor properties have played (and others likely will play in the future) an important role in the HCP process as mitigation sites. Landowners, desiring to remove red-cockadeds from their lands via an HCP, are provided a list of acceptable mitigation sites by USFWS (see U.S. FWS 2003 for details). These third-party mitigation sites inform the HCP applicant of the amount of funding required to take on the responsibilities of harboring the "replacement" woodpecker group. Typically, neither USFWS nor the state wildlife agency becomes significantly involved in these pricing negotiations. Their primary interest remains in ensuring that the administrative, permitting, monitoring, and conservation aspects of the HCP are completed and meet acceptable standards. Following USFWS protocols and guidance, mitigation sites are required to provide a minimum amount of acreage and habitat (i.e., timber) (with appropriate management) to establish each new woodpecker group. There are both short-term direct management costs and long-term indirect opportunity costs to the third-party landowner of taking on this mitigation responsibility.

Valuing these costs by individual HCP varies greatly and, based on results of previous HCPs, appears to primarily be a function of ownership class; for example, conservation lands, like The Nature Conservancy (TNC) or state lands—which are acceptable sites, unlike federal lands—versus industrial forestlands. Conservation lands, already harboring red-cockaded woodpeckers and planning to expand their extant population, have a very low opportunity cost related to timber resources, since they are not in the timber business. On the other hand, industrial forest landowners have a high opportunity cost because the alternative land use to providing woodpecker habitat is short rotation forestry, with guaranteed incomes at approximately 15–30-year intervals. Due to this pricing disparity, a direct result of ownership class and landowner objectives, mitigation costs for establishing new woodpecker groups with their associated habitat for the 15 HCP permits approved to date have ranged from $9,000 to $100,000.

Successful implementation during the past decade of the Service's private land red-cockaded woodpecker conservation strategy, and safe harbor specifically, has substantially improved the status and trend of private land woodpecker populations. As of January 2008, 43 percent of all known private land red-cockaded groups were in state or USFWS conservation partnerships. With the 2006 approval of the remaining three statewide safe harbor initiatives (Alabama, Florida, and North Carolina outside the Sandhills), significant opportunities exist to enroll additional landowners into safe harbor partnerships. One profound benefit of safe harbor, perhaps too often forgotten, is that under SHMAs occupied habitat for existing groups (the baseline) is actively managed for the species' benefit with prescribed burning and timber thinning. This is in stark contrast to private landowners' normal responsibilities under the ESA, which rewards passive behavior and not active habitat management for listed species. This benefit to existing groups, the fact that new woodpecker groups are being established, and that additional "beyond safe harbor" conservation initiatives have spun off from safe harbor partnerships have all contributed to improvement of the woodpecker's status on private land.

Regardless of the scale of implementation, whether at the individual property owner or statewide level, USFWS's relationship with private property owners regarding red-cockadeds has improved substantially since the early 1990s. The old paradigm of conflict has been replaced with a new paradigm of cooperation. Hopefully, a direct outcome of this paradigm shift will be that a significant percentage of previously concerned private landowners no longer view the ESA as onerous, inflexible, and rife with regulatory burdens.

We believe that the creation of safe harbor under the ESA, which is silent on the concept, clearly illustrates the possibilities and potential for creative solutions to seemingly complex issues related to conservation of listed species on private lands. Indeed, the initial success of safe harbor for red-cockaded woodpeckers led to its national implementation for other species. The species and landowners enrolled to date on a national level are a testament to the program's flexibility and forward-looking approach to conservation of rare fauna on nonfederal lands. The variety of species covered in SHAs confirms the program's conservation value: some examples include, birds (Attwater's prairie chicken, northern spotted owl), mammals (San Joaquin kit fox), amphibians (Chiricahua leopard frog, Houston toad), reptiles (gopher tortoise), fish (Oregon chub, desert pupfish), and arthropods (Cheeca Lodge Schaus swallowtail butterfly). Further solutions will be forthcoming for this and other listed species issues. However, such solutions will require partnerships of affected and interested constituents, founded in trust and sound science, to identify the challenges and then use the ESA's flexibility to discover and implement workable solutions. In our opinion, to be workable, solutions must be ecologically sound, economically prudent, socially and culturally acceptable, and politically, legally, and administratively feasible. Safe harbor satisfies all of these requirements and, at the end of the day, is a common-sense approach to helping save rare flora and fauna on private lands.

Acknowledgments

We offer a very special thanks to the following "red-cockaded woodpecker" biologists who implement safe harbor programs for providing accomplishment data by state: Susan Miller (North Carolina Sandhills), Ricky Maxey (Texas), Nicole Hawkins ([formerly Chadwick] South Carolina), Phil Spivey (Georgia), Brian van Eerden (Virginia), and Eric Baka (Louisiana). Thanks also to Sarah Lauerman, Clemson University graduate student and office assistant for the Clemson Field Office of the Service, for designing and constructing our tables, and to Nicole Edwards, a long-time colleague and former employee (currently a volunteer) of the Service's Clemson Field Office, for assembling our literature-cited section.

References

Bean, M. J., J. P. Jenny, and B. van Eerden. 2001. "Safe Harbor Agreements: Carving Out a New Role for NGOs." *Conservation Biology in Practice* 2: 8–16.

Bonnie, R. 1997a. "Safe Harbor for the Red-cockaded Woodpecker." *Journal of Forestry* 95:17–22.

———. 1997b. "Strategies for Conservation of the Endangered Red-cockaded Woodpecker on Private Lands." *Endangered Species Update* 14:45–47.

———. 1997c. "A Market-based Approach to Conservation of the Red-cockaded Woodpecker on Private Lands." In *Proceedings of the Symposium on the Economics of Wildlife Resources on Private Lands*, R. Johnson ed., 102–110. Auburn, AL: Auburn University.

Bonnie, R., and M. Bean. 1996. "Habitat Trading for Red-cockaded Woodpeckers: Enhancing Recovery, Reducing Conflicts." *Endangered Species Update* 13: 7–9.

Campbell, P. V., S.C. Belfit, A. E. Bivings, E. Wolters and F. A. Annand. 2004. "The North Carolina Sandhills Conservation Partnership: A Collaborative Approach to Red-cockaded Woodpecker Recovery." In *Red-cockaded Woodpecker: Road to Recovery*, R. Costa and S. J. Daniels ed. Blaine, WA: Hancock House Publishers.

Cone, B. 1994. "My Red-cockaded Woodpecker Dilemma." *Forest Farmer* 53: 22–23.

Costa, R. 1992. *Red-cockaded Woodpecker Procedures Manual for Private Lands*. Atlanta, GA: U.S. Fish and Wildlife Service.

———. 1995. "Red-cockaded Woodpecker Recovery and Private Lands: A Conservation Strategy Responsive to the Issues." In *Red-cockaded Woodpecker: Recovery, Ecology and Management*, D. L. Kulhavy, R. G. Hooper, and R. Costa eds., 67–74. Nacogdoches, TX: Center for Applied Studies in Forestry, College of Forestry, Stephen F. Austin State University.

———. 1997. "The U.S. Fish and Wildlife's Red-cockaded Woodpecker Private Lands Conservation Strategy: An Evaluation." *Endangered Species Update* 14: 40–44.

———. 1999. "Safe Harbor: A Private Lands Conservation Strategy for Longleaf Pine Habitat and Red-cockaded Woodpeckers." In *Proceedings of the 2nd Longleaf Alliance Conference*. Longleaf Alliance Report Number 4, 42–44.

Costa, R., and R. S. DeLotelle. 2006. "Reintroduction of Fauna to Longleaf Pine Ecosystems: Opportunities and Challenges." In *Longleaf Pine Ecosystems: Ecology,*

Management, and Restoration, J. Shibu, E. Jokela, and D. Miller eds. New York: Springer-Verlag.

Costa, R., and J. W. Edwards. 1997. "Cooperative Conservation Agreements for Managing Red-cockaded Woodpeckers on Industrial Forest Lands: What Are the Motivations?" In *Proceedings of the Symposium on the Economics of Wildlife Resources on Private Lands*, R. Johnson ed., 111–124. Auburn, AL: Auburn University.

Costa, R., and E. T. Kennedy. 1996. "An Incentive Program to Enhance Conservation of Longleaf Pine and Red-cockaded Woodpeckers on Private Land: The Case of Safe Harbor." In *Proceedings of the First Longleaf Alliance Conference: Longleaf Pine: A Regional Perspective of Challenges and Opportunities*, compiled by J. S. Kush. Auburn, AL: The Longleaf Alliance, 30–33.

Costa, R., S. L. Miller, and S. M. Lohr. 2001. "A Common Sense Approach to Conserving Rare Species and Their Habitats on Private Lands: Safe Harbor and Red-cockaded Woodpeckers—A Case Study." *Transactions of the North American Wildlife and Natural Resources Conference* 66: 498–510.

Drake, D. 2000. "Non-industrial Private Forestlands, Red-cockaded Woodpeckers, and a Safe Harbor." Dissertation. Raleigh, NC: North Carolina State University.

Edwards, N., and S. L. Miller. 2004. "Red-cockaded Woodpecker Population Data Collection and Dissemination in the Information Age." In *Red-cockaded Woodpecker: Road to Recovery*, R. Costa and S. J. Daniels eds. Blaine, WA: Hancock House Publishers.

Environmental Defense. 1999. *Safe Harbor: Helping Landowners Help Endangered Species*. Washington, D.C.: Environmental Defense.

Hagan, G., and R. Costa. 2001. "Rare Woodpeckers Reintroduced to North Florida." *Endangered Species Bulletin* 26: 30–31.

Kennedy, E. T., R. Costa, and W. M. Smathers, Jr. 1996. "New Directions for Red-cockaded Woodpecker Habitat Conservation: Economic Incentives." *Journal of Forestry* 94: 22–26.

Lohr, S. M. 2000. "The South Carolina Safe Harbor Program: Protecting Landowners and Preserving the Federally Endangered Red-cockaded Woodpecker." *American Bar Association, Section of the Environment, Energy, and Resources* 1: 643–648.

Michael, J. A. 1999. "The Endangered Species Act and Private Landowner Incentives." Dissertation. Raleigh, NC: North Carolina State University.

Miller, S. L., P. Campbell, and M. A. Cantrell. 2004. "North Carolina Sandhills Red-cockaded Woodpecker Safe Harbor Program: Current Status and Lessons Learned." In *Red-cockaded Woodpecker: Road to Recovery*, R. Costa and S. J. Daniels eds. Blaine, WA: Hancock House Publishers.

Ozier, J. C., W. W. Baker, and P. B. Spivey. 2004. "Status and Management of the Red-cockaded Woodpecker in Georgia: Emphasis on Private Lands." In *Red-cockaded Woodpecker: Road to Recovery*, R. Costa and S. J. Daniels eds. Blaine, WA: Hancock House Publishers.

Saenz, D., K. A. Baum, R. N. Connor, D. C. Rudolph, and R. Costa. 2002. "Large-Scale Translocation Strategies for Reintroduced Red-cockaded Woodpeckers." *Journal of Wildlife Management* 66: 212–221.

U.S. FWS (U.S. Fish and Wildlife Service). 2003. *Red-cockaded Woodpecker* (Picoides borealis) *Recovery Plan: Second Revision.* Atlanta, GA: U.S. Fish and Wildlife Service.

Vandersteen, C. A. 1992. "Time Running out for Landowners." *Forests and People* 42: 7.

Endnotes

1 As discussed later, one outgrowth of the Costa (1995) manuscript resulted in USFWS recognition that the most effective approach for engaging private landowners in red-cockaded woodpecker conservation was to outline the science-based regulatory standard for compliance with the ESA.
2 A red-cockaded woodpecker group is the social unit—that is, "family" of birds—which occupies and defends a territory. Groups can consist of a solitary bird (usually a male waiting for a dispersing female) or a potential breeding pair with zero to four helpers, typically male offspring from previous breeding seasons.
3 The selected strategy of USFWS issuing the permit to itself created a lively and animated policy discussion. In the end, however, the Service's Regional Directorate felt that the potential benefits associated with getting this program off the ground, by using a creative application of its authority provided by the ESA, outweighed the potential risks.
4 While this process is underway, individual landowners seeking to pursue an SHA have only one option—that is, to develop an individual application and seek USFWS (and state) approval. In some cases, this has happened. In most cases, interested landowners are waiting (and have been waiting) for this process to be completed and the statewide programmatic(s) to go "live."
5 In the case of red-cockaded woodpeckers, the recovery coordinator has served that role. The previous recovery coordinator was in place during the period from 1991 to 2007, wherein he initiated the concept of statewide red-cockaded woodpecker conservation plans, was part of the founding team that created the safe harbor policy, and held the nation's first safe harbor permit.
6 Transaction costs in the present sense are the creation of the documents, costs of consultants or others to write the SHA, and other time-consuming administrative elements of approving and issuing the permit.

CHAPTER 9

Saving *Salmo*: Federalism and the Conservation of Maine's Atlantic Salmon

ALISON RIESER

> *The State of Maine continues to object to the listing of salmon on the seven Down East rivers in the strongest possible terms. It should be clearly understood that this Administration has worked with the Services in good faith to develop the (Conservation) Plan as an alternative to listing. Should listing occur, however, all cooperation with the Services will cease, implementation of the Plan will be suspended, and we will pursue all available avenues, including litigation and legislative solutions to prevent this misapplication of the Act.*—Angus S. King, Governor of Maine (Oct. 10, 1996 Letter to USFWS and NMFS, quoted in Baum 1997).

In the last decade of the twentieth century, state and federal officials reluctantly acknowledged that restoring wild salmon would take more than making more fish. The anadromous Atlantic salmon (*Salmo salar*) had once flourished in river systems throughout New England, but the economies of the nineteenth century unwittingly reduced the salmon's range to a few river systems in Maine. In 2000, the remnant populations that return to eight of the minor coastal river systems of eastern Maine were listed as endangered under the U.S. Endangered Species Act (ESA).

The bitter fight that broke out between Maine and federal officials over the listing represents probably a new low in environmental federalism. Combatants pit the tenuous prospects of a much-revered biological relic, facing a changing climate regime, against the bright promise of economic revival of the boarded-up fishing and farming towns of Down East Maine. Far from the traditional, low-tech industries of that region that wax and wane seasonally, foreign investors in the new salmon-farming venture were determined to use technology, economies of scale, and intensive production methods to overcome ecological constraints and thereby ensure global competitiveness. A newly elected governor, who was independent of political party affiliation, was determined that "common sense" and higher economic aspirations would prevail over environmental fear-mongering and NIMBY-ism.

In an attempt at compromise, federal officials mustered whatever dexterity they could under the ESA. They used newly-minted ESA policies to avoid dealing with the hydropower dams on the salmon's largest remaining riverine habitat and to maintain the state's primacy in devising a conservation strategy. In the end, this flexibility was insufficient to bridge the differences between state goals and federal responsibilities. A century-old partnership turned into a brawl over the interpretation of genetic data and a rhetorical spat over the difference between a salmon in Maine and a Maine salmon. A political atmosphere that encouraged anti-federal grandstanding found a convenient whipping boy in the proposed listing, despite the flexibility its federal administrators had shown. Accommodation had turned to anger in the space of less than two years.

The salmon farming industry's resistance to changing their increasingly intensive and risk-prone husbandry practice had undermined the federally endorsed state conservation plan. After one year of implementation, it was clear to federal officials that the plan was underfunded and not tough enough on the growing risks that aquaculture posed to the meager numbers of returning salmon, and that it was unlikely to be strengthened. When conservation groups sued federal ESA administrators, the listing proposal was reinstated. This time, the status was "endangered," with no plan to rely on state, local, and voluntary measures in lieu of federal restrictions. When the listing was final, the State of Maine challenged it in court, faulting its underlying science and its unwarranted intrusion on state sovereign interests. The federal court upheld the listing in 2003.

The election of a new governor eventually laid the legal battle to rest. A victory in court for the federal listing decision helped the state come to terms with the ESA. This victory was assisted by an independent scientific report vindicating the view that the Maine salmon was a genetically distinct and significant population segment (DPS) entitled to recognition and protection as a "species" under the Act.

In the final analysis, however, it took a citizen suit ruling, under the federal Clean Water Act (CWA), to demonstrate to the state and the aquaculture industry that, without a doubt, federal environmental law controlled. The federal judge showed he was willing to deal with industry intransigence. If the regulators would not, he would order the salmon farms to cease stocking nonnative strains of Atlantic salmon immediately or hold them in contempt of court, even if this imposed costs the industry had hoped it could avoid under a state conservation regime.

Recovery planning for the Maine Atlantic salmon then began, with both the state and the industry promising to take a cooperative approach. Meanwhile, a second report from the same independent scientific panel that supported the DPS determination dropped the proverbial second shoe, making clear that recovery activities that focused too narrowly on the eight rivers of the DPS would not be adequate. Notwithstanding the federal listing agencies' victory on the definition of what a Maine salmon is, the panel concluded that rehabilitating the species in Maine must include helping the populations whose habitat is diminished by dams.

Independent of the ESA listing and recovery efforts, private and nongovernmental groups began to tackle the dams, brokering a multiparty agreement to restore the mighty Penobscot, the river to which most Atlantic salmon in Maine return. For the sake of the salmon, three of the worst-offending dams will be bought from their power-company owner and pulled down, while other dams, less damaging to habitat, will increase their power output. As her predecessor Bruce Babbitt had done in 1999, at the historic breaching of the Edwards Dam on the Kennebec (McPhee 1999), Interior Secretary Gail Norton found herself, in 2004, at a summer photo opportunity on the banks of a Maine river, extolling the virtue of cooperation in regaining our common natural heritage (*Bangor Daily News* 2004). Despite her surprise appearance to sign the agreement personally, she did not bring news of any federal funds to help meet the multimillion-dollar purchase price for the dams. That news was to take another four years and the intervention of many more and new players, in the saga of the Atlantic salmon listing. Finally, the listing process came full circle when salmon in Maine's four largest, industrialized rivers were added to the endangered listing of Gulf of Maine salmon, along with an extensive determination of its critical habitat—virtually the entire watersheds of all significant salmon rivers in Maine.

The Natural History of the Atlantic Salmon

Edward Baum, the chief salmon biologist for the Maine Atlantic salmon commission for over 30 years, published a book at the time of the proposed listing entitled *Maine's Atlantic Salmon: A National Treasure.* (This case study draws heavily from this authoritative account.) Baum (1997) described the basic natural history features of the Atlantic salmon that make its conservation such a challenge to biologists and politicians alike. Atlantic salmon were native to most major river systems north of the Hudson River, but by 1865, salmon had vanished from all southern New England rivers due to overfishing, pollution, and dams. By 1900, no wild salmon spawned in the Connecticut, Merrimack, or Androscoggin rivers; by 1870, only seven or eight rivers in Maine contained populations, down from between 28 to 34 rivers (Figure 9-1).

One hundred years later, despite this severe restriction in its habitat, approximately 1.5 million Atlantic salmon returned to the rivers of North America. By 2003, however, this number was down to fewer than 500,000, and by 2004, to fewer than 100,000, with the majority returning to salmon rivers of eastern Canada. This level of return was less than half the number needed to meet conservation targets for the rivers of North America (Porter, G. 2003).

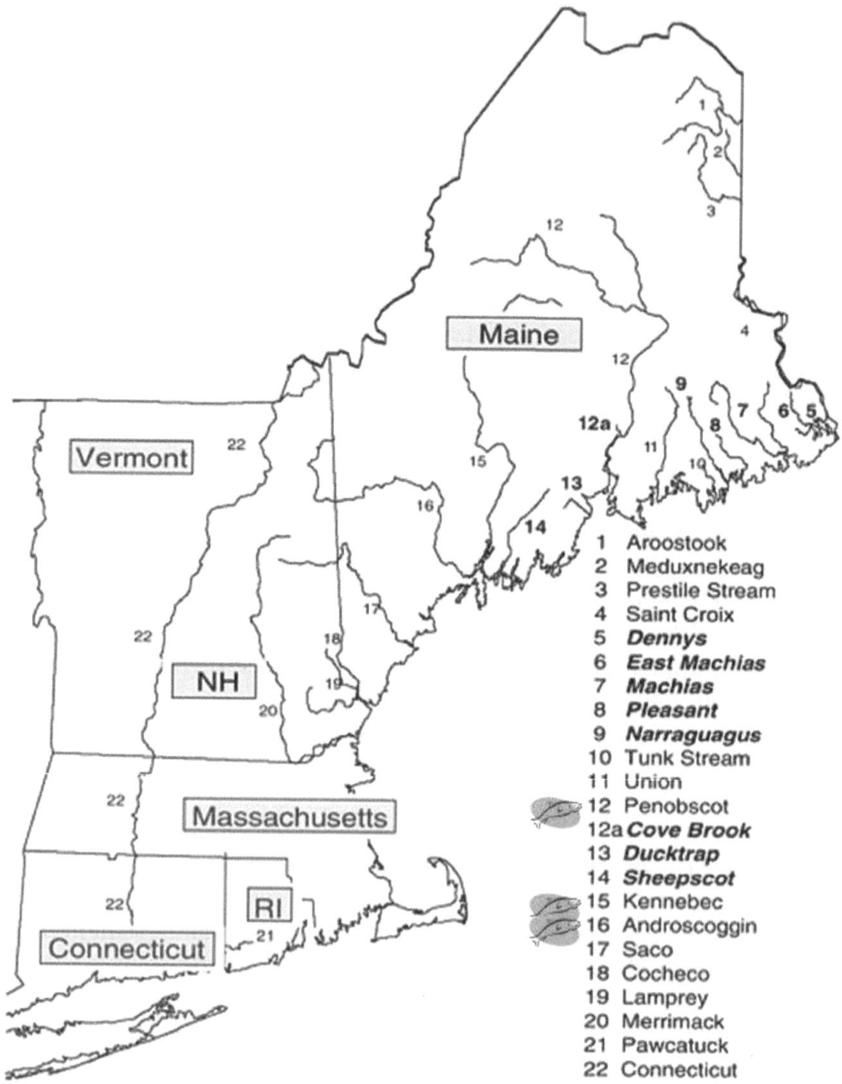

FIGURE 9-1. Historic Atlantic Salmon Rivers in U.S.

Note: The eight DPS rivers listed in 2000 are in bold; the fish icons mark the three rivers added to the DPS in 2009. Modified from NAS (2004).

Fewer than 862 adult salmon returned to Maine rivers to spawn in 2002, down from an estimated 940 in 2001. Most of those fish returned to the Penobscot (782 in 2002, and 786 in 2001). In 2003, the number returning to U.S. rivers was less than 5 percent of spawning requirements. In the eight rivers in Maine where salmon were initially listed as endangered, only 72 salmon returned in 2003, up from 33 in 2002 (NAS 2004).

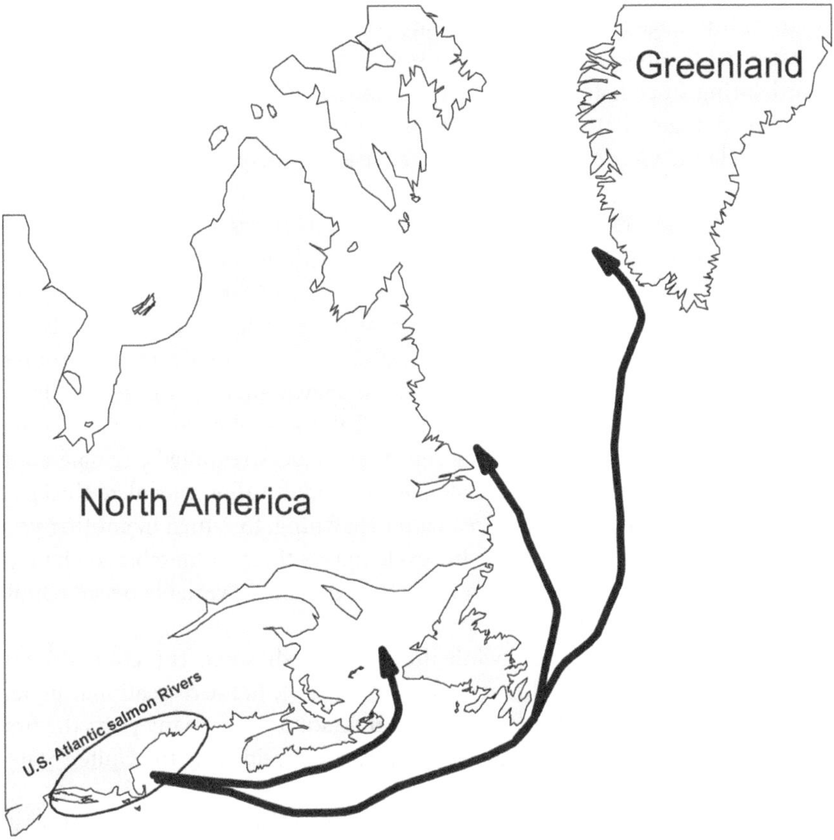

FIGURE 9-2. Marine Migration Routes of U.S.-Origin Atlantic Salmon

Source: Fay et al. (2006).

Each life stage of the Atlantic salmon has a distinctive name, and each has specific habitat requirements. Salmon are anadromous—meaning that as adults, after spending one or more winters in the open ocean, they return in the spring to the rivers in which they were born. They spend up to five months in their natal waters, making their way to stretches of the river with just the right combination of temperature, water flow, and bottom type. As autumn daylight begins to fade and water temperatures cool, spawning females deposit their eggs into one or two groups of gravel pits that are collectively called a *redd*. The eggs hatch five or six months later, in March or April. The eyed eggs hatch into *alevins* and remain buried in the gravel for another six weeks. After they absorb their yolk sacs, the *fry* are free-swimming, leaving the gravel beds to begin to feed, first on plankton and then on insect larvae and insects. As they grow larger, the youngest juvenile salmon become *parr*, so named for the bar-like markings on their sides that serve as camouflage. Most salmon spend two years at this stage (about 20 percent are parr for another year), and the rivers must provide nursery habitat for three year-classes of the young fish. In their third spring since hatching, salmon become *smolts*, undergoing major physiological changes that prepare them for the journey down river and into the North Atlantic Ocean, which takes place from mid-April to mid-June (Baum 1997).

After the seven-inch salmon smolts leave the rivers, they spend from one to three winters in the North Atlantic Ocean, feeding in the rich waters off the coasts of Labrador, Newfoundland, and west and east Greenland (Figure 9-2). Maine salmon have been found in waters as far away as the Faroe Islands and above the Arctic Circle, off Greenland. After retracing their migration routes, some salmon return to their natal rivers to spawn after only one sea winter, and others after two or three years at sea. Salmon that return after only one winter are called *grisle*. After they find their natal streambeds, female adult salmon create redds, and then male parr and adult salmon fertilize the eggs. Some adults will return to the ocean after spawning, to return in another year to spawn again. This spectacular life cycle makes them vulnerable to changes in their river habitat, as well as to numerous predators, variable ocean conditions, and ocean fisheries.

Adult salmon do not feed while they are in fresh water. This could explain why catching an Atlantic salmon is a major challenge to the recreational angler, earning it the moniker, "the king of fish." By tradition, for many years the first salmon caught in the spring would be sent to the President of the United States to served at the White House (Baum 1997).

When they are in rivers, salmon require gravel beds and deep, cool pools and eddies of moving water. These conditions were once abundant in the rivers of Maine, but the timber economy of the nineteenth century took a huge toll on the salmon's habitat. Salmon were once present in somewhere between 28 to 34 rivers in the state, but the mountains of sawdust and other lumbering debris discarded by sawmills wreaked havoc on the salmon's migration routes. Although Maine lawmakers outlawed this practice in 1834 in the Kennebec and later in other rivers, the damage had been done. Extensive timber cutting had also altered riparian habitat and affected the rivers, through the practice of driving logs down rivers to the sawmills. A successful drive required loggers to widen the rivers, clear away rock obstructions, then to create huge plumes of water released from behind the drive dams (Jenkins 2003).

These re-engineered rivers became scoured-out transportation corridors for timber, the exact opposite of what a salmon needs for migration and spawning. At the same time, thousands of mill dams were constructed, powering tanneries, paper mills, textile mills, and other factories. By 1872, only eight rivers in Maine were known to have salmon populations (Jenkins 2003). The twentieth century then brought hydropower dams to the salmon's remaining rivers.

Conservation Efforts Prior to the 1993 Listing Petition

By the late 1880s, it was apparent to many that the loss of river habitat was largely responsible for Atlantic salmon disappearing from New England's rivers. Instead of placing regulatory controls on the lumber industry, mills, and dams, however, New England states, with federal assistance, chose to restock the depleted rivers with fish made in hatcheries. Salmon were by this time highly prized as recreational fishing quarry. It was an easy call for these governments to put to work new knowledge of salmon biology and fish culture methods; then, however, hatchery-based restocking took on a life of its own. For over a century, fish-making through hatcheries was the basis of a cooperative relationship between federal and state fish and wildlife agencies. Salmon returns continued to fluctuate in response to other factors, but the state–federal relationship based upon a mutual faith in restocking never faltered. This state of affairs was to change drastically when an endangered species listing loomed on the horizon. That extensive restocking had gone on for so many years became a point of division between federal and state officials.

The Era of Making Salmon

Science and technology have been at the heart of the federal–state salmon relationship since the late nineteenth century. In an 1874 report to the U.S. Fish Commission, Charles G. Atkins, Maine's fisheries commissioner, expressed optimism that knowledge of how to make salmon in hatcheries, together with improvements in passages for migration and federal–state cooperation, could restore salmon to self-sustaining levels (Jenkins 2003). The states began restocking on their own, creating the first fish hatcheries in the 1860s—first in New Hampshire, then in Maine, Vermont, and Massachusetts. Federal help in fish stocking came after Congress created the U.S. Fish Commission in 1871. Its first director, Spencer Baird, believed strongly that hatcheries were the answer to dwindling stocks of food fish, and the states received federal monies to build them. In the late 1880s, a hefty trade was underway, as fish eggs and fingerlings from species in the eastern United States were introduced into western waters, and vice versa.

The belief in and reliance upon fish culture and stocking continued throughout most of the twentieth century. In Maine, this occurred pursuant to cooperative agreements between the State of Maine and the United States Fish and Wildlife Service (USFWS) and its predecessors, the Fish Commission, and the Bureau of Sport Fisheries. By the end of the twentieth century, almost 100 million young salmon had been stocked in Maine rivers (Baum 1997). The problem was that fish eggs, juveniles, and adult fish were being brought in from distant ecosystems, even as biologists began to recognize fidelity to the natal stream (the stream of birth) as a key adaptation of the salmon species. It was not until 1991 that stocking programs in Maine began using river-specific strains, rather than hatchery stock derived from Canadian or Penobscot River-returning adults, or from hybrids of the two. This long history was later to play a major role in the State of Maine's challenge to the scientific validity of the listing decision.

Maine has long known that salmon in its rivers are at risk. In 1945, the Maine legislature created the Atlantic Sea Run Commission, whose job it was to identify ways to strengthen salmon runs in Maine. In 1947, the Commission reported that less than 10 percent of the original habitat was accessible to returning adult salmon due to dams and other obstructions in Maine's rivers. Natural runs of salmon were all but extinct, except for the small remaining run in the Penobscot and in the restocked rivers of eastern Maine. But the Commission also had some good news to report: wild salmon still spawned in Maine rivers, and conditions in the rivers had improved. If pollution could be abated, fishways constructed, and water diversions screened, it might be possible to restore the salmon. In the late 1960s, Maine's salmon commission developed a cooperative plan with USFWS and the Maine inland fish and game department to restore salmon runs, largely for the benefit of the recreational angler.

The cooperation of the two Maine agencies with USFWS proved insufficient, however, to deal with all the private land management activities that were eating away at the remaining salmon habitat. Neither agency had sufficient legal authority or political clout to coordinate land and river management in the manner needed for salmon restoration. This was especially true in the watersheds where timber and pulpwood harvesting practices included clear-cutting, a practice that causes erosion and sedimentation and destabilizes water flows, which creates water shortages and inappropriate temperatures.

Another land use emerged that Maine agencies were also loath to burden with regulation. Blueberry production began in some of the watersheds of the remaining salmon rivers in eastern Maine in the 1980s. Blueberry cultivation entails water withdrawal for irrigation during the growing season and for protection from frost and the application of pesticides and herbicides to the barrens, all of which created further problems for salmon. The Maine agencies adopted a series of river-specific plans to address these issues and increase the likelihood that naturally spawning and stocked fish would survive and return to the rivers.

The 1980s were apparently a period of good winter survival at sea. By the mid-1980s, salmon populations were relatively high; the number of fish returning from the sea to spawn numbered in the thousands statewide (6,000 to 10,000). The recreational fishery thrived, landing annually between 1,000 and 2,000 fish (Baum 1997; Jenkins 2003). The early to mid-1980s was also a period of intensive and expanded restoration efforts, enthusiastically supported by busy anglers, who were happy with the good salmon runs and wanted the chance to experience them statewide. Maine's salmon commission developed a new strategic plan for Atlantic salmon in 1984 and sent its biologists to Canada and Greenland to learn more about the winter lives of salmon.

Salmon Farming Enters the Picture

Farming of hatchery-bred Atlantic salmon began in Maine's coastal waters in the 1980s. The state–federal restocking program for the recreational angler, in fact, helped launch the salmon aquaculture industry in Maine, by diverting smolts from the federal salmon hatchery to the private growers. After this "jumpstart," the industry began relying on its own broodstock, derived in part from the European cousins of the Maine Atlantic salmon. Placing pens in Maine's coastal waters to grow salmon smolts to market size was authorized by the Department of Marine Resources, under Maine's aquaculture leasing law—a law written largely to encourage shellfish aquaculture, especially the small-scale blue mussel farms that had cropped up in the coastal rivers of mid-coast Maine. As the Maine salmon farming industry began to expand, it also grew in Canada, Norway, Scotland, and other countries that had wild salmon rivers. Later in the 1980s, Maine's aquaculture companies repaid the jumpstart loan by giving USFWS and the Maine salmon commission fry and parr they had reared in their now-booming private hatchery facilities.

The transfer of salmon smolts from the publicly funded hatchery to the new farming industry angered the angling community. Sports fishing lobbyists convinced the Maine legislature to increase public representation on the state salmon commission—presumably to prevent the commission from forgetting that its mission was to restore salmon runs for the benefit of recreational fisheries (Baum 1997). But the expansion of the salmon commission in 1987 eroded the effectiveness of its statewide program. By increasing the number of public members from one to three, factions were allowed to emerge, representing different regions and groups of anglers. The warring factions competed for the limited program funds and blamed each other when the runs dropped. The commission tried to maintain its focus and prevent politically driven priorities by adopting a strategic plan.

However, the political infighting continued. When salmon runs and angler landings dropped precipitously in the mid-1980s, due largely to poor marine survival of wild and hatchery smolts, statewide returns dropped from 6,000–10,000 to 1,500–3,000, even with increased stocking. By the end of the 1980s, the populations again declined, as marine survival fell by an estimated 80 percent (Jenkins 2003).

A severe state budget crisis in the late 1980s, accompanied by reduction in staff at all government agencies, curtailed the restocking program, dooming the salmon commission's statewide restoration and management plan. State funding was withdrawn for the biologists at the salmon commission. The shortfall in revenues from salmon fishing licenses required the commission to appeal to USFWS and the National Marine Fisheries Service (NMFS) to support a scaled-back restoration program, focusing on the seven rivers that still had wild salmon.

The 1992 Prelisting Recovery Plan

In response to the precipitous drop in numbers of returning adults, in November 1991, USFWS listed five of the seven Maine rivers with wild runs (Dennys, East Machias, Machias, Pleasant, and Narraguagus) as Category 2 under the ESA. This listing category no longer exists, but at the time it signified sufficient concern that a threatened or endangered listing might become necessary in the future. In the meantime, biological information would be collected, and the species would be monitored even more closely than under the cooperative agreements with the states (U.S. FWS 1991). In 1994, two other rivers with wild runs, the Ducktrap and Sheepscot—inadvertently omitted in 1991—were added to the Category 2 list.

The 1991 listing prompted a much more focused restoration effort by the state salmon commission, in cooperation with USFWS. Instead of trying to satisfy everyone and restore 16 rivers, the agencies focused on the rivers that still had wild runs. This focus involved limiting the restocking to only river-specific strains of salmon. Monitoring returning tagged salmon had shown that wild salmon survive at higher rates than hatchery salmon. The decision was made to increase the rivers' own production of wild smolts by saturating any vacant or underused habitat with fry, so that when conditions allowed greater winter survival, there would be many fish to take full advantage of the improved conditions at sea. The two agencies created a "prelisting recovery plan" for wild Maine salmon populations in 1992. It was based on the assumption that listing as endangered or threatened could be avoided by concerted conservation efforts to boost the populations' status. On the other hand, if listing eventually happened, the plan would serve as the basis for a full recovery plan (Baum 1997).

The prelisting recovery plan had six elements, the first of which focused on development of river-specific brood stocks to support fry stocking, which would in turn maximize each river's production of wild smolts. Research on the genetic characteristics of all salmon stocks in Maine rivers would help focus future recovery efforts. By installing weirs to trap fish on existing runs, fish could be trapped and used for broodstock and to collect data on the strength of the runs. A complete inventory of all salmon habitat and the identification of threats to the salmon and their elimination were also in the plan. USFWS constructed a new state-of-the-art hatchery in Maine to accommodate the river-specific strains. It was a big operation; each river had its own room in the hatchery.

The river-specific stocking program released more than 1.2 million fry into five of the seven rivers in the years between 1992 and 1996; the use of

Penobscot River-origin salmon was stopped completely in 1991. Senator George Mitchell helped to secure most of the funding to carry out the pre-listing plan from congressional appropriations. Along with Senator William Cohen, Mitchell restored the federal funding through an amendment to the Interior Department's 1993 appropriations bill, sending $550,000 to provide funds for restoration and research on the Down East rivers. Adult salmon returned in low numbers, however, reaching new lows in the early 1990s.

In 1995, the Maine salmon commission adopted a statewide salmon restoration and management plan to guide the preparation and implementation of river-specific plans for the next five years. It also adopted a statewide regulation limiting salmon angling to a catch-and-release fishery. Although during the previous decade recreational fishermen, especially those from Down East, had vehemently opposed proposed reductions in the season catch limit, the catch-and-release rule drew very little opposition (Baum 1997).

Actions Regarding the Ocean Fisheries for Atlantic Salmon

An ocean fishery for Atlantic salmon had been carried out off the coast of Greenland since the early 1960s, after a big concentration of fish was discovered in the Davis Straits (Berrill 1997). Tagging returns revealed that this fishery caught mixed stocks, including salmon from Maine as well as European rivers (Baum 1997).

A U.S. nongovernmental organization (NGO) called Trout Unlimited, through a campaign led by retired businessman Richard A. Buck, brought pressure to bear on the U.S. government. Buck sought an end to the rapidly escalating intercept fishery, which caught Atlantic salmon in their winter feeding grounds off the coast of West Greenland. Buck created the Committee on the Atlantic Salmon Emergency (CASE) in 1968 and enlisted a number of American celebrities who were fans of Atlantic salmon—including Bing Crosby, Ted Williams, and Curt Gowdy—to help publicize the salmon's plight and the need to end the high-seas fishery. Buck sought and received an audience with the Danish foreign minister in 1971 (Baum 1997). CASE lobbied Congress to pass the Pelly Amendment to the Fishermen's Protection Act, giving the Executive Branch the power to bar Danish goods from the U.S. market if Denmark did not curtail overharvesting off western Greenland.

After passage of the Pelly Amendment and the threat of an embargo, the Danish government met with U.S. officials. The result was the U.S.-Danish Fisheries Agreement, signed on February 22, 1972, to phase out, by 1976, the catching of salmon in waters beyond the territorial sea of Greenland. The West Greenland fishery was managed by catch quotas from 1972 onward (Baum 1997). Buck then organized a new entity, Restoration of Atlantic Salmon in America, Inc., to press the U.S. State Department to negotiate an international agreement to protect salmon from all fishing during its ocean migration. His participation in the negotiations resulted in a treaty forming the North

Atlantic Salmon Conservation Organization (NASCO) in 1983; Buck then served as one of the three U.S. commissioners appointed by the President, from 1983 until 1991.

The International Council for Exploration of the Seas (ICES) provides NASCO with scientific advice in the form of abundance estimates and escapement requirements for the 600 stocks of Atlantic salmon in North America and the 1,500 stocks in the Northeast Atlantic. Its job includes research and providing catch advice, via the Advisory Committee on Fishery Management, for the two high-seas mixed stock fisheries off western Greenland, a fishery that takes mainly North American and southern European stocks (ICES 2002). In the early 1990s, ICES recommended a zero quota. Eventually, U.S. (and Canadian) NGOs, with some government support, bought the fishing rights the Greenland fishers had under the NASCO quotas.

In the late 1980s, the New England Fishery Management Council was asked to develop a fishery management plan to prohibit fishing for and possession of Atlantic salmon in the U.S. 200-mile Exclusive Economic Zone (EEZ), presumably to meet the U.S.'s international obligations under NASCO (NEFMC 1987). But the U.S. government apparently balked at closing down high-seas fishing for salmon by U.S. vessels (Jenkins 2003). The government of Canada closed the Newfoundland salmon fishery for five years beginning in 1992. This was followed by the two-year buyout of the Greenland fishers' quota. Curtailing these fisheries was essential to providing escapement of adult salmon to return to spawn in Maine's rivers. The West Greenland fishery was stopped in 2004 (NASCO 2004).

International Attention to Aquaculture Impacts on Wild Salmon

In 1994, responding to concerns about the growing salmon farming industry's impact on wild salmon, nation members of NASCO, which include the seven largest producers of farmed salmon, adopted the so-called Oslo resolution. The parties agreed to adopt national controls, including criteria for siting pens away from wild salmon rivers, measures to prevent escapes of farmed salmon from the pens, and controls on disease and parasites, with an annual reporting commitment. Signatories of the Oslo resolution are the U.S., Canada, Norway, Scotland, Iceland, and the Faroe Islands.

By 2002, many observers were disappointed by the slow progress under the Oslo resolution. An NGO report on national compliance with the agreement gave the United States the second worst score (0.50 out of a possible 10 points) (Porter, G. 2003), due largely to the failure of government authorities to issue CWA discharge permits to the salmon farms that would limit genetic, disease, and other risks to wild salmon, but also for failing to apply siting criteria to locate pens safely away from the wild salmon rivers and to require farms to report major fish escapes (Goode 2003).

In a similar vein, ICES developed a Code of Practice on the Introduction and Transfer of Marine Organisms in 1984 and 1988. In 1994 the code was updated and sent to the United Nations (U.N.) Food and Agriculture Organization for inclusion as part of its guidelines under the Code of Conduct for Responsible Fisheries (UN FAO 1996). NASCO and ICES convened an international symposium in Bergen, Norway, in 2005 to examine the science underlying concerns regarding the interaction of salmon aquaculture and wild salmon.

The Proposed Listing as Threatened and the Special Section 4(d) Rule Deal

The relationship between Maine and the federal fish and wildlife agencies changed dramatically after USFWS and NMFS received petitions in 1993 to list the Atlantic salmon as endangered. The petitions came at a time when the ESA was under siege, triggered in large part by the Services' interpretation that "indirect takes" through significant habitat degradation were prohibited by the Act. Advisors to Interior Secretary Babbitt were very active in seeking ways to reduce political opposition to the Act while preserving its protective essence.

Out of this search came some new concepts and interpretations, including the "distinct population segment" (DPS) and habitat conservation plans (U.S. FWS 1996). Some of these tools were used in the Atlantic salmon listing—first to keep the State of Maine in the lead in defining which private actions constituted a take. One commentator suggested that the Services used the DPS concept to try to limit the potential application of the take prohibition to constituencies that were less politically powerful than the dam-operating power companies and their customers (Siegel 1997). Or at least that was the hope. By the time the Services were able to make a deal with the state about the listing, a new industry had developed in Maine that posed new threats to wild salmon. Although much smaller than the timber companies and hydropower consumers of the Pacific Northwest that NMFS had had to contend with, the salmon farming companies would prove to be formidable opponents of federal regulation. A new governor was also on the scene, one who was willing to take on federal authority in order to protect both old and new jobs in Maine.

The 1993 Petitions

USFWS received a petition to list the Atlantic salmon under the ESA on October 1, 1993, from a new organization called RESTORE: The North Woods, which is based in Concord, Massachusetts. Organized to help promote the creation of a new national park in the remaining forestlands of northern New England, RESTORE asked USFWS to list the salmon throughout its historical range in the U.S. Co-petitioners were the Biodiversity Legal Foundation and Jeffrey Elliott, a New Hampshire resident.

Apparently, USFWS's first response was to panic; all the controversial Pacific salmon listing petitions had been submitted to NMFS, leaving USFWS off that particular hook. Aware of the century-long restocking effort, USFWS created a genetics working group at its lab in West Virginia to collect new samples and analyze archived samples to determine whether any of Maine's seven wild salmon populations even qualified for listing as a species.

An identical petition was sent one month later to NMFS, triggering something of a turf battle with USFWS over which agency had jurisdiction. The agencies finally resolved the issue, with a memorandum of understanding between their northeast regional directors. They agreed to review the petition jointly, in view of the fact that Atlantic salmon spend half their lives in fresh water and the other half in marine waters.

The Services published a notice in the Federal Register that the petitions had been received and then invited public comment until April 1994 (NMFS and U.S. FWS 1995a). After the close of the comment period, they created a biological review team composed of six members, three from each of the two agencies, who then spent the remainder of the year gathering information and reviewing materials that had been submitted.

Business interests in Maine, as well as departments of the state's government, sent dozens of letters, all arguing that listing would be an economic disaster that was not warranted. The Maine forest industry promptly formed a committee to fight the petition. Later in the year, a coalition of private industry, state and federal agencies, and private individuals created a nonprofit organization called SHARE—for Salmon Habitat and River Enhancement—with the purpose of conserving and enhancing salmon habitat in the five most eastern salmon rivers, through voluntary action and cooperation.

In January 1995, when no action on the petitions appeared imminent, the petitioners filed suit to compel the Services to act. The newly inaugurated Governor of Maine, Angus F. King, Jr., immediately spoke out against the potential listing. On March 17, 1995, the Services announced that the biological status review was finished. They had concluded that the best available biological evidence showed that listing as endangered throughout the salmon's historic range was not warranted, largely because indigenous salmon in rivers south of the Kennebec had been extirpated in the nineteenth century. They had, however, "determined that sufficient information was available to support appropriate listing actions for the Distinct Population Segment (DPS) that consists of populations in the Sheepscot, Ducktrap, Narraguagus, Pleasant, Machias, East Machias, and Dennys rivers." The status of the populations in the lower Kennebec River, the Penobscot River, Tunk Stream, and the lower St. Croix River was uncertain and required further study. The Services requested comments on whether any native, naturally reproducing populations still existed in these rivers and tributaries. Meanwhile, work would begin immediately on a proposed rule to list the DPS.

The 1995 Proposed Listing as Threatened

The Services were as good as their word, for on September 29, 1995, they published a proposed listing of the seven Maine rivers as a threatened DPS. The proposed rule contained a special feature. The State of Maine was invited to submit a conservation plan that, under a special section 4(d) rule, would provide regulations in lieu of federal controls, perhaps obviating the need for a listing altogether. A section 4(d) rule is a mechanism developed by the Services in the early 1990s to reduce the risk that private parties will be subject to citizen suits for allowing an indirect taking of a species listed as threatened. Under the rule, an approved state conservation plan defines the range of actions that would constitute acceptable indirect takes, as long as efforts are made to conserve and restore the species (Bean and Rowland 1997, 267–70).

In the proposed listing, the Services identified three major threats to salmon: poaching, low natural survival at sea, and the potential impacts from salmon aquaculture and hatcheries through disease transmission and loss of genetic integrity via escaped salmon. The state conservation plan the Services were inviting would presumably minimize those threats (NMFS and U.S. FWS 1995b).

The Services' discussion of ESA listing factor D—whether existing regulatory mechanisms were inadequate—made this expectation clear. The Services reported that the status review had noted that more stringent implementation and enforcement of existing regulations would strengthen their effectiveness. They specifically mentioned the need for strengthening regulations of salmon farming—first, in relation to the genetic threat posed by escapees from pens located within 20 kilometers of five of the seven rivers in the proposed DPS and, second, in regard to the growing risk that penned salmon could transmit disease to wild salmon migrating nearby. The Services also suggested that possible new measures could require changes in broodstock and prohibit the use of European strains for broodstock in the aquaculture companies' hatcheries (NMFS and U.S. FWS 1995b, 50532–33).

The date of the proposed rule's publication, September 29, 1995, turned out to be the last day before a Congressional moratorium took effect on ESA listing actions. When the moratorium expired in April 1996, the Services reopened the comment period. The State of Maine's comments were accompanied by a letter from Governor Angus King stating that he opposed the listing in the strongest possible terms, proposing instead that the Services enter into a cooperative agreement with the state. At some point while the proposed listing was pending, Senator William Cohen sent a letter to Secretary Babbitt. The letter contained a thinly veiled threat that the senator would support an amendment to the ESA that would require the Services to consider social and economic factors in listing decisions (Siegel 1997).

State of Maine's Response to the 1995 Proposed Listing

After the proposed listing was published, the governor's office took control of making salmon policy for the state. Earlier in 1995, the Maine legislature had enacted a bill replacing the Atlantic Sea Run Salmon Commission with a new entity, the Atlantic Salmon Authority. The bill had been a compromise between those who wanted to abolish the commission and those seeking a more diverse membership and a more independent body that would be free of political interference from the governor's office and executive departments. The Atlantic Salmon Authority was to have "sole authority" over salmon in all waters of the state, and the administrators of the two government departments would be outnumbered by public and tribal members of the governing board.

On the date the new Authority was to have come into being, the Services proposed the listing. The governor delayed appointing the public members of the authority's board until late in the next legislative session, in 1996, when he got a bill quickly passed, without public debate, to delay giving the Authority "sole authority" over salmon in Maine until July 1, 1997 (Baum 1997).

Less than one month after the listing was proposed, on October 20, 1995, Governor King issued an executive order creating a Maine Atlantic Salmon Task Force, whose job was to advise the governor on how to respond to the proposed listing of salmon in the seven rivers; to develop a conservation plan for the recovery of salmon and its habitat on the seven rivers; and to weigh in on whether the populations in the other Maine rivers were native and naturally reproducing. The task force was chaired by the governor's appointed Commissioner of Inland Fisheries and Wildlife, who in turn created six technical working groups to tackle the job. Members of the task force included representatives of the timber companies, salmon farmers, blueberry growers, environmental advocacy groups, and state agencies.

Using materials from the working groups, the governor sent a response to the Services on December 27, 1995, stating:

> *the State of Maine is strongly opposed to the proposed threatened species listing on the seven rivers on the grounds that the stocks in the seven rivers do not meet the criteria for listing under the Act and that listing would be counterproductive to the superior protection afforded the species under the existing Maine regulatory mechanism, as enhanced by voluntary private/public partnership to conserve and restore salmon runs.*

The governor's letter asked the Services instead to enter into a cooperative agreement for the implementation of the conservation plan under development by the Task Force. He presumably thought that a cooperative agreement would be superior to the proposed section 4(d) rule, because there would be no listing.

The Maine Conservation Plan of 1997

The Task Force submitted the first draft of the conservation plan in November 1996, and it was sent to the Services for an informal review. After receiving their comments two months later, the Task Force revised the draft, and the governor's office released it for public comment in early March 1997. The Services then published a notice in the Federal Register inviting public comment on the plan for 30 days.

It was not until the Task Force had nearly completed the first draft of the plan that the governor appointed the citizen members of the board for the new Atlantic Salmon Authority. The board's first task was to submit a report to the legislature on its plan for the management of the state's salmon fishery. But, because the governor had given the Task Force the job of devising a plan for salmon in the seven Down East rivers—and the Services were still funding most of the conservation activities in the state—few believed that sufficient budget and staff would be given to implement any plan the board could devise for the remaining nine rivers with salmon populations.

Meanwhile, the Services were revisiting the definition of a DPS. On February 7, 1996, the USFWS published a new interpretation. Three criteria would apply: whether the population was 1) discrete in relation to the remainder of the species to which it was a part; 2) significant to the species to which it belonged; and 3) met the standards for listing as either threatened or endangered (U.S. FWS 1996). Underlying the revised definition was a belief that new techniques of genetic analysis would allow these criteria to be applied objectively.

The Task Force took 16 months to write a conservation plan that could withstand review: it was heavily assisted by the Services. NMFS contributed $60,000 so Maine could hire someone to help pull the plan together (Baum 1997). In the plan, the Task Force noted its consensus view that forces beyond Maine's control are responsible for the status of the Atlantic salmon in Maine. In its view, these forces include cyclical stock fluctuation, strongly influenced by low marine survival beyond Maine's state waters, and overfishing on the high seas. Because these factors will ultimately determine the fate of salmon runs, "[t]he Conservation Plan is designed to assure that Maine has taken all reasonable steps to assure successful restoration if and when the international commercial fishing and ocean temperature conditions improve."

When it was completed, the Maine Plan evidently contained enough to satisfy the Services, who were probably anxious at this point to allow the state to retain primary control over salmon restoration efforts. The plan contained a voluntary agreement among the various industries whose practices affect salmon habitat to improve fish management techniques, restore degraded habitat and protect habitat integrity, provide more comprehensive protection of salmon, develop new public education programs, and effectively enforce existing regulations (Jenkins 2003). The Atlantic Salmon Federation criticized the plan for failing to address adequately threats posed by aquaculture through interbreeding by escaped farmed fish, the industry's use of nonnative strains, and disease. The plan's effectiveness regarding these threats depended upon the industry's strategy to prevent escapes through stronger containment nets, vaccination against disease, and harvesting penned fish before they were reproductively mature (MASTF 1997).

The Services then reopened the public comment period on the proposed listing, to solicit public comment on the Maine plan as well as on new information on the salmon, including the latest adult returns, redd counts, fry stocking, habitat assessments, and new commercial fishing agreements and management measures (NMFS and U.S. FWS 1997a).

Withdrawal of the Proposed Listing

In December 1997, Interior Secretary Bruce Babbitt and Commerce Under Secretary for Oceans Terry Garcia, attended a meeting at the Maine State Capitol in Augusta, Maine, to discuss the proposed listing. At a public ceremony later that day, Secretary Babbitt praised the Maine Conservation Plan and the cooperation that had led to its approval. He called it "a new chapter in conservation history" that would "stand as a model for the nation," and he praised Governor Angus King for "show[ing] great leadership in forging this collaboration, which will continue to enhance the ecology and economy of the state for years to come." He had come not only to praise the state, but also to announce that federal agencies were withdrawing the proposed listing under the ESA (Jenkins 2003). He gave some hint at the orchestration that had led to the moment when he noted:

> *The announcement today is short and sweet by joint agreement of the National Marine Fisheries Service, in the Department of Commerce, and the Fish and Wildlife Service, in the Department of Interior. We are here gratefully and happily to say to the people of the great State of Maine the petition that lists the Atlantic Salmon is hereby withdrawn. Yeah, I kinda thought that would be a crowd pleaser! But it didn't just happen. This happy event today is the combination of a lot of work by some very determined people led by your good Governor, who several years ago sat down with Molly Beattie, the Director of the Fish and Wildlife Service, and the people from the National Marine Fisheries and made a simple point, and that was that the protection of the Atlantic salmon ought to be worked out on the ground under the Governor's leadership by the affected people in the state agencies and the conservation organizations of the State of Maine. The Governor's pitch to us then and now was very simple. He says, 'Rather than setting up the inevitably antagonistic form of federal regulation we can, working together, buy all of the stakeholders into a plan to protect this fish because the people of Maine have a deep and abiding love for their*

> *land and their resources and this salmon,' and that of course is what leads us to the work product today, which is the Conservation Agreement, which has been put together under the Governor's direction, which is the substitute for the regulatory action of listing and which substitutes precisely because by its terms it removes the threat that could've caused the listing. And I would say in conclusion that this is a big win for the people of Maine.* (Jenkins 2003, 862–63)

To explain why the proposed listing was being withdrawn, the Services emphasized that the ESA required them to consider whether the state and local efforts to conserve are sufficient to prevent a species' further decline or to recover a species that is in peril. They said they had considered the species' current status and had taken into account efforts such as the development of the Maine Conservation Plan and its implementation to date, as well as private and federal actions and international efforts to control ocean harvest through NASCO. Based on this review, the DPS was not likely to become extinct in the foreseeable future and therefore the listing was not warranted. They noted that the Distinct Population Segment would be renamed the "Gulf of Maine DPS," in recognition that additional populations can be added in the future if they are found to be naturally reproducing with historic, river-specific characteristics (NMFS and U.S. FWS 1995c).

Although the exact timing is unclear, sometime after the listing petitions were filed in 1993—probably in the months before the proposed listing was announced—negotiations between the Services and Maine officials resulted in a deal. The terms were that the wild salmon in eastern rivers of Maine would be recognized as a "distinct population segment" that qualified as a species under the ESA, but the listing would be withdrawn because Maine would develop a state conservation plan. They also agreed, at some point, that the Services would fund an analysis of the available genetics data to see if the salmon runs were, in fact, distinct from the hundreds of thousands of Atlantic salmon that lived in the aquaculture pens in the salmon farms now dotting the Down East coast (Malakoff 1998).

In their public notice explaining why the listing was being withdrawn, the Services said they believed the most significant threats posed by escaped farmed salmon through interbreeding with wild fish would be alleviated by full implementation of a code of containment and the construction of weirs at which farmed fish could be removed (NMFS and U.S. FWS 1997b). They seemed to signal that the state was not being given carte blanche to make only desultory efforts at recovery. The Services promised the public they would report annually on what progress was being made under the Maine plan and to make the report available for public comment. They suggested that the listing could be reinstated if certain circumstances arose, including a significant deterioration of the DPS's biological status.

Breakdown of the Deal and the Final Listing as Endangered

Governor King directed the state agencies to use their authorities to carry out the conservation plan and gave the executive branch's Land and Water Resources Council (LWRC) oversight responsibility for the plan's implementation. The LWRC is chaired by the State Planning Office director and includes commissioners of all the state departments with any responsibility for the environment, natural resources, economic development, and infrastructure. The LWRC created an Atlantic Salmon Committee that included the department heads, as well as the chair of the new Atlantic Salmon Authority and a representative of each of the local watershed councils (MASTF 1997; Jenkins 2003).

By 1999, Maine had spent roughly $1 million to carry out the plan, and another $1 million had been designated for continuing efforts. The legislature appropriated money to hire a coordinator for the plan and a biologist for the Atlantic Salmon Authority. Challenge grants were issued for local habitat restoration efforts in the seven rivers. By all accounts, quite a lot of work was completed: fish weirs were built to trap and count returning adult fish and to collect any escaped farmed salmon; a multi-year, river-specific fry stocking plan was developed; habitat assessments were done for the most important spawning and juvenile nursery areas; over 100 beaver dams and other obstructions were removed; and water withdrawal management plans were completed for the Narraguagus, Machias, and Pleasant rivers, with money from the federal government and the blueberry growers. All recreational fishing was banned, including catch-and-release (MASC 2000).

Despite some skepticism regarding the commitments for voluntary actions made by members of the Task Force in drafting the plan, the signs were strong that a good faith effort had begun. The Maine Aquaculture Association, the salmon farmers' and shellfish growers' trade association, developed a voluntary code of conduct to reduce the risk that salmon would escape from the pens, as well as a disease-prevention protocol. For each of the seven rivers, a local watershed council was established. A training program for code enforcement officers, emphasizing the salmon conservation plan, was developed. And the private and public hatchery operators began to follow new protocols to prevent the introduction of diseases to the wild salmon populations.

The 1999 Annual Report on the Maine Conservation Plan

As promised, the Services announced the availability of the first annual report on progress under the Maine Conservation Plan and invited public comment, asking specifically whether the existing measures remained adequate in light of current knowledge (NMFS and U.S. FWS 1999a). The Services prepared their own review of the report and sent these comments to the state. After receiving the comments, the state made a series of amendments to the plan. The Services said that after a final version of the report was submitted, they would reconvene the biological review team to conduct a new status review of the DPS, then decide whether to re-initiate the listing. The new status review would take account of the genetics study by the United States Geological Survey (USGS) that the Services funded in 1998 after the proposed listing was withdrawn.

One week after availability of the annual report on the Maine plan was announced, the groups that had submitted the original listing petitions in 1993 expressed their dissatisfaction with the whole state plan arrangement. The Defenders of Wildlife sued the Services in January 1999 in federal court in Washington, D.C., challenging their decision to withdraw the proposed listing. The plaintiffs argued it had been improper to consider the Maine plan so heavily in the decision to withdraw the proposed listing. A couple of months later, Trout Unlimited, an organization of recreational fishers long active in Atlantic salmon politics, also filed suit, claiming the proposed listing was improperly withdrawn. The State of Maine sought to intervene as a defendant in these suits but failed, then threatened to sue the Services if the listing was re-initiated, as the plaintiffs demanded.

Six months earlier, a federal district court in Oregon had ruled that NMFS had improperly relied on a State of Oregon conservation plan composed largely of voluntary actions and commitments that could not be enforced by law. NMFS had used the state plan to explain why the Oregon coho salmon did not warrant listing, not long after publishing a proposed rule indicating that the evolutionarily significant unit of coho should be listed (U.S. District Court 1998). The Oregon plan had been developed under a memorandum of agreement between NMFS and the governor of Oregon, just days before a court-ordered deadline to complete a pending listing decision on the Oregon coho salmon. The court interpreted the ESA listing factor D, on the adequacy of existing regulatory mechanisms, as allowing NMFS only to consider currently operational conservation measures, not those planned for the future and based on voluntary actions (U.S. District Court 1998, 1154–55).

The State of Maine later claimed that it was the Services' fear of the Atlantic salmon lawsuit and this legal precedent, rather than any legitimate consideration under the ESA, which led them to re-initiate the listing. At least one observer concluded, however, that by January 1999, the Services had been quietly preparing to re-initiate the listing for some time (Jenkins 2003).

The 1999 Status Review and Re-initiation of the Proposed Listing

When they received the first annual report on the Maine plan, the Services decided to prepare a new biological status review and reconvened the biological review team. The new status review they released in July 1999 was highly critical of current conservation efforts; it must have seemed like a bombshell to those who thought the state plan was working. The status review levied most of its criticism at the underregulation of the rapidly expanding salmon aquaculture industry by both the Army Corps of Engineers and the state (Colligan et al. 1999).

The 1999 status review concluded that, notwithstanding all the efforts of the past two years, the Gulf of Maine Distinct Population Segment was in danger of extinction, due to continued low levels of spawning stocks, low juvenile survival, and increased threats of disease and loss of genetic integrity through the escape of farmed salmon. It noted some progress on reducing threats: habitat enhancement and protection actions by the watershed councils were evident, and water withdrawal from aquifers near spawning streams was being dealt with through management plans and regulations. Also, ocean harvesting was now restricted, and recreational fishing in Maine had been closed, with no significant poaching problem during 1997–1999. The greatly expanded salmon farming industry in the vicinity of the DPS rivers, however, posed a major threat to recovery—the most serious threat being to the genetic integrity of wild salmon from escapees. In particular, the Services noted they had been unsuccessful in convincing the Army Corps of Engineers to ensure that salmon farmers complied with their Rivers and Harbors Act section 10 permit conditions, requiring that they stock the pens only with native strains of Atlantic salmon. Similarly, state regulations on fish and egg imports had failed to restrict aquaculture operators from expanding their use of hybrids using European strains. The Review stated:

> *The … Team believes that current aquaculture practices have the potential to disrupt, displace and genetically contaminate the DPS through redd superimposition, hybridization, disease transfer, and competition. Although discussions with environmental regulators in Maine are ongoing, to date the Services' efforts to obtain state and industry agreement to address aquaculture use of European stocks, crosses between European and North American stocks, and the importation of European milt have so far been unsuccessful. In fact, the use of hybrids by the industry appears to be increasing. In 1998, six million Atlantic salmon were raised in sea cages near the mouths of many DPS rivers and other Atlantic salmon rivers, and escapees have been documented in several of these rivers. Additionally, the escape of juvenile salmon from aquaculture hatcheries within DPS watersheds creates further concern because they compete with indigenous stock in fresh water, and identification is more difficult. The level of these threats is being elevated due to low spawner abundance and industry expansion.… [While discussions are ongoing,] comprehensive protective measures to address the threats are not in place, and aquaculture practices continue to pose a serious threat to the genetic integrity of the Gulf of Maine DPS. Action to address the use of pure European strains and hybrids is necessary. Additionally, weirs on several of the rivers scheduled for construction since 1996*

> *must be completed to help protect stocks from this threat. ... This status review acknowledges the considerable efforts being put forth by the State of Maine and public and private sectors partners to protect Atlantic salmon. The fact remains, however, that under current circumstances, it is the opinion of the [Team] that the Gulf of Maine DPS of Atlantic salmon is in danger of extinction.*

Federal and state agencies were cited as failing to take the genetic risk seriously:

> *The risk inherent in wild stock interacting with escapees has increased significantly from what it was believed to have been three years ago [in 1996] when certain restrictions on the importation and use of foreign salmon stocks were believed ... to be in place and enforced. Regulations governing import and placement of aquaculture fish fall short on two counts: 1) the Maine state law ... regulating import fails to restrict European milt from entering the state as it does fish or eggs, thus enabling expansion of the use of hybrids ... and 2) the Corps of Engineers continues not to enforce permit conditions under §10 of the Rivers and Harbors Act, which prohibits placing European strain or hybrids in sea cages. Failure to adequately enforce certain existing regulations or correct deficiencies identified in others significantly increases risks to the survival of severely depressed existing wild salmon populations. The [Team] concludes that the Gulf of Maine DPS is in danger of extinction.*

Most significantly, the Services announced they had begun to prepare a new proposed listing; this time, however, the listing would be as endangered.

The 2000 Listing as Endangered

After publishing the 1999 status review, the Services, as promised, then published a proposed listing of the salmon in eight rivers of the DPS as "endangered." With the breakdown of the compromise with Maine, there was no reason to re-propose a special section 4(d) rule and threatened listing. Without a listing, there would also be no way of compelling the Army Corps of Engineers to enforce permit conditions on salmon strains. A tributary of the Penobscot River, Cove Brook, was added to the previously proposed seven rivers (NMFS and U.S. FWS 1999b). The notice of the proposed listing began the one-year timetable for final action.

Almost immediately, the State of Maine filed a Freedom of Information Act (FOIA) request for all the documents pertaining to the 1999 decision to list, soon followed by a second request, this time for all the data and documents regarding the listing. The state filed suit under FOIA in February 2000, before the listing was final, to get access to data used in the 1999 status review. Governor King stated publicly that there was not enough scientific data to show that wild salmon in Maine's rivers were genetically distinct from other salmon. In December 2000, one month after the final listing, the court hearing the FOIA case allowed USFWS to withhold some documents, but ordered that it disclose all others—rejecting the claim that the documents were exempt as attorney work products or under the attorney-client privilege.

The proposed listing also stopped efforts by state agencies to make safer aquaculture practices a matter of regulation. In the second annual report on the Maine Conservation Plan, the LWRC reported that the proposed listing stopped the move toward regulation dead in its tracks:

> *The Aquaculture Industry operated its facilities under the second year of the industry's voluntary code of containment practices. In response to concerns about the effectiveness of voluntary management practices, the Department of Marine Resources is developing draft rules codifying the industry's containment code of practices. The Land and Water Resource Council approved this measure as an amendment to the ASCP on March 18th [1999]. DMR suspended the proposed rule-making on containment standards and procedures in light of the federal government's listing proposal under the Endangered Species Act, and is currently considering an appropriate regulatory mechanism.* (MLWRC 1999)

Meanwhile, the conservation plaintiffs agreed to stay their litigation as long as the Services did not extend the review period and kept to the one-year timetable.

(Maine later cited the agencies' agreement to this condition as evidence they were somehow conspiring with the listing petitioners, having made up their minds to follow through with the listing, regardless of what new information came in during the comment period.)

When the final decision to list was published in November 2000, the Services cited several factors related to aquaculture as the major reason for their change of heart, including the emergence of new disease and genetic threats, as well as continued concern over the intensive stocking practices using hybrids by the salmon farms (NMFS and U.S. FWS 2000). The State of Maine made good on its promise and filed suit one month later, challenging the scientific basis for the determination that the Maine salmon were a DPS. The conservation NGOs sought to intervene on behalf of the Services, and business groups (the blueberry growers and salmon farming companies) on behalf of the state. The conservation groups' motion to intervene was denied, and later, the business plaintiffs' motions were dismissed for lack of standing. The court took judicial notice of the state's sovereign interests as a basis for its standing (Sanders 2003).

As if to demonstrate the need for federal oversight, a massive escape of penned salmon occurred, adjacent to two of the listed salmon rivers, during an ocean storm in Maine in February 2001; state officials failed to report it to the Services for three weeks (Clancy 2001).

While Maine's legal challenge was pending, further developments lessened the chance that Maine would be able to make the case that the listing was unwarranted and unnecessary. At the urging of Maine's Senator Olympia Snowe, Congress had asked the National Academy of Sciences to convene a study committee to look into the scientific basis of the listing, including the contested genetics study. It asked the academy to report on these findings in an interim report and to then submit a final report on the most effective means of restoring Atlantic salmon (NAS 2004).

In January 2002, the interim report found that wild salmon in the eight rivers had remained genetically distinct from other wild populations despite over 100 years of stocking. It also found that adult returns in 2000, of 75 to 100 fish, had been the lowest in 10 years and was 50 percent below the average returns of the last decade. The following month, the U.S. Department of Agriculture ordered all salmon to be removed from the pens in Cobscook Bay, adjacent to Canadian waters, where there was an outbreak of infectious salmon anemia, which led to the destruction of all the salmon in the pens.

Ultimately, the federal court rejected Maine's challenge to the listing, finding that the DPS determination was supported in the record, was based upon the best scientific evidence, and was not an abuse of discretion (U.S. District Court 2003a).

Three months after Judge Carter's ruling in *Maine v. Norton*, the new governor of Maine, John E. Baldacci—a former state representative and member of Congress—announced that the state would not appeal and would drop its opposition to the listing. He noted that the state had signed a settlement agreement with the Services that "ensures that the state and federal governments will work cooperatively to protect and restore this important part of Maine's heritage" (Carrier 2003). The agreement committed the federal agencies to work with Maine in developing and implementing a recovery plan. The federal agencies also promised to complete a new and comprehensive status review and a full listing process before any additional Maine rivers or runs were added to the DPS.

The Impact of the Clean Water Act's Cooperative Federalism

By the time the listing case was decided, Judge Carter was very familiar with the issues surrounding salmon farming and the risks that it posed to wild salmon. Judge Carter had previously ruled that two of the salmon farms were point sources under the CWA and in violation for discharging fish, feed, and other materials into U.S. waters without National Pollutant Discharge Elimination System (NPDES) permits. The citizen suit had been filed in late 2000, one year after Maine had submitted an application to the U.S. Environmental Protection Agency (EPA) for delegation of the authority to issue NPDES permits (U.S. District Court 2003b). Judge Carter agreed that it was regulatory negligence of the EPA not to have issued the permits to those companies that had applied for them in 1990; nevertheless, the farms were in violation. He ordered a ban on the use of nonnative strains, in view of the threat that escaped salmon of nonnative origin posed to wild salmon.

Judge Carter's order has been described as accomplishing in one paragraph what the Services had been trying to do for 15 years through negotiations with the Army Corps, EPA, and the state, but had been unable to achieve due to heavy industry lobbying. Industry pressure had thwarted recommendations that had been made repeatedly at the staff level on this and other issues (Goode 2003). In September 2002, EPA at last proposed effluent limitation guidelines for wastewater discharges from fish farming operations in U.S. waters, establishing minimum standards for state water quality regulation.

Did the salmon farming companies take this particular federal official, the judge, more seriously than they did the agencies? To make sure they did, in February 2003, Judge Carter ordered them not to stock fish in their pens in the spring. Two months later, Fjord Seafood, the Norwegian parent company of Atlantic Salmon of Maine, petitioned the Services to release the genetics data they relied on for the ESA-based permit restrictions on use of European strains in their broodstock. They sought to take advantage of the new procedures of the Data Quality Act of 2001, aimed at ensuring the integrity of data and information used in federal regulatory policies (Porter, A. 2003). Conservation groups immediately petitioned in opposition to the request. Fjord's aim was to once again reevaluate the USGS genetics study regarding the genetic isolation of Gulf of Maine DPS from all other Atlantic salmon of North American origin—the same data that the State of Maine had questioned vehemently but then accepted.

When Atlantic Salmon of Maine, Inc. went ahead with the stocking, Judge Carter held them in contempt for discharging a new year-class of fish into their pens without a permit. In June 2003, after Judge Carter's contempt order, Fjord Seafood announced it was adopting a new cooperative attitude toward government involvement in its business decisions. Not long afterward, it announced that Atlantic Salmon of Maine was for sale.

For many years, the EPA and Maine had negotiated whether the state was eligible for delegation of the NPDES program from EPA and disagreed over the manner in which the state would regulate the discharges of the salmon farming industry (Fleming 2001–2002; Price 2004). After the Gulf of Maine wild salmon DPS was listed, it seems the logjam was broken on several pending actions. Formal interagency consultation took place under section 7 of the ESA, concerning the NPDES delegation as well as on the Army Corps of Engineers' permit conditions for salmon pens under the Rivers and Harbors Act (NMFS and U.S. FWS 2003). Also, in 2001, EPA and the Services released a final coordination memorandum of agreement regarding the CWA and the ESA at the same time that they released a final biological opinion on the Maine NPDES delegation. The opinion concluded that conditions must be included in NPDES permits for salmon farms to prevent escapes or accidental releases. EPA approved the Maine program on February 28, 2001 (EPA 2001) and adopted the Services' recommendation that to avoid jeopardy, EPA should object to any Maine-issued permits for salmon farms that did not protect endangered wild salmon.

Recovery Planning for the Maine Salmon: State–Federal Cooperation at Last?

The National Academy of Sciences released its final report, *Atlantic Salmon in Maine*, in January 2004, making a number of recommendations for the conservation and recovery of Atlantic salmon. After all the focus on the risks posed by salmon farming, the report ranked dams and development as the highest risk factors for salmon. It concluded that much more attention should focus on the salmon returning to the Penobscot River and suggested that decision-analytic tools be used to craft an effective recovery plan.

The Services released their Recovery Plan for the Gulf of Maine DPS of Atlantic Salmon in December 2005, after public hearings and independent peer review (NMFS and U.S. FWS 2005). The plan had been prepared in consultation with Maine's Atlantic Salmon Commission, but the Services described the draft plan as more comprehensive than the Maine Conservation Plan. The plan proposed actions that complement the Maine plan but, in the Services' view, addressed the threat of salmon farming in more detail and tackled rangewide threats, beyond the State of Maine's jurisdiction, including low ocean survival. The plan itemized over 100 action items that are necessary for the Atlantic salmon's conservation and recovery, but it did not include any demographic reclassification and recovery criteria. These were deferred until the plan could be reviewed and revised, after three years. A total cost for the entire recovery program was not estimated, but a figure of over $33 million was given for the first three years. It concluded that it would take 12 to 15 years to determine if the species' decline had halted. The plan included stocking, population research, stream flow studies, and fish passage improvements.

But what about those dams? Independent of the ESA listing and interagency recovery efforts, private groups, NGOs, and the Penobscot Indian Nation began to tackle the dams. In anticipation of their pending relicensing, they brokered a multiparty agreement to restore the Penobscot, the river to which most Atlantic salmon in Maine return. The deal they struck would allow the coalition for the river's restoration to buy three of the dams on the Penobscot from the private electric utility, in exchange for an agreement not to contest an application to the Federal Energy Regulatory Commission to increase power production at two other dams. The agreement follows in the spirit of the pact to remove the Edwards Dam on the Kennebec, an event Secretary Bruce Babbitt had attended and heralded as a new era in humanity's relationship with nature (McPhee 1999). This time, however, the coalition had to raise the estimated $25 million purchase price. Interior Secretary Gail Norton attended the signing of the Penobscot River restoration agreement but did not promise to obtain federal funds to carry out the ambitious plan (*Bangor Daily News* 2004).

In February 2008, the Penobscot Indian Nation and its partners announced they had succeeded in raising $25 million for the first phase of restoring the Penobscot, their ancestral river and watershed (Toensing 2008)—including a $10 million appropriation from Congress (Daley 2007) that will allow them to buy and tear down the two dams closest to the ocean, build fishways at others, restore Atlantic salmon and other native fishes, and perhaps give real substance to the Penobscot Nation's tribal fishing rights.

Conclusion

In this case study, we see that cooperation with the State of Maine broke down to the point of litigation, despite the federal agencies' efforts to exercise as much flexibility as they could plausibly read into the ESA. Maine regulatory officials, guarding the state's agriculture and fearing the loss of a new industry in an economically challenged county, were unable to impose on the resistant industry the strict conditions made necessary by federal determination that raising salmon in net-pens, adjacent to the river mouths, posed a risk to wild salmon running in those rivers. After the state failed to disprove the genetic distinctiveness of the remaining wild salmon populations running in its rivers and the endangered listing was upheld, the CWA's cooperative federalism mechanisms became the focus.

But Maine's effort to maintain state control over discharge conditions, through federal delegation of the NPDES permit program, also failed, stopped by a CWA citizen suit. By ruling that the salmon farmers were discharging pollutants by stocking their net-pens with nonnative strains of Atlantic salmon—and requiring them to empty their pens, under threat of contempt citations while they obtained an NPDES permit—the court exercised power, under the CWA, to resolve the issue that had stymied cooperation under the ESA.

The court's intervention, in turn, cleared the way for a resumption of federal–state cooperation in developing the Atlantic salmon recovery plan. Prospects for recovery through incremental improvements in habitat, by this point, were rapidly dwindling, as fish numbers continued to decline. Bold new actions were needed. A cooperative effort, via a privately negotiated agreement among hydropower companies, conservation groups, and the Penobscot Indian Nation, now held the greatest promise. Their agreement brought the federal and state agencies together, in a multiparty pact, to remove the worst offending dams within the watershed and allow ecological restoration important to salmon and other fish species, whose company in the river might be essential for the salmon.

Epilogue

A citizens' petition to revise the DPS to include salmon in the Kennebec River prompted the federal agencies to convene a new Biological Review Team to review newly available genetic data and other information necessary to evaluate the salmon populations that had been excluded from the 2000 DPS listing. The Team's Status Review was unequivocal: the best available data indicated that that all Atlantic salmon in Maine should be in the same population segment, including the populations in the three large rivers that had been excluded from the DPS in 2000. The inclusion of these salmon, which included fish spawned in the federal conservation hatcheries, did not change the DPS's conservation status. The numbers were so low that a new population viability analysis indicated a very high risk of extinction that ranged from 19 percent to 74 percent within the next 100 years (Fay et al. 2006).

In 2009, the Services finally listed the salmon inhabiting the three largest rivers in Maine—the industrialized Androscoggin and Kennebec, and the many-dammed Penobscot—as part of an expanded Gulf of Maine DPS (NMFS and U.S. FWS 2009a). A determination of the DPS's critical habitat as including the entire geographic range of wild salmon in Maine followed on the heels of the new listing (NMFS and U.S. FWS 2009b). The State of Maine and its congressional delegation urged the Secretaries to classify the expanded DPS as threatened but exclude the Androscoggin (into which the largest remaining paper mill discharges its effluent) and again rely on state regulation to implement the recovery plan (NMFS and U.S. FWS 2009a, 29385; State of Maine 2008). The Services remained unconvinced, however, of the state's commitment to making the hard decisions needed to promote salmon recovery. The state had recently eliminated the Atlantic Salmon Commission and merged it into a marine resources department that had never been enthusiastic about dealing with the salmon's plight (Baum 2008). The Services were encouraged by news that the Penobscot Restoration Agreement partners had secured $25 million to buy three of the dams and retire them and to modify another with fish passages. But the benefits of this voluntary conservation agreement, however, were not yet certain enough to decrease the predicted risk of extinction. In the final analysis, the conservation plan and restoration agreement were not enough to change the risk assessment under the criteria of the Services' joint "Policy for Evaluation of Conservation Efforts When Making Listing Decisions" (NMFS and U.S. FWS 2009a, 29377 citing 68 Fed. Reg. 15100, Mar. 28, 2003). The Gulf of Maine Atlantic salmon was clearly endangered, and it would be a shared federal responsibility to manage its recovery. The Services no doubt agreed with the comments of Baum, who had been the state's chief salmon biologist for decades and now supported the expansion of the DPS, that the history of state and federal cooperation showed that "Maine Atlantic salmon are not as important to the State as they are to the rest of the nation" (Baum 2008, 2).

Appendix 9-1 lists the chronology of the Atlantic salmon endangered species listing.

APPENDIX 9-1. Atlantic Salmon Endangered Species Listing Chronology

1991	USFWS designates Atlantic salmon in five Maine rivers as candidate species Maine Legislature enacts P.L. 1991, c. 381, banning import and introduction into state waters of salmon eggs from Icelandic or European stocks, effective Jan 1, 1995, but authorizing administrative exemptions for legitimate research.
1992	USFWS and Maine complete a prelisting recovery plan.
1993	USFWS and NMFS receive petitions to list as threatened all naturally spawning Atlantic salmon throughout its historic U.S. range.
1994	Services find that listing might be warranted and forms biological review team.
1995	Petitioning groups file suit to compel action on the listing petitions. Governor Angus King speaks out against the potential listing.
1995	Services publish Status Review and determine that listing is not warranted throughout the entire U.S. range, but that salmon in seven Maine rivers could constitute a Distinct Population Segment (DPS) that warrants listing as threatened.
1995	Maine Legislature amends P.L. 1991, banning nonnative fish and eggs, to allow administrative exemptions for "legitimate aquaculture projects."
1995	Services propose to list the seven rivers' DPS as threatened due to low natural survival at sea, poaching, and potential impacts from salmon aquaculture and hatcheries on disease and genetic integrity. State of Maine is invited to submit a conservation plan as a special section 4(d) rule, in lieu of federal protective regulations. Governor of Maine appoints a task force to prepare the plan.
1996	After congressional moratorium ends on processing listings, public comment is invited on proposed listing as threatened. State of Maine files comments objecting to the proposed listing "in the strongest possible terms" and proposes a cooperative agreement as an alternative to listing.
1997	Services receive the Maine Atlantic Salmon Conservation Plan (ASCP) and invite the public to comment on it and on new information on stock status and conservation efforts.
1997	Services withdraw the proposed listing of the DPS as not warranted, in light of the Maine conservation plan and conservation efforts underway. Services rename it the Gulf of Maine DPS, acknowledging the possible addition of other populations if found to be naturally reproducing with historic, river-specific characteristics. Services also promise to make annual reports on progress, under the Maine plan, available for public review. Circumstances that would warrant reinstatement of listing include deterioration of DPS's biological status.
1999	Services invite public comment on the first annual report on the Maine plan and whether existing measures remain adequate, in light of current knowledge. Services send comments to State of Maine. Defenders of Wildlife et al. file suit in federal court challenging December 1997 withdrawal of the proposed listing, alleging improper consideration of Maine plan in decision to withdraw.
1999	Maine submits amendments to its conservation plan in light of Services' and public comments on the first annual report. Maine Legislature creates the Atlantic Salmon Commission.
1999	Atlantic Salmon Federation and Trout Unlimited file suit, challenging the withdrawal of the proposed listing.

1999	Services publish a new biological Status Review, which concludes the DPS is in danger of extinction—notwithstanding Maine's efforts under the conservation plan—due to continued low levels of spawning stocks, low juvenile survival, and increased threats of disease and genetic integrity through escapes from net-pens. Services begin preparing a new proposed listing. Status Review notes Services are unsuccessful convincing the Army Corps to ensure compliance with section 10 permit conditions, requiring sea farm stocking with native strains only, and that state regulations on fish and egg imports fail to restrict aquaculture operators from expanding their use of hybrids from European strains.
1999	Services publish a proposed listing of Gulf of Maine DPS in eight rivers as endangered—citing risk of extinction due to disease, inadequate regulation of aquaculture and water withdrawals, and escapes from fish pens and hatcheries. Maine files an FOIA request for all documents pertaining to the proposed listing.
2000	Services publish final rule listing the Gulf of Maine DPS as endangered. State of Maine and several business groups, including agriculture and aquaculture companies, file suit challenging the scientific basis of the DPS determination and the basis for the listing, claiming that fear of litigation rather than science led to listing. Conservation groups are denied permission to intervene on behalf of Services.
2002	In a citizen suit, Federal District of Maine Judge Carter finds two Maine salmon farming companies in violation of the CWA for operating their net-pen farms without NPDES permits; he orders farms to remove all nonnative-strain fish the companies stocked in their pens while the remedy phase was under review.
2003	Judge Carter rejects Maine's challenge to the merits of the listing, finding that DPS policy was properly applied, DPS determination was supported in the record, and listing was based on the best scientific evidence and not an abuse of discretion.
2003	Governor Baldacci decides not to appeal the decision upholding the listing. In settlement agreement, Services and state agree to cooperate on recovery planning and activities and to complete a new status review before any other rivers are added to the DPS.
2003	Judge Carter orders salmon farms not to stock fish in pens and then finds them in contempt of court for stocking a new year-class despite absence of an NPDES permit.
2003	NMFS releases the final biological opinion on proposed modifications to Army Corps section 10 permit conditions for pen salmon farming in Maine, including special conditions regarding genetic strains that may be used.
2003	First Circuit Court of Appeals upholds Judge Carter's ruling under the CWA. Governor Baldacci appoints a new task force to make recommendations to address social, environmental, and economic issues of aquaculture in Maine.
2004	National Academy of Sciences releases its final report, *Atlantic Salmon in Maine*, ranking dams and development as the highest risk factors for salmon, recommending expansion of the DPS to include Maine's three largest rivers.
2005	Services release the final Recovery Plan.
2006	New biological Status Review concludes that available genetic and ecological data support listing as endangered all naturally spawned and conservation hatchery populations of Atlantic salmon whose freshwater range occurs in the watersheds from Androscoggin River north to the Dennys River.
2008	State of Maine eliminates independent Atlantic Salmon Commission; Services propose listing expanded DPS and critical habitat; 2,100 adult salmon return to Penobscot River.
2009	Services publish final rule expanding DPS and designating critical habitat.

Endnote

This chapter was published originally in the University of Maine School of Law *Ocean and Coastal Law Journal* 16(1): 135–179, Fall 2010, and is reprinted with permission.

References

Bangor Daily News. 2004. "Saving Salmon." June 22, A8.

Baum, Edward. 1997. *Maine's Atlantic Salmon: A National Treasure*. Hermon, ME: Atlantic Salmon Unlimited.

———. 2008. "Comments on Proposed Endangered Status for the Gulf of Maine Distinct Population Segment of Atlantic Salmon." December 2 (on file with the author).

Bean, M. J. and M. J. Rowland. 1997. *National Wildlife Law*. 3rd ed. Westport, CT: Praeger Publishers.

Berrill, Michael. 1997. "Salmon the Greatest Challenge." In *The Plundered Seas: Can the World's Fish Be Saved?* 153–172 San Francisco, CA: Sierra Club Books.

Carrier, Paul. 2003. "State Ends Atlantic Salmon Fight." *Portland Press Herald*, July 25.

Clancy, M. A. 2001. "100,000 Salmon Escape from Pens; Aquaculture Moratorium Sought." *Bangor Daily News*, February 23, A1.

Colligan, M. A., J. F. Kocik, D. C. Kimball, G. Marancik, J. F. McKeon, and P. R. Nickerson. 1999. *Review of the Status of Anadromous Atlantic Salmon (*Salmo salar*) under the U.S. Endangered Species Act. National Oceanic and Atmospheric Administration and U.S. Fish and Wildlife Service*. Gloucester, MA.

Daley, Beth. 2007. "US Aid May Preserve Huge Maine Tract." *Boston Globe*, December 23.

EPA (Environmental Protection Agency). 2001. "Approval of Application by Maine to Administer the NPDES Program; Maine." *Federal Register* 66: 12791–95, February 28.

Fay, C., M. Bartron, S. Craig, A. Hecht, J. Pruden, R. Saunders, T. Sheehan, and J. Trial. 2006. "Status Review for Anadromous Atlantic Salmon (*Salmo salar*) in the United States." Report to the National Marine Fisheries Service and U.S. Fish and Wildlife Service, Gloucester, MA.

Fleming, Roger. 2001–2002. "Does the Clean Water Act Protect Endangered Species? The Case of Maine's Wild Atlantic Salmon." *Ocean and Coastal Law Journal* 7(2): 259–327.

Goode, Andrew. 2003. "Editorial: Task Force to Help Plan for Aquaculture's Future." *Bangor Daily News*, June 10.

ICES (International Council for the Exploration of the Seas). 2002. "What's Happening to Atlantic Salmon?" *ICES CIEM Newsletter*, 39: 8–11.

Jenkins, David. 2003. "Atlantic Salmon, Endangered Species, and the Failure of Environmental Policies." *Comparative Studies in Society and History* 45(4): 843–872.

Malakoff, David. 1998. "Atlantic Salmon Spawn Fight Over Species Protection." *Science* 279(5352): 800. February 6.

MASC (Maine Atlantic Salmon Commission). 2000. "Atlantic Salmon Conservation Plan for Seven Maine Rivers." 2000 Annual Progress Report.

MASTF (Maine Atlantic Salmon Task Force). 1997. "Atlantic Salmon Conservation Plan for Seven Maine Rivers." March.

McPhee, John. 1999. "The Control of Nature: Farewell to the Nineteenth Century—The Breaching of the Edwards Dam." *The New Yorker* 75(28): 44–53, September 27.

MLWRC (Maine Land and Water Resources Council). 1999. "Second Annual Progress Report for 1998 on the Maine Atlantic Salmon Conservation Plan." Augusta, ME: State Planning Office.

NAS (National Academy of Sciences). 2004. *Atlantic Salmon in Maine.* Washington, DC: National Academy Press.

NASCO (North Atlantic Salmon Conservation Organization). 2004. "Report on the 21st Annual Meeting of the Council."

NEFMC (New England Regional Fishery Management Council). 1987. "Fishery Management Plan for Atlantic Salmon: Incorporating an Environmental Assessment and Regulatory Impact Review/Initial Regulatory Flexibility Analysis." Saugus, MA: New England Fishery Management Council.

NMFS and U.S. FWS (United States Department of Commerce, National Oceanic and Atmospheric Administration, National Marine Fisheries Service and United States Department of Interior, Fish and Wildlife Service). 1995a. "12-Month Finding for a Petition to List the Anadromous Atlantic Salmon (*Salmo salar*) Populations in the U.S. as Endangered or Threatened." *Federal Register* 60: 14410–12. March 17.

———. 1995b. "Proposed Rule, Proposed Threatened Status for a Distinct Population Segment of Anadromous Atlantic Salmon (*Salmo salar*) in Seven Maine Rivers." *Federal Register* 60: 50530–39. September 29.

———. 1995c. "Withdrawal of a Petition to List a Distinct Population Segment of Atlantic Salmon (*Salmo salar*) as Threatened." *Federal Register* 62: 66325–38. December 17.

———. 1997a. "Reopening of Comment Period on Proposed Threatened Status for a Distinct Population Segment of Anadromous Atlantic Salmon (*Salmo salar*) in the Seven Rivers." *Federal Register* 62: 28413–15. May 23.

———. 1997b. "Withdrawal of Proposed Rule to List a Distinct Population Segment of Atlantic Salmon (*Salmo salar*) as Threatened." *Federal Register* 62: 66325–38, December 18.

———. 1999a. "Availability of an Annual Report on Implementation of the Conservation Plan for Atlantic Salmon in Seven Maine Rivers." *Federal Register* 64: 3067–68. January 20.

———. 1999b. "Proposed Rule. Proposed Endangered Status for a Distinct Population Segment of Anadromous Atlantic Salmon (*Salmo salar*) in the Gulf of Maine." *Federal Register* 64: 62627. November 17.

———. 2000. "Final Endangered Status of a Distinct Population Segment of Atlantic Salmon (*Salmo salar*) in the Gulf of Maine; Final Rule." *Federal Register* 65: 69459–83. November 17.

———. 2003. "Biological Opinion on Existing Army Corps of Engineers Section 10 Permits for Salmon Aquaculture in Maine Waters."

———. 2005. "Notice of Availability for the Final Recovery Plan for the Distinct Population Segment of Atlantic Salmon (*Salmo salar*) in the Gulf of Maine." *Federal Register* 70: 75473–78. December 20.

———. 2009a. "Determination of Endangered Status for the Gulf of Maine Distinct Population Segment of Atlantic Salmon (*Salmo salar*); Final Rule." *Federal Register* 70: 75473–78. June 19.

———. 2009b. "Designation of Critical Habitat for the Gulf of Maine Distinct Population Segment of Atlantic Salmon (*Salmo salar*) Gulf of Maine Distinct Population Segment; Final Rule." *Federal Register* 70: 75473–78. June 19.

Porter, Aaron. 2003. "Fjord Challenges Federal Salmon Genetics Data, Abstracts and Reviews." *Center for Regulatory Effectiveness Newsletter*. April 28.

Porter, Gareth. 2003. *Protecting Wild Atlantic Salmon from Impacts of Salmon Aquaculture: A Country-by-Country Progress Report.* St. Andrews, N.B., Canada and Washington, D.C.: Atlantic Salmon Federation and World Wildlife Fund.

Price, Brian R., II. 2004. "Note, Maine Aquaculture, Atlantic Salmon, and Inertia: What is the Future for Maine's Net Pen Salmon Industry?" *Boston College Environmental Affairs Law Review* 31(3): 683–722.

Sanders, Hanna. 2003. "Casenote, *State of Maine v. Norton*: Assessing the Role of Judicial Notice." *Ocean and Coastal Law Journal* 9(1): 125–137.

Siegel, Kirk G. 1997. "Comment, Challenging the 'Distinct Population Segment' Definition of Atlantic Salmon Under the Endangered Species Act," *Ocean and Coastal Law Journal* 2: 341–380.

State of Maine. 2008. "Comments on Proposed Expanded Distinct Population Segment for Atlantic Salmon." Augusta, ME. December 2.

Toensing, Gale C. 2008. "The Fish Run Through It." *Indian Country Today*. January 17.

UN FAO (United Nations Food and Agriculture Organization). 1996. *Precautionary Approach to Capture Fisheries and Species Introductions, FAO Technical Guidelines for Responsible Fisheries*, v. 2. Rome: FAO.

U.S. District Court (U.S. District Court for the District of Oregon). 1998. *Oregon Natural Resources Council v. Daley*, 6 F.Supp.2d 1139–1161, June 1.

U.S. District Court (U.S. District Court for the District of Maine). 2003a. *State of Maine v. Norton*, 257 F. Supp.2d 357–407, April 24.

———. 2003b. *U.S. Public Interest Research Group v. Atlantic Salmon of Maine, LLC*, 257 F.Supp.2d 407–36, May 28.

U.S. FWS (U.S., Fish and Wildlife Service). 1991. "Notice of Review: Animal Candidate Review for Listing as Endangered or Threatened Species, 56." *Federal Register*. 58804. November 21.

———. 1996. "Policy Regarding the Recognition of Distinct Vertebrate Population Segments Under the Endangered Species Act." *Federal Register* 61: 4722–25. February 7.

CHAPTER 10

Oregon Coastal Coho Restoration and the Endangered Species Act

Gail L. Achterman and Julia Doermann

Pacific salmon define the place some call "Salmon Nation" (Wolf and Zuckerman 1999). Yet in 1991, in spite of decades-long restoration efforts, 101 out of 214 native, naturally spawning runs were at high risk of extinction (Nehlson et al. 1991, 4–21). State officials, federal agencies, and business, community, and environmental leaders sought an alternative to listing salmon under the Endangered Species Act (ESA)—one focused on restoring salmon runs, not simply avoiding listing. This case study examines the Oregon Coastal Salmon Restoration Initiative (OCSRI), which later became the Oregon Plan, a state strategy for conserving salmon and watersheds throughout Oregon.[1]

The Oregon Plan seeks to restore salmon and watersheds while meeting Clean Water Act (CWA) and ESA requirements. The state effort is designed to take the lead in salmon recovery, salmon management, and watershed restoration, without the additional federal regulations that come with a listing. Appendix 10-1 at the end of this chapter provides a chronology of major events at the state and federal levels and in the courts concerning coho salmon conservation in Oregon.

Oregon Coastal Coho

Oregon coastal coho begin their lives in the upper watersheds of coastal streams. Figure 10-1 shows the Oregon coastal coho evolutionarily significant unit (ESU).[2] They rear their young in areas sheltered from high flows. Coho migrate to the ocean in the spring, where they live for two to three years, growing to 8 to 12 pounds. Most return to fresh water in the fall of their third year and spawn in the late fall and winter on gravel bars in upper watersheds. Their life cycle depends on good water quality; spawning, rearing, and feeding habitat; stream flow; ocean conditions; and passable streams. Their diverse life histories, genetic diversity, and metapopulation organization help them cope with natural environmental variation (Lichatowich 1999).

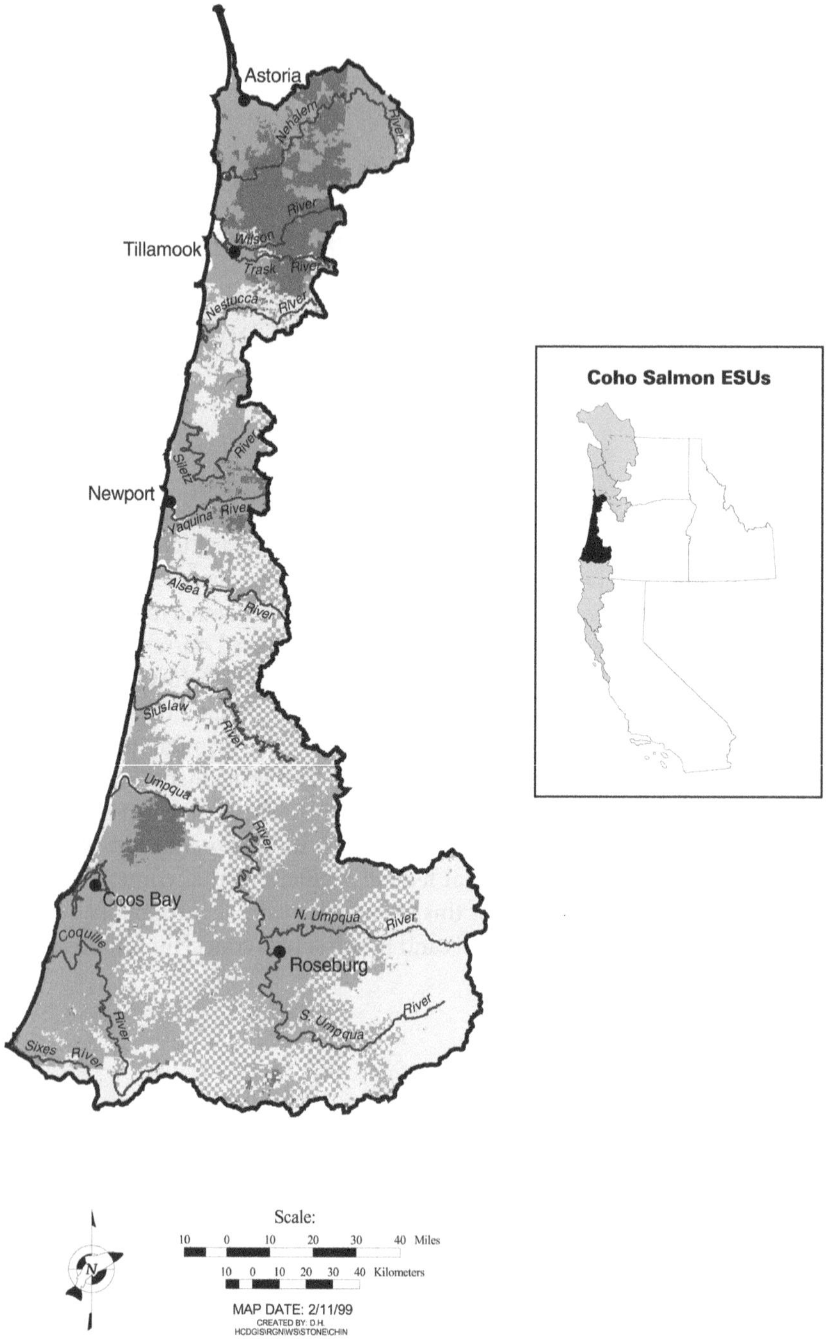

FIGURE 10-1. Oregon Coastal Coho Evolutionary Significant Unit

Source: National Marine Fisheries Service.

Coho declines reflect the direct impacts and cumulative effects of varied human activities. The factors associated with coho decline are generally referred to as the "four Hs": habitat (destruction of habitat by urban development, farming and forestry practices, channelization, irrigation and other water withdrawals, mining, pollution, and other human activities); harvest; hatcheries; and hydropower (Lichatowich 1999; NMFS 2004b). All have taken their toll on coho over decades. Nonnative commercial fish harvesting began when canneries were built on virtually every coastal stream, beginning in the mid-1800s (Taylor 1999, 64–65, 133–65). Commercial and sport fishing caught over two million coho in 1923, before overfishing caused coho populations to plummet in the 1940s and 1950s. Hatchery construction helped increase runs in the 1960s. Coho harvest reached a peak of over four million fish in the mid-1970s (Lichatowich 1999, 208). Natural forces and fluctuations, such as cycling ocean conditions and predators, also affect salmon populations and interplay with human-caused factors of decline (Stouder et al. 1997, 331–352). When ocean survival is poor, high-quality freshwater habitat is essential, and 60 percent to 70 percent of onshore habitat is on private land (NMFS 2004c).

Historical Context

The Oregon Plan grew from decades of failed efforts to save salmon runs, the nascent community-based watershed restoration movement, CWA implementation, and Oregon's experience with northern spotted owl protection under the ESA.

Traditional Salmon Conservation

Calls to stop the decline of Oregon salmon runs date from as early as 1875 (Taylor 1999, 3–4). Because salmon are anadromous, they integrate human activities and ecological processes across the landscape and reflect them in their health, individually and as a species (Stouder et al. 1997, 13–16). The process of identifying and agreeing on the activities that hurt the fish, combined with their complex life cycle, make conservation challenging. It is too easy to avoid responsibility for salmon conservation by blaming someone or something else. Until recently, government responded to calls to save salmon by adopting technical or regulatory fixes. Fish managers believed that salmon runs could be restored by engineering artificial substitutes for functioning natural systems, rather than restoring and protecting natural processes (Taylor 1999, Chapter 3). State laws attempted to save the fish by regulating harvest gear and seasons, requiring fish passage on dams, and prohibiting dumping waste in salmon streams. State officials embraced hatcheries as a panacea for salmon declines (Montgomery 2003, 149–176). But by the 1980s, it was clear that the hatchery panacea and technical fixes did not work (Lichatowich 1999, 52–80).

The first comprehensive effort to restore Pacific salmon runs occurred in the Columbia Basin, where precipitous declines of Columbia Basin runs—combined with needed reform of power marketing legislation—resulted in the creation of the Northwest Power Planning Council in 1980 (Blumm 2002). The Columbia Basin fish restoration effort set the stage for the Oregon Plan, stimulating voluntary fish and watershed restoration work throughout the Pacific Northwest.

Watershed Restoration and Water Quality

In the 1980s, Oregon developed incentives to stimulate voluntary salmon and watershed restoration efforts to supplement traditional regulatory approaches and hatcheries. A volunteer salmon and trout enhancement program encouraged local groups to improve stream habitat and install small hatch boxes to restore runs. The Governor's Watershed Enhancement Board (GWEB) was created to fund watershed restoration projects. GWEB investments generated considerable enthusiasm—resulting in the Oregon legislature urging counties to form local watershed councils to assess watershed conditions and to develop and implement action plans, aimed at achieving "sustainable watershed health." By the mid-1990s, watershed councils were working in all coastal subbasins.

Local watershed restoration work was also prompted by CWA requirements. By the mid-1990s, only 11.6 percent of assessed streams in the range of the coho met all state water quality standards (OCSRI 1997, 17:B-1), which necessitated development of total maximum daily load allocations (TMDLs). The only effective way to meet the standards was to address nonpoint pollution sources. TMDL preparation engaged citizens with state and federal agencies in assessing their watersheds.

Northern Spotted Owls and the Northwest Forest Plan

After the northern spotted owl was listed as endangered under the ESA in June 1990, Oregonians experienced the community polarization that ESA implementation can create. That experience and the subsequent adoption of the Northwest Forest Plan, in April 1994, shaped Oregon's response to salmon conservation. A key component of the Northwest Forest Plan was the Aquatic Conservation Strategy (the ACS), which is designed to maintain and restore the ecological health of watersheds in federal forests. The ACS imposed binding standards and guidelines restricting certain activities within riparian reserves and key watersheds. Proposed federal timber sales and other projects must meet these ACS requirements.

Listing Petition and State Response

On July 21, 1993, several environmental groups petitioned the National Marine Fisheries Service (NMFS) to list five stocks of coho salmon as threatened or endangered under the ESA and to designate critical habitat needed to recover the species.[3] The productivity of wild coho salmon had fallen to less than five percent of historic levels (Cone 1995, 282–3). NMFS reviewed the petition through a comprehensive status review of coho in Oregon, Washington, and California, focusing on the relationship between wild naturally spawning populations and hatchery fish (NMFS 1993). A key issue was whether hatchery fish should count in the listing decision.

While no one was sure about when or whether listing would occur, there was widespread interest in protecting and restoring coho. The question was how to do so most effectively and quickly. The Oregon Department of Fish and Wildlife (ODFW) curtailed harvest of wild coho within three miles offshore and all Oregon fresh water. ODFW also started marking all hatchery fish, so that sport and commercial fishermen could catch them but release wild coho. The Pacific Salmon Task Force was created to optimize use of federal expertise and resources to restore salmon runs while NMFS completed the status review (OCSRI 1997, 17:D-2).

Based on their past experience with ESA implementation—in particular, the case of the northern spotted owl—many public and private entities in the state were concerned that listing would bring bureaucratic procedures and plans with debatable benefits for salmon restoration. John Kitzhaber took office as governor in January 1995. He immediately made recovering coastal coho salmon a priority—engaging legislative leaders, the forest industry, agricultural and ranching interests, environmental organizations, federal land management agencies, and state government to build a plan designed to align state laws and management practices for coho restoration.

On July 25, 1995, NMFS proposed listing two ESUs of coastal coho as "threatened," including Oregon coastal coho. The ESU included both hatchery and natural spawning coho—which were genetically indistinguishable—but only "naturally reproducing populations" were proposed for listing. Up to 62 percent of the returning fish were hatchery fish in subbasins heavily influenced by hatcheries (NMFS 1995).

In the proposed listing, NMFS addressed the adequacy of existing state regulatory mechanisms to reverse the identified factors for decline. NMFS acknowledged the existence of new local, state, and federal efforts, such as the ACS and watershed council programs, but it was unable to determine the conservation effect of these initiatives and found them inadequate to preclude listing the species. The threat of listing stimulated development of the OCSRI. State officials wanted to know what they needed to do to create regulatory mechanisms adequate to meet ESA requirements.

Building the OCSRI

OCSRI emphasized three key elements: 1) harvest and hatchery reform; 2) community-based watershed councils; and 3) interagency and intergovernmental coordination (OCSRI 1997, 1:10). Harvest and hatchery reform built on ODFW's earlier efforts. ODFW's wild fish policy was changed to reduce straying of hatchery fish and hatchery releases. GWEB, local watershed councils, and soil and water conservation districts worked to motivate private landowners to make voluntary commitments to watershed restoration. The GWEB grant program was strengthened, focusing on completing watershed assessments and implementing high-priority restoration projects.

Four regional teams of state field managers provided direct assistance to watershed councils, coordinated projects and reviewed grant applications. Often, the regional team gatherings were the first time that state and federal officials responsible for various aspects of watershed management had ever met or worked together. The major new element needed was state and federal agency coordination at all levels. To succeed, agency budgets and programs needed to be integrated, and programs needed to be aligned across jurisdictions in order to support restoration actions. Formal memoranda of understanding were negotiated and signed with participating federal agencies to harmonize programs and manage for healthy and sustainable watersheds.

A seven-member Independent Multidisciplinary Science Team (IMST) was created to provide scientific review and impartial oversight and advice. Other groups worked on specific issues, such as the Coastal Salmon Restoration and Production Task Force (harvest and hatchery policy), the Forest Practice Advisory Committee (reviewing forest practice rules), S.B. 1010 planning groups (agricultural water quality management plans), and TMDL groups.

Between a reenergized GWEB, the new interagency/intergovernmental teams developing the OCSRI, local technical support teams, and the federal agency memoranda of understanding, an unprecedented level of coordination was achieved, resulting in an integrated coho salmon restoration program. Traditional state fishery management changed, TMDL preparation accelerated, and Forest Practices Act regulations were revised to protect core coho habitat areas. Oregon's comprehensive land-use planning regulations were revised to strengthen requirements for local governments to protect natural resources. Work began on habitat conservation plans for large state forests in the Coast Range.

By August 1996, the draft OCSRI was available for public review, even though NMFS did not release formal guidance for the state to use in developing its strategy until September 1996. A panel of scientists peer reviewed the Oregon Plan. In February 1997, a revised and updated draft was presented to the state legislature. It contained 16 lengthy chapters, including evaluation of the causes for decline; a review of past failed restoration efforts; mission, goals, and strategies; funding methods; and specific agency actions (OCSRI

1997). Oregon's response to a complex system was complex itself (Arha et al. 2003, 27).

The final OCSRI was released on March 10, 1997. The legislature passed a bill on March 25, 1997, adopting the plan and creating a legislative oversight framework (Oregon Legislature 1997a). The Oregon legislature, governor, state natural resource agencies, and a good portion of Oregon citizens viewed the Oregon Plan as a model of cooperative resource management, simultaneously addressing CWA and ESA objectives. They believed it met the ESA standard for avoiding listing on the grounds of "adequate existing state regulatory mechanisms."

The 1997 Oregon legislature appropriated $30 million for the 1997–1999 biennium to support OCSRI implementation. Of these funds, $20 million provided grant money for 1) on-the-ground watershed restoration and enhancement projects; 2) watershed capacity building, assessment, and monitoring; 3) industry compliance (primarily agriculture and forestry); and 4) outreach and education. The remaining $10 million funded 58 new positions to assist watershed councils in their on-the-ground coho conservation work and to assist in developing TMDLs. The timber industry supported the Oregon Plan—agreeing to an increase in the timber severance tax to raise $14.6 million of the new funds the legislature dedicated to OCSRI. If the Oregon coastal coho ESU were listed under the ESA, the legislature agreed to void the increase in the timber severance tax. The industry would only commit to funding recovery if coho were not listed. The industry supplemented the severance taxes with an additional voluntary restoration investment on private timberlands of over $130 million (Oregon Legislature 1997b).

Memorandum of Agreement Between Oregon and NMFS

The OCSRI was developed and funded between January 1995 and March 1997 to create a comprehensive state plan to restore coastal coho. State officials and citizens wanted to avoid listing wild coastal coho under the ESA. A "no list" decision in April 1997, acknowledging the OCSRI's sufficiency, would affirm the Oregon Plan.[4] The key question was whether NMFS would find the OCSRI sufficient. NMFS had not prioritized any of the 30 factors of decline and insisted that, to the extent possible, all factors should be addressed by the Oregon Plan. The OCSRI drafters tried to cover all factors, but several issues were not addressed to the satisfaction of NMFS, particularly the adequacy of habitat protection measures over the long term, and the adequacy of Oregon's forest practices rules. The rules, which apply to state and private lands, differed significantly from the federal ACS.

To address unresolved issues related to habitat protection, NMFS and the state decided to supplement the OCSRI with a memorandum of agreement (MOA). The MOA also tried to address the standard for deciding not to list the species. The questions that arose were: Would the OCSRI have to 1) ensure a self-sustaining population (sufficiently protected not to warrant listing); 2) avoid jeopardy (making sure the species did not drop below the self-sustaining level); or 3) actually recover the species (increasing population numbers and viability)? Oregon believed that if Oregon coastal coho were self-sustaining, they should not be listed. During the negotiations, NMFS pressed for a "no jeopardy" standard as a condition not to list. Yet if the fish were listed, the state and private landowners would be subject only to a "no take" standard. Essentially, NMFS wanted the state to assume the burden of applying a higher standard to state forest management and private land regulation than NMFS could require if the species were listed.

In the end, Under Secretary of Commerce Terry Garcia negotiated the final MOA with Governor Kitzhaber on April 22, 1997. It committed the state to implementing the OCSRI, specifying that the coho would be listed if the IMST reported that OCSRI implementation "is not effective in improving and sustaining Oregon coastal coho." It set a timeline for forest practices rule revisions and required habitat conservation plans (HCPs) for coastal state forests (State of Oregon and NMFS 1997).

Political tension surrounding the MOA and the listing decision was extremely high. NMFS staff urged listing in spite of the OCSRI and issued reports, even after the MOA was signed, concluding that the OCSRI would not successfully halt or reverse the downward habitat trend.

The "No List" Decision

On April 27, 1997, Oregon Plan supporters were elated when NMFS decided Oregon coastal coho did not warrant listing. NMFS reviewed the OCSRI in detail, commenting that the OCSRI's "greatest contribution" was that it "provides a comprehensive framework for integrating habitat protection and restoration efforts by all entities, public and private" (NMFS 1997, 24604). NMFS explained its decision not to list Oregon coastal coho as threatened, as follows:

> *On the basis of the harvest and hatchery improvements together with the habitat protections in the NFP and given the improving trends in escapement, the Oregon Coast coho is not likely to become endangered in the interval between this decision and the adoption of improved habitat measures by the State of Oregon* (NMFS 1997, 24607).

The finding was made after NMFS made a detailed response to public comments on the adequacy of the OCSRI. These NMFS responses later undercut the agency's final decision not to list the species. In its responses, NMFS stated, "[t]he Oregon Forest Practices Act ... does not have implementing rules that adequately protect coho salmon habitat" (NMFS 1997, 24596). NMFS was critical of the state dredge-and-fill program's lack of cumulative effects analysis. NMFS acknowledged that TMDLs would be significant in the long term, but found that their efficacy could not be determined in the short term. State agricultural water quality management plans were noted, but NMFS found it "open to question" whether they would "adequately address salmonid habitat factors" (NMFS 1997, 24597). NMFS found Oregon's land-use planning program wanting, because it did not require sufficient protection of riparian areas.

After the "no list" decision, efforts to implement the MOA escalated tensions between NMFS and the state. In February 1998, three panels of NMFS scientists found the proposed forest practices revisions inadequate, citing inadequate riparian buffers, insufficient protection of unstable slopes, and short timber rotations. NMFS went on to offer a model draft of state forest practices regulation it would prefer. The Oregon Department of Forestry and the state timber industry were affronted, pointing out that NMFS had a poor understanding of forestry practices in general and of existing state forest practices regulations in particular. The timber industry was furious, believing it was singled out for heavy regulation. Governor Kitzhaber protested to the Secretary of Commerce, faulting NMFS for chilling state salmon restoration efforts (Kitzhaber 1998).

Initial Listing Decision

In spite of the MOA, the Oregon Natural Resources Council (ONRC) and several other environmental and fisheries groups challenged the "no list" decision. The district court ruled that NMFS's reliance on the OCSRI violated the ESA and constituted an arbitrary and capricious action (*ONRC v. Daley* 1998). It held that NMFS applied an improper legal standard because, instead of finding that the species would not likely become endangered "in the foreseeable future," it found that coho were not likely to become endangered before the adoption of improved habitat measures under the OCSRI (*ONRC v. Daley* 1998, 1160–61). The court did not decide how long the "foreseeable future" is, but the two-year period when Oregon forest practices regulations were to be revised was, in the court's opinion, too short to meet the statutory requirement of ESA (*ONRC v. Daley* 1998, 1151).

The court also found that NMFS relied on impermissible factors by using the voluntary and future efforts promised in the OCSRI and MOA (*ONRC v. Daley* 1998, 1156–57). The OCSRI, the MOA, and the actions taken under them were not "existing regulatory mechanisms" as required by section 4(a)(1)(D) of the ESA, since they were not all in place. Section 4(b)(1)(B) allows consideration of broader measures than section 4(a)(1)(D) does, but the court found that even it did not permit consideration of future actions. The court concluded that the Secretary of Commerce "may not rely on plans for future actions to reduce threats and protect a species as a basis for deciding that listing is not currently warranted" (*ONRC v. Daley* 1998, 1154). The MOA provided no further assurances, because either party could terminate it. The court concluded, "This court declines to tie the fate of the Oregon Coast ESU to the whim of politics and promises of future state conservation actions that may be years or decades away from implementation" (*ONRC v. Daley* 1998, 1158).

Without any legislative history to guide it, the court found that voluntary measures are "speculative," providing no assurance of species protection. Voluntary measures "should be given no weight" at the listing stage (*ONRC v. Daley* 1998, 1155). It also found that there was no certainty to the funding sources and that conservation efforts by individuals would count only if they are mandated, as part of a federal agency's regulatory requirements, and already in place at the time of the listing decision.

The opinion reflects a curious distinction between the certainty of future federal actions and the certainty of state actions. For example, the court concluded that NMFS could take federal land management plans (the Northwest Forest Plan and its ACS) into account when making listing decisions, even though they would also take time to implement, were subject to change, and rely on annual federal appropriations. It is unclear how state agency plans, as embodied in the OCSRI, are significantly different from federal land management agency plans in terms of the likelihood of future implementation. The court seemed to assume that conservation action would occur if Oregon coastal coho were listed, even though the only obligation of state and private landowners, under the ESA, is to avoid take, not to restore the species.

Throughout the opinion, the judge pointed to statements in the administrative record that contradicted NMFS's decision not to list. OCSRI drafters and supporters suspected that some NMFS staff intentionally included harmful information in the record to undermine any decision not to list. When the court relied on such information in its decision, it created deep distrust between state officials and some NMFS staff.

In 1998, Oregon coastal coho were listed under the ESA (NMFS 1998). Even so, OCSRI implementation continued. But the listing decision created regulatory uncertainty that OCSRI was developed to alleviate.

Oregon Plan Implementation

After the listing, the state focused on implementing the Oregon Plan and obtaining regulatory assurances under the ESA for landowners participating in the plan. OCSRI implementation was aided, in November 1998, when a ballot measure dedicated 15 percent of net lottery proceeds to parks, salmon, and watershed restoration. The ballot measure assured large, stable funding for the Oregon Plan. Oregon's congressional delegation sought greater federal assistance and more flexibility in the use of federal appropriations. The Congress authorized the U.S. Forest Service to undertake high-priority restoration and enhancement work on private land, if doing so would provide a higher net benefit to the overall watershed or ecosystem health than investments solely on public lands. The Secure Rural School and Community Self Determination Act of 2000 directed a portion of timber receipts to ecosystem and watershed restoration. The governor also worked with other Pacific Coast states to create a federal Pacific Coastal Salmon Recovery Fund (PCSRF) to help recovery efforts. Pursuant to a memorandum of understanding (MOU) with NMFS, the Oregon Watershed Enhancement Board (OWEB) distributes PCSRF funds.

In January 1999, Governor Kitzhaber signed Executive Order No. 99-01, to replace and broaden the 1996 MOA with NMFS to protect and restore all salmonids and improve water quality in Oregon. He chose to direct specific state agencies' actions himself to improve and strengthen the Oregon Plan, rather than following the detailed prescriptions in the MOA. State agencies were directed to use their authority to "not appreciably reduce the likelihood of the survival and recovery of salmonids in the wild" (Kitzhaber 1999).

With the governor's direct involvement, state agencies redoubled their efforts, using their management and regulatory authority to help restore coho and other salmon populations and watersheds. These efforts ranged from changing state water appropriation rules to adoption of new state forest practice rules. State officials continued to press their case with NMFS, pointing out that rigid application of ESA prohibitions would not address the major factors affecting coho recovery, since they only regulate future actions and cannot remedy historical habitat changes. In their opinion, the federal ESA was good at prohibiting activities on the landscape that could harm coho salmon, but very poor at encouraging management activities that would actively benefit and restore coho salmon populations. OCSRI, in contrast, harnessed the resources of federal, state, and local agencies, landowners, and other private groups to implement needed conservation actions.

Federal, state, and local agencies responsible for land management embraced the Oregon Plan. The U.S. Forest Service and the Bureau of Land Management, under the Northwest Forest Plan and its ACS strategy, have worked very closely with OWEB, ODFW, and local watershed councils. The Army Corps of Engineers, the Environmental Protection Agency (EPA), the Bureau of Reclamation, the Natural Resources Conservation Service, and the United States Fish and Wildlife Service (USFWS) also have continued to collaborate with OWEB in applying their respective programs to support the Oregon Plan. These partnerships have resulted in additional federal funding for salmon and watershed restoration efforts under OCSRI.

In contrast to the friction between ODFW and NMFS over the regulatory actions under ESA, the two agencies have cooperated in administering fish hatchery and harvest programs offshore and onshore. State hatchery managers work with the harvest managers. The Oregon Plan called for substantial reforms in fish harvest and hatchery management, and the coho population upturn indicates the reforms have been effective.

At the local government level, the Oregon Plan has enabled OWEB to partner with the local watershed councils and soil and water conservation districts and to coordinate a wide variety of federal, state, and local programs to better implement specific salmon and watershed restoration projects. This has resulted in an unprecedented level of close cooperation and program integration on the local level.

The state legislature continued to support the Oregon Plan, in spite of the listing. The 1999 session convened shortly after the effective date of the listing and passage of the ballot measure dedicating lottery funds to salmon and watershed restoration. The measure required the legislature to designate one agency to distribute the restoration funds. The result was House Bill 3225, which reconstituted GWEB as the agency to administer the salmon and watershed restoration funding. GWEB was renamed OWEB (the Oregon Watershed Enhancement Board). Its responsibilities were expanded to include setting statewide and regional restoration goals and priorities for funding decisions; developing guidance and protocols for watershed assessments and watershed action plans; and developing and implementing a standardized data collection, management, and reporting system. Even though the OCSRI did not prevent listing of Oregon coastal coho, by the end of the 1999 legislature, the state had renewed and strengthened the Oregon Plan.

The Oregon legislature, governor, state and local agencies, and a large portion of affected residents were beginning to view restoring salmon and watersheds as a part of life. Nevertheless, they all wanted to know what the conservation goal was. When could they declare their efforts a success and a conservation victory? What would be needed to delist coho? When NMFS was not forthcoming with any measurable salmon restoration goals, the frustrated state legislature decided to address these concerns. The Oregon legislature passed a law defining "recovery," "listed unit," "native fish," "naturally produced," "population," and "self-sustaining" (ch. 452, 2003 Oregon Laws).

Conservation Planning

Ted Kulongoski took office as governor in 2003. He announced his intent to focus on removing barriers to salmon recovery by prioritizing the factors for its decline and identifying recovery "bottlenecks," rather than the "across-the-board, do whatever seems like a good idea approach" of the plan's early years (Byler 2004). The first step in this effort was preparation of a comprehensive assessment of the Oregon Coast ESU (ODFW 2005) and a new cooperative agreement with NMFS to work together on recovery planning (Kulongoski 2003). The Coho Assessment evaluated the performance of the Oregon Plan, identified the current limiting factors to salmon restoration, and provided a guide for future restoration efforts.[5] The assessment also addressed NMFS's new Policy for Evaluation of Conservation Efforts as a part of the cooperative agreement with NMFS (NMFS 2003). The assessment concluded that Oregon coastal coho are currently viable and that measures taken under the Oregon Plan have stopped, if not reversed, the deterioration of coho habitat. NMFS later praised the assessment as "an unprecedented, rigorous analysis" (NMFS 2006). NMFS submitted detailed comments on the draft assessment, including reviews by its Northwest Fisheries Science Center. NMFS deferred a final decision on relisting in order to analyze the final assessment (NMFS 2005).

Following completion of the assessment, ODFW prepared a comprehensive conservation plan for Oregon coastal coho under the state Native Fish Conservation Policy (OAR 635-007-0502 to 0509). The plan built on all of the efforts taken under the Oregon Plan. The Plan was completed in February 2007 following a two-year collaborative public process. It establishes specific conservation goals and actions based upon analysis of limiting factors and threats (ODFW 2007). The Conservation Plan's key goal is to provide greater support to local community-based organizations engaged in habitat restoration—building upon the fundamental premise of the Oregon Plan—that habitat improvement on private land is most likely to occur through incentive-based cooperative partnerships with landowners. The plan contains most of the elements of a federal ESA recovery plan. The biological objectives, however, are higher than those required to meet federal recovery standards.

Results

The Coho Assessment evaluated the biological viability of the ESU, the key risk factors, and the status and trends of all restoration work (ODFW 2005). A summary of the results for the entire ESU is shown in Figure 10-2 (ODFW 2010). As noted above, the assessment concludes that the ESU is now viable and strong enough to sustain itself under the foreseeable range of future environmental conditions.

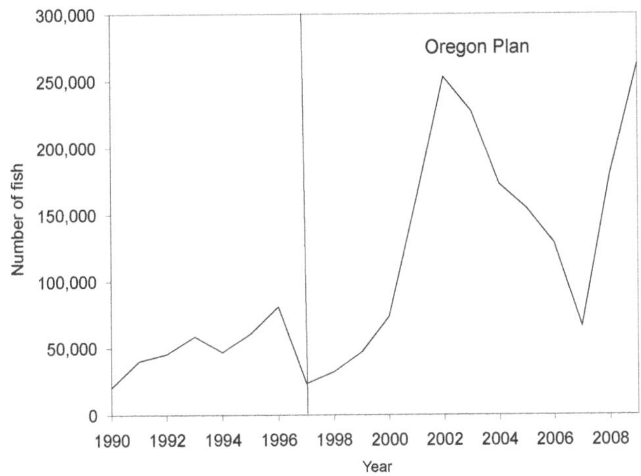

FIGURE 10-2. Spawning Population of Wild Adult Coastal Coho, 1990–2009

Source: ODFW Coastal Salmonid Inventory Project, http://oregonstate.edu/Dept/ODFW/spawn/index.htm (accessed July 9, 2010).

Coho returns since the late 1980s had increased nearly twenty-fold, from the record low of 14,068 wild coho in 1997 to nearly 300,000 in 2009 (ODFW 2010). Historical land, water, and fish management activities that contributed to declines appeared to have been arrested. Fish harvest rates are at a historically low level. Genetic management programs have been drafted for all 10 coastal hatcheries. Abundance and density of coho spawners throughout the ESU increased since 1998 to the highest average level observed in five decades (OWEB 2005). Since completion of the assessment in 2005, state and federal agencies have gathered more status and trend information. Mean population abundance from 2002–2006 is approximately 152,960, well above the 1992–1996 average of 52,845 (NMFS 2008). Concerns about the status of the ESU continue, however, because of a decline in productivity (recruits per spawner) observed from 2003 until 2007.

It is too soon to link these increased returns with watershed restoration efforts, but the assessment concluded that the primary habitat-related threats to coho viability are being addressed through ongoing conservation efforts, although the effects of ocean conditions and climate change appear to be adversely affecting coho. Since 1995, within the ESU, watershed restoration projects have included upgrading 6,200 road miles and retiring 575 miles; treating nearly 2,000 in-stream miles to reduce sediment and increase stream complexity; and treating 2,800 miles of riparian zone, 554 miles of which have been fenced. In addition, a considerable effort has been dedicated to restoring wetlands and upland habitats, including improving flow through nearly 30,000 culverts and road crossings in the ESU (Oregon Explorer 2010).

The unanswered question remains whether freshwater habitat can support and sustain the high numbers of natural spawners necessary to sustain healthy, viable salmon populations. Three key remaining risk factors are marine habitat, stream complexity, and climate change. The Coho Assessment's ability to narrow the key environmental factors that affect salmon populations stands in stark contrast to the many high-risk factors identified by NMFS while listing coho salmon as threatened. If the Coho Assessment is credible, then its findings support the conclusion that the risks to coho salmon from fish harvest, hatcheries, fish passage, and water quality have been substantially reduced from what they were in 1997.

ESA Implementation

NMFS deferred critical habitat designation when it listed Oregon coastal coho. It was proposed on May 10, 1999 and became effective on March 17, 2000. Critical habitat designation refers to range of habitat that is deemed necessary to recover the listed species. The final rule included all river reaches accessible to Oregon coastal coho and "the area adjacent to a stream that provides the following functions: shade, sediment transport, nutrient or chemical regulation, streambank stability, and input of large woody debris or organic matter" (NMFS 2000a). What this meant, in practice, was that anyone proposing activities within riparian areas would have to prepare detailed studies on how the activity affected the specified functions. The proposal raised significant concerns among riparian landowners about economic impact on their property, because of the breadth of the proposed riparian buffer and the number of designated streams. In its final rule, NMFS acknowledged the broad economic impact that the designation would have on private activities in industries such as fishing, agriculture, forestry, transportation and utilities, land development, mining, recreation, and tourism. Local governments "could face significant changes" in their approaches to planning, zoning, erosion and sediment control, water supply, storm water, and wastewater discharges.

The designation of coastal coho critical habitat raised another political furor in Oregon. NMFS attempted to calm public fears by noting that the designation does not prohibit private landowners from modifying streams or riparian areas, especially if no federal funding or permits are involved. Yet many state, local, and private activities do require federal permits, and added costs of ESA consultations were bound to increase with the critical habitat designation. The National Association of Homebuilders challenged the designations. A consent decree was negotiated, and NMFS vacated the designations on April 30, 2002 (*National Association of Homebuilders v. Evans* 2002). The Pacific Coast Federation of Fishermen's Associations then sued NMFS for failing to designate critical habitat in a timely manner. On December 14, 2004, the agency again proposed to designate critical habitat. The new proposal, if adopted, would have changed policy on the most controversial aspects of the 2000 designation. First, it would have limited designation to identified stream reaches where coho have actually been observed, or where local biologists determine them to be, and not to currently unoccupied reaches. Second, instead of designating general areas serving riparian functions, it would limit the designation to "the width of the stream area defined by the ordinary high-water line" (NMFS 2004c).

The 2004 critical habitat designation was withdrawn on January 19, 2006, when NMFS decided that listing Oregon coastal coho was not warranted (NMFS 2006). However, as discussed below, this decision started another round of litigation resulting in relisting in 2008. Critical habitat was designated concurrently with that listing, using the same principles as the 2004 listing—only designating habitat below the "high-water line" and excluding unoccupied reaches (NMFS 2008). The current (2010) NMFS proposal to retain the coho listing will also retain the 2008 critical habitat designation (NMFS 2010). This latest proposal is a result of the settlement between NMFS and Douglas County in *Douglas County v. Balsiger*. Comments on the latest proposed listing were due July 26, 2010.

Section 4(d) Take Limits and Guidance

The state sought assurances from the federal government that, if landowners were complying with the provisions of the Oregon Plan, they would be protected from liability for take under the ESA. In response, NMFS made concerted efforts to use its ESA authority to support the Oregon Plan. NMFS issued innovative rules prohibiting take of coho but "limiting" application of the take prohibition to activities governed by specific programs that were designed to minimize impacts to the species (NMFS 2000b). NMFS took care to avoid imposing federal prescriptive rules. Instead, it reviewed a variety of state and local programs to see if they were sufficient to avoid jeopardy; if they were, NMFS agreed to not apply the take prohibition. Federal agencies such as USFWS and NMFS enjoy considerable discretion under section 4(d) of the Act to tailor appropriate compliance and protective measures for conserving threatened (not endangered) species.

According to NMFS, to qualify a state or other program under section 4(d) required at a minimum ensuring that management activities would not jeopardize the continued existence of the species or "appreciably reduce the likelihood of survival and recovery of any of the ESUs in the wild" (NMFS 2000b, 42433). In the case of coastal coho, however, NMFS called for a stricter standard for select activities with a history of "particularly destructive" impacts on salmon populations. In order to qualify for a 4(d) limit, such programs would have to show something more than "no jeopardy"; they would have to make specific, positive contributions to recovery.

The July 2000 rule also tried to alleviate public concerns by providing "take guidance." This guidance clarified how NMFS would approach ESA enforcement. NMFS planned to target flagrant or particularly damaging violations of the take prohibition. It did not plan to pursue enforcement action against private parties acting in accordance with other state and federal regulations. While this enforcement approach attempted to address landowner concerns, landowners criticized it, in part because of what many viewed as a double standard. NMFS continued to allow ocean harvest of other salmon that resulted in by-catch and associated mortality of coastal coho. Landowners believed they were being asked to bear a disproportionate burden to produce a few more young fish that might well be killed by ocean fisheries. The take prohibition did not become effective until January 8, 2001, over two years after the listing. Although potential civil and criminal liability worried landowners and nonfederal government agencies, NMFS has taken few enforcement actions since the listing. Even investigations have been minimal.

The primary practical effect of the listing and take prohibition has been on the ability of third parties to bring citizen suits. Litigation is a very real concern for large landowners and state and local governments. Their concerns were realized by litigation against the Oregon Department of Forestry in 2003. The Pacific Rivers Council and other environmental organizations sued the State

Forester to enjoin certain logging practices as illegal "take." When the court refused to dismiss the suit, the Board of Forestry and the legislature changed the law and rules under the Forest Practices Act. Instead of requiring the State Forester to affirmatively sign off on forest operations plans, they now required "reviewing and commenting" only, expressly to avoid state liability for indirect take (*Pacific Rivers Council v. Brown* 2003).

Alsea Valley and Delisting

When NMFS listed Oregon coastal coho as threatened in 1998, it only listed "naturally spawned" fish, not hatchery-spawned fish (NMFS 1998). ODFW kills hatchery-spawned fish when they return to the hatcheries, in order to protect naturally spawning populations. A citizen filmed ODFW employees killing the hatchery fish in 1999, just when landowners were worried about the impact of proposed critical habitat designations. Landowners feared that if listed species were in streams on their property, they would be required to provide riparian buffers, taking land out of production for farming and forestry, or to restrict water withdrawals for irrigation. Some commercial and sports fishermen, whose fishing was restricted following listing, asked why coastal coho were being killed at hatcheries if they were threatened with extinction. Ultimately, a group led by the Alsea Valley Alliance formed to sue NMFS, contending that the distinction NMFS made between naturally spawned coho and hatchery-spawned coho was arbitrary and capricious under the Administrative Procedures Act.

The district court agreed with the plaintiffs. Although it upheld the NMFS policy of treating ESUs as distinct population segments (DPSs), based on genetic and geographic range (NMFS 1991), the court found no basis for listing distinctions below the subspecies or DPS level. The court agreed that naturally spawned coho could have been listed as a DPS, but the facts in the record did not support this. The judge agreed that genetic diversity is an important factor for the long-term success of a threatened species, but he did not think genetic considerations could justify a listing that, on its face, contradicted the statutory definition of "species." A stay protected coho status until the Ninth Circuit dismissed all appeals on February 24, 2004, at which point Oregon coastal coho were delisted (*Alsea Valley Alliance v. Evans* 2001). In response to the *Alsea* decision and a series of resulting listing and delisting petitions, NMFS updated the Oregon coastal coho status review (NMFS 2002).

On June 14, 2004, NMFS proposed to relist Oregon coastal coho as threatened, based upon the review done by the Pacific Salmonid Biological Review Team (BRT). The BRT was closely divided about whether Oregon coastal coho were threatened, but the majority supported relisting, primarily because of their uncertainty about future ocean conditions. Under NMFS's new hatchery listing policy, if the genetic divergence between the local natural population and the hatchery stocks is no more than would be expected between closely related populations within the ESU, the hatchery stocks are considered in the listing decision (NMFS 2005). This new policy reinforced NMFS's continued commitment to give special consideration to naturally spawning "wild" populations (Brown 2005). Oregon coastal coho meet this standard. The proposed listing noted that NMFS expected additional viability information from the Oregon Coastal Coho Assessment to inform its final decision (NMFS 2004a).[6]

The 2005 hatchery policy was challenged from both sides, by both environmental groups and the original Alsea Valley Alliance and Building Industry Association of Washington. The Ninth Circuit ultimately upheld the hatchery policy against all challenges, overturning the district court of Western Washington. The Ninth Circuit found that the Alsea decision did not require equal treatment of hatchery and natural fish and that the policy permissibly "considers both the positive and the negative effects of hatchery fish on the viability of natural populations" (*Trout Unlimited v. Lohn* 2009).[7] Application of the ESA to state hatchery programs is likely to remain contentious as hatcheries attempt to balance their dual role of increasing salmon available for fishing and preventing natural salmon from becoming extinct (Stein 2010).

Policy for Evaluation of Conservation Efforts

Oregon and NMFS began collaborating to assess the status of conservation of Oregon coastal coho in 2003 using the "Policy for Evaluation of Conservation Efforts when Making Listing Decisions" (the PECE process) to identify criteria that NMFS would use to determine whether conservation efforts contribute to making listing unnecessary. The primary objectives of the collaborative effort were to 1) assess the Oregon Plan efforts to conserve and rebuild coastal coho populations and 2) use the assessment a) to inform NMFS's status review listing determination, b) as a basis to seek legal assurances for local participants, and c) as a foundation for a coho recovery. As discussed above, Oregon's Coho Assessment is now completed (ODFW 2005).

The standard applied under the PECE process is "whether the conservation effort improves the status of the listed species" (NMFS 2003, 15114). In practice, according to Oregon officials, the standard is higher than the standards for federal recovery planning because states must ensure implementation. Certainty of implementation and effectiveness are not criteria applied to the review of federal recovery plans under section 6. The PECE process might yet build on the history of the last decade—recognizing the conservation effectiveness of the Oregon Plan and determining that it satisfies the conditions to preclude the need to list the coastal coho in Oregon under the ESA or adopting the state's Conservation Plan as the federal ESA recovery plan.

Trout Unlimited and Relisting

On January 19, 2006, NMFS withdrew the proposed listing of coastal coho and the proposed critical habitat designation (NMFS 2006), after consideration of the Oregon Assessment. NMFS noted the extremely encouraging increases in returning fish. It acknowledged that its BRT disagreed with the assessment about whether the increases could be sustained in the future, due to uncertainties in freshwater habitat and ocean conditions. Yet, NMFS ultimately accepted Oregon's conclusion that Oregon coastal coho have the resiliency to survive periods of poor ocean conditions and remain viable. NMFS decided not to relist.

In June 2006, Trout Unlimited and seven other organizations challenged NMFS's decision not to list Oregon coastal coho. On July 13, 2007, the district court again ordered NMFS to list the ESU as threatened. The court found that NMFS's 2006 decision not to list was not supported by the best available science. The court specifically rejected reliance on Oregon's Assessment because of its assumption of resilience at low abundance, lack of findings on the adequacy of habitat, and uncertainty about the adequacy of harvest and hatchery reforms (*Trout Unlimited v. Lohn* 2007). NMFS complied with the order and relisted Oregon coastal coho on February 4, 2008 (NMFS 2008).

Of particular note for purposes of this case study is the court's discussion of the Oregon Plan. In withdrawing the proposed listing, NMFS cited the assessment's finding that measures undertaken under the Oregon Plan "make it unlikely that habitat conditions will be degraded further in the future" (NMFS 2008). The court pointed out that the assessment's viability analysis was based on the biological performance of coho, not on the effectiveness of actions under the Oregon Plan. It also noted lack of data documenting implementation of Oregon Plan measures and their effects. The court again rejected reliance on the effectiveness of the Oregon Plan as a basis for not listing Oregon coastal coho (*Trout Unlimited v. Lohn* 2007).

NMFS again reviewed available scientific information when it relisted Oregon coastal coho in 2008. Given the 2003–2006 abundance and marine survival information and preliminary spawning surveys from 2007, NMFS concluded that the ESU is likely to become endangered in the foreseeable future throughout all or a significant portion of its range. Nonetheless, NMFS noted that, had it been given time to do a more thorough analysis, it might have reached a different conclusion. Simultaneous with the listing decision, NMFS both designated critical habitat and also applied the 4(d) rules codified at 50 CFR 223.203(b) to Oregon coastal coho (NMFS 2008). NMFS and state officials anticipate that the Oregon Coastal Coho Conservation Plan will provide the starting point for federal recovery plan development.

Following the initiation of a 2009 status review and continuing litigation, in 2010 NMFS formally proposed to retain the current coastal coho listing as threatened. The proposed listing notice analyzes the continuing efforts of the State of Oregon and state and federal laws affecting salmon, but makes no mention of the Oregon Coastal Coho Conservation Plan. Ultimately NMFS concludes that existing regulatory mechanisms are inadequate to conclude that coastal coho salmon areas are no longer threatened. The critical habitat identified in the 2008 listing will remain in effect (NMFS 2010).

Lessons for Recovery Efforts

The Oregon coastal coho conservation effort provides several broad lessons that could facilitate cooperative federalism for future species conservation. Coho are far-ranging species spending key portions of their lives at sea and traversing multiple state and federal jurisdictions. Consequently, close cooperation across state and federal jurisdictions is essential for its restoration and recovery. No state or federal entity on its own can ensure coho restoration.

For any partnership to be effective there must be 1) a shared vision; 2) allocation of responsibilities to match expertise; and 3) good working relationships. The Oregon Plan was able to build new partnerships across state government, with unprecedented levels of coordination and collaboration, to develop a broad-based conservation strategy for restoring salmon populations and watershed. Most importantly, the plan invited affected private industries and landowners, notably the forest industry, to participate in the planning and execution of the conservation strategy.

The plan also enabled strong partnership among state, local, and federal agencies. There were instances during the planning and execution of the plan, however, when the cooperation between state agencies and NMFS on ESA regulatory issues left much to be desired. The factors leading to collaboration, and lack thereof, between NMFS and state agencies in sharing a common vision, allocating responsibilities, and sustaining good working relationships are discussed below.

Common Vision

Arguably, Oregon natural resources agencies and NMFS share a common vision of healthy watersheds supporting viable salmon populations. It was extremely difficult, however, for ODFW and NMFS to jointly determine a measurable conservation target of specific salmon numbers and associated habitat. Throughout the development of the Oregon Plan, state agencies repeatedly asked NMFS for recovery targets. For many years, NMFS did not set precise, measurable recovery goals for coastal coho populations. Absence of transparent recovery goals often presents the highest hurdle to credible and effective state–federal collaboration in recovering threatened and endangered species. The state agencies are invited to marshal the resources and implement a species conservation strategy, while the federal agency reserves the right to unilaterally determine, at some undisclosed future time, what in its opinion constitutes the goal line for recovery. And there are no guarantees that the goal line, once stated, will not be unilaterally moved again by the federal agency. Such vagaries breed resentment and distrust.

Often, not enough is known to determine precise recovery goals. Most field biologists have learned to live with these inherent uncertainties of species conservation. What is perplexing is the seeming unwillingness of state and federal biologists to agree to address these uncertainties jointly through a measured and prudent conservation strategy. In the absence of a joint federal-state process for developing specific species recovery targets, states often embark on their own efforts, like the Oregon Assessment and Conservation Plan, with federal agencies reviewing and commenting as critics.

Oregon agencies suffered similar frustrations with the refusal of NMFS to prioritize factors for coastal coho decline and resulting restoration actions. By declining to prioritize threats faced by coastal coho, NMFS seemed to recommend a conservation strategy that was to "do it all" wherever, whenever, and however possible. No management agency with scarce resources has that luxury. Once again, with no answers from NMFS to its queries, the state initiated its own comprehensive Coho Assessment to identify and rank limiting factors for coho restoration at the subbasin scale and allocate funds to restoration projects accordingly. In the absence of federal prioritization, affected private groups often suspect that they are being asked to bear a disproportionate share of the restoration burden.

Roles and Responsibilities

The contrasting approaches of NMFS, administering the ESA, and state agencies, implementing the Oregon Plan, reflect their respective competencies. NMFS is tasked with administering a powerful federal regulatory statute with specific mechanisms for species conservation including critical habitat designation and take prohibitions. The state agencies in the Oregon Plan are implementing a comprehensive and collaborative state conservation strategy to restore salmon populations and watershed health. Governor Kitzhaber analogized the difference between the federal regulatory approach and the Oregon Plan approach to the difference between controlling point sources and nonpoint sources of pollution. The Oregon Plan focuses on individual, diffuse actions and tries to address these many small coho needs across the natural landscape. It is these "nonpoint" factors of decline that must be addressed to restore salmon runs and watershed health simultaneously.

Rigid regulatory application of the ESA by federal agencies, with what at times seems to be a disproportionate focus on prohibiting land-use activities, is ill suited to promoting myriad conservation actions across large and complex landscapes. NMFS is best at providing funding, developing the national policy guidance needed for a species that crosses state and international boundaries, providing the "safety net" of enforcement action and regulations if state programs do not result in species recovery, and working with large organizations to regulate their activities. The states are better suited to implementing grassroots programs on the ground because of their better understanding of local communities and resources, their financial efficiency, and their flexibility. They can raise matching funds and in-kind support, work with local partners, and integrate programs across agencies. A key to success, under the Oregon

Plan, has been using state and federal technical assistance and funds to promote voluntary landowner action. The MOU between NMFS and OWEB on managing PCSRF funds is a good example of this leveraging of multiple funding sources, resulting in efficient and effective investment.

Working Relationship

From the Oregon Plan's inception, there has been a high level of NMFS involvement, both formally and informally, in plan development and execution. State officials have also participated in many of NMFS's technical review teams. In the course of these interactions, tensions frequently surfaced between the federal and state agencies—originating largely from their respective regulatory and managerial approaches to species conservation. Although NMFS frequently refers to the state as a "co-manager," in practice NMFS remains the sole regulatory power under the ESA. A few high-profile incidents—including the February 1998 NMFS letter proposing new forest practice rules, and the staff memoranda in NMFS files that were relied on in *ONRC v. Daley*—created mistrust among state agencies and NMFS. On the other hand, NMFS worked diligently with Oregon agencies to develop section 4(d) rules for the state. In addition, ODFW and NMFS have enjoyed an effective working relationship in managing fish hatcheries and harvests and working with OWEB on the PCSRF.

There are always opportunities for state and federal agencies that share a common goal of species conservation and wildlife management to work well together. Good working relationships require high trust—attained through high involvement, marked by frequent good-faith dealings. Effective working relationships also depend on intrinsic and extrinsic political factors ranging from leadership and agency culture to available resources and political support (Scheberle 2004). The Oregon Plan, championed by the state governor and legislature, has achieved extraordinary collaboration among state, federal, and local agencies within the state. The relationship between NMFS and Oregon agencies concerning coastal coho restoration has been very contentious on occasion but has improved over time through better communication and new cooperative processes such as those associated with recovery planning.

Conclusion

Two broad themes emerge from the Oregon Plan for salmon and watershed restoration. The first is the sheer complexity associated with conservation of an anadromous fish species. Coho cross multiple landownerships, jurisdictions, lands, and oceans in the cycle of their lives. Conserving such a species poses complex ecological, economic, political, and regulatory challenges. The second theme is appropriate governance structure to address the multifaceted conservation of a species such as coastal coho. The administration of the ESA by NMFS offers one example of governance, and the Oregon Plan, to restore salmon and watershed, another.

Conservation of anadromous fish species, such as coho, is substantially more complex than that of most terrestrial species. Anadromous fish are affected by human activities and quality of habitat in both fresh water and oceans. Our understanding of ocean and freshwater ecosystems and the effects of human activities upon them is rather rudimentary. In the case of coastal coho, these activities include forestry practices, farming, urban runoff, and other activities carried out by an overlapping array of local, state, federal, and private entities. No single state, federal, or local agency has the needed jurisdictional reach or resources to conserve coho.

The ESA regulatory approach has one major advantage, given the complexity of coho conservation. An ESA listing provides strong regulatory control to dissuade management activities that may "take" the species. However, takings prohibitions do not encourage the activities actually needed to recover the species from its threatened condition. The Oregon Plan statewide conservation strategy attempts to identify and address the complex and interconnected range of activities that affects state watersheds and salmon populations, promoting voluntary restoration work that will never be done by government alone.

The ESA approach suffers from the rigidity of the regulatory process and the huge associated costs borne by increasingly disgruntled private land managers. The Oregon Plan approach is collaborative, comprehensive, and nuanced: the state agencies offer regulatory relief and financial assistance to encourage conservation actions across the landscapes and to restore salmon populations and watershed health. These two approaches to salmon recovery have often collided, but have begun to coexist and complement each other, even as litigation continues. The possibility of listing coastal coho as threatened under the ESA was a catalyst in the development of the Oregon Plan. Participants in the Oregon Plan now enjoy regulatory relief under the ESA section 4(d) rules developed by NMFS, with cooperation of the state agencies. Without the Oregon Plan, salmon runs and watershed health in the state would be substantially diminished.

The case of coastal coho presents a window into the future: species conservation issues immersed in multifaceted complexity that existing governance structure is ill equipped to address. A strong regulatory framework and enforcement are needed, as characterized by current ESA administration. Reliance on listing and regulation alone, however, will not recover threatened species and restore their habitats. It is imperative that federal agencies administering the ESA recognize the limitations of the present approach and dedicate their efforts to actively encouraging and supporting restorative conservation strategies such as the Oregon Plan.

The challenge before us is how to modify the existing statutory framework to encourage and support a complementary collaborative approach to effective species conservation. The Oregon Plan experience highlights areas where federal statutory and regulatory reform can facilitate better state–federal collaboration. One broad area of reform would be clarification of the standards affecting state–federal partnerships. The Congress or federal agencies should develop clear standards for state species conservation efforts that would preclude the listing of the species. This was the biggest bone of contention between Oregon and NMFS—the latter's inability to articulate a sure standard that the state had to meet to preclude the listing of coastal coho. The PECE policy is a worthy but unsatisfying first step, as demonstrated by its lack of application.

The Congress and federal agencies should also articulate clear standards for state species conservation programs and landowner actions that qualify for section 4(d) taking limits for threatened species. Little-used section 4(d) could become a robust conservation tool in the ESA toolbox. By emphasizing a programmatic approach, section 4(d) could be used to delegate enforcement to a state through recognition of state conservation plans such as the Oregon Plan (Fischman 2004, 479).

Administration of the ESA should emphasize recovery of threatened and endangered species. States and other federal and nonfederal parties should be encouraged to actively engage and lead recovery efforts. Federal agencies administering the ESA lack resources and expertise to undertake recovery on their own, and they need partners. An inclusive state conservation strategy like the Oregon Plan—when approved by NMFS and USFWS—should be incorporated into management plans for federal land management agencies and implemented accordingly. NMFS and USFWS should explore ways to modify section 4(f) and section 7(a)(1) of the ESA to make this happen (Rohlf 2004, 542–546). Actions consistent with an approved recovery plan should not require consultations under section 7 of the ESA. This would substantially reduce public resources spent on redundant bureaucratic processes.

Regulatory reform should encourage integration of compliance among multiple federal and state environmental statutes. The Oregon Plan represents a state conservation initiative unmatched in its breadth. There are not enough resources available to muster such state efforts for all listed species individually, or to selectively satisfy the proscriptions of all state and federal environmental statutes. Such conservation efforts should capitalize on available energy and resources to address the conservation needs of multiple species and meet regulatory compliance requirements of related environmental statutes. The Oregon Plan attempted to do just that, by focusing on watershed health and attempting to meet both ESA and CWA requirements for salmon restoration and water quality. Improved state–federal collaboration on species conservation would be greatly aided by concerted state and federal efforts to align compliance requirements under various state and federal laws.

In conclusion, the greatest lesson of the Oregon Plan for other states in the field of species conservation is to "just do it." The Oregon Plan, prodded by

APPENDIX 10-1. Timeline of Key Events

1987	
Oregon	Governor's Watershed Enhancement Board (GWEB) created and funded.
1993	
Private	July 21—Listing petition filed for West Coast coho populations.
Oregon	GWEB and Strategic Water Management Group initiate watershed management, based on local watershed councils.
1995	
Judicial	June 1—Oregon Natural Resources Council (ONRC) files suit in the Northern District of California for failure to act within one year on listing petition.
NMFS	July 25—NMFS proposes listing the Oregon Coast coho ESU as threatened.
Oregon	October—Governor initiates Coastal Salmon Restoration Initiative (OCSRI).
1996	
Oregon	August—Draft OCSRI released for review.
1997	
Oregon	March 25—Legislature approves the final OCSRI, to become The Oregon Plan for Salmon and Watersheds.
Oregon/NMFS	April 25—Oregon and NMFS sign Memorandum of Agreement.
NMFS	May 6—NMFS determines that the Oregon coho ESU does not warrant listing based on state conservation measures.
1998	
Judicial	June 1—the U.S. District Court of Oregon rules in *ONRC v. Daley* that the NMFS final rule regarding the Oregon Coast coho ESU was arbitrary and capricious.
NMFS	August 10—NMFS lists Oregon coastal coho ESU as threatened under the ESA.
Oregon	November—Ballot Measure 66 passes to provide dedicated funds for salmon and watershed restoration.
1999	
Oregon	January 8—Governor issues Executive Order to implement the Oregon Plan.
NMFS	May 10—NMFS proposes critical habitat for Oregon coastal coho ESU.
2000	
NMFS	February 16—NMFS designates critical habitat for Oregon coastal coho.
NMFS	July 10—NMFS issues take prohibitions for Oregon coastal coho with 4(d) limits.
2001	
Judicial	September 10—*Alsea Valley Alliance v. Evans* invalidates the Oregon Coast coho listing because of how NMFS defined the "distinct population segment" to exclude hatchery fish.
2002	
NMFS	February 11—NMFS reinitiates status reviews for 15 ESUs of salmon and steelhead in response to the *Alsea* ruling.
2003	
Oregon	Legislature defines recovery targets for coho.

APPENDIX 10-1. Timeline of Key Events (*continued*)

2004	
Oregon	Oregon coastal coho assessment and stakeholder team meetings begin.
Judicial	February 24—Ninth Circuit dissolves stay of District Court decision in *Alsea*, resulting in delisting.
NMFS	June 14—NMFS proposes listing of Oregon coastal coho as threatened.
NMFS	December 14—NMFS proposes critical habitat designation for Oregon coastal coho.
2005	
Oregon	May—ODFW releases Final Coastal Coho Assessment Report concluding that coho are viable and likely to persist into the foreseeable future.
NMFS	June 28—NMFS extends final listing determination for Oregon coastal coho in order to consider Assessment results and finalizes new policy on treatment of hatchery fish.
2006	
NMFS	January 19—NMFS withdraws proposed listing of Oregon coastal coho as unwarranted, based partially on Coho Assessment.
Private	June—Trout Unlimited and seven other organizations file suits challenging NMFS's salmon regulations, including its decision not to list Oregon coastal coho.
2007	
Oregon	February 26—ODFW issues final Conservation Plan for Oregon coastal coho. Implementing rules adopted March 2007.
Judicial	October 5—U.S. District Court for Oregon in *Trout Unlimited v. Lohn* invalidates the NMFS decision not to list, finding it to be "arbitrary, capricious, contrary to the best available evidence."
2008	
NMFS	February 4—NMFS lists Oregon coastal coho as threatened, designates critical habitat and adopts 4(d) rules.
2009	
Judicial	March 16—The Ninth Circuit upholds NMFS 2005 hatchery policy against challenges from fishing, business, and environmental organizations in *Trout Unlimited v. Lohn* and *Alsea Valley Alliance v. Lautenbacher*.
2010	
NMFS	May 26—NMFS proposes to retain ESA listing for Oregon coastal coho.

the potential listing of coastal coho, represents a trailblazing state initiative addressing watershed health and salmon conservation. The state did not receive the cooperation and support it expected from NMFS; motivated by the gravity of the task, however, it forged ahead. Salmon and watershed health are important to Oregon's economy and culture. The state did not use the perceived lack of clear direction and standards from NMFS as an excuse for inaction. Instead, when appropriate federal standards and guidance were not forthcoming, the state developed its own. In the long run, by committing itself to a proactive conservation strategy in the Oregon Plan, Oregon was better able to negotiate with the federal agencies and address the conservation and economic development aspirations of its residents. Daniel Kemmis, former mayor of Missoula, Montana, suggests we are entering a new era where community cooperation and collaboration will require a new brand of cooperative federalism adjusted to natural landforms, with each level of organization sovereign in its own domain. Regulatory authority would be vested in the lowest level at which it can be competently exercised (Kemmis 2001). The Oregon Plan exemplifies this new approach to governance.

The conception and execution of the Oregon Plan represents a historic chapter in the story of states taking the lead in species conservation. The greatest accomplishment of the Oregon Plan might be the reconfiguration of state laws and institutions to emphasize watershed restoration amid economic and community development. In so doing, the state has aligned its governance apparatus to match the complexity of the natural systems it is attempting to conserve. The new state governance mechanism developed in the Oregon Plan creates a strong community, better prepared to effectively address the complex species conservation challenges of tomorrow. NMFS and USFWS should develop new rules and regulations under the ESA to encourage and reward state initiatives such as the Oregon Plan with appropriate regulatory relief and financial assistance in order to better achieve species conservation on the ground.

References

Alsea Valley Alliance v. Evans. 2001. 161 F. Supp. 2d 1154 (D. Or. 2001). Appeal dismissed for lack of jurisdiction, 358 F.3d 1181 (9th Cir. 2004).

Alsea Valley Alliance v. Lautenbacher. 2007. WL 2344927; 2007 U.S. Dist. LEXIS 60203 (D. Or. August 14, 2007), upheld in memorandum opinion released concurrently with *Trout Unlimited v. Lohn*, 2009 WL 690078 (9th Cir. March 16, 2009).

Arha, Kaush, Hal Salwasser, and Gail Achterman. 2003. "The Oregon Plan for Salmon and Watersheds: A Perspective." Policy Paper 2003-03. Corvallis, OR: Institute for Natural Resources.

Blumm, Michael C. 2002. *Sacrificing the Salmon: A Legal and Policy History of the Decline of Columbia Basin Salmon*. Lake Mary, FL: BookWorld Publications.

Brown, Joseph A. 2005. "Commercially Harvesting Endangered Salmon? Rethinking the Interaction of Salmon Hatcheries and the Endangered Species Act after *Alsea Valley Alliance v. Evans*." *Georgia Law Review* 40(1): 253.

Byler, Thomas. 2004. Personal interview with Thomas Byler, the Director of the Oregon Watershed Enhancement Board.

Cone, Joseph. 1995. *A Common Fate*. Corvallis, OR: Oregon State University Press.

Fischman, Robert L. 2004. "Predictions and Prescriptions for the Endangered Species Act." *Journal of Environmental Law* 34(2): 4451–4481.

Kemmis, Daniel. 2001. *This Sovereign Land*. Washington, D.C.: Island Press.

Kitzhaber, John A. 1998. Letter to Secretary of National Marine Fisheries Service William Daley. February 26.

———. 1999. Executive Order No. EO 99-01: The Oregon Plan for Salmon and Watersheds. January 8.

Kulongoski, Ted. 2003. "State of Oregon and Federal Government Developing Joint Plan to Restore Coastal Coho." Press Release. June 13. http://governor.oregon.gov/Gov/p2003/press_061303.shtml (accessed Sept. 9, 2010).

Lichatowich, James A. 1999. *Salmon without Rivers: A History of the Pacific Salmon Crisis*. Washington, D.C.: Island Press.

Montgomery, David R. 2003. *King of Fish: The Thousand-Year Run of Salmon*. Boulder, CO: Westview Press.

National Association of Homebuilders v. Evans. 2002. WL 1205743, No. 00-CV-2799 (D.D.C. 2002).

Nehlson, Willa, Jack E. Williams, and James A. Lichatowich. 1991. "Pacific Salmon at the Crossroads: Stocks at Risk from California, Oregon, Idaho, and Washington." *Fisheries* 16: 4–21.

NMFS (U.S. Department of Commerce National Oceanic and Atmospheric Administration National Marine Fisheries Service). 1991. "Policy on Applying the Definition of Species under the Endangered Species Act to Pacific Salmon." *Federal Register* 56: 58613, November 20.

———. 1993. "Listing Endangered and Threatened Species and Designating Critical Habitat: Petition to List Five Stocks of Oregon Coho Salmon." *Federal Register* 58: 57770-01, October 27.

———. 1995. "Proposed Threatened Status for Three Contiguous ESUs of Coho Salmon Ranging from Oregon through Central California." *Federal Register* 60: 30011, July 25.

———. 1997. "Threatened Status for Southern Oregon/Northern California Evolutionary Significant Unit (ESU) of Coho Salmon." *Federal Register* 62: 24588, May 6.

———. 1998. "Endangered and Threatened Species; Threatened Status for the Oregon Coast Evolutionarily Significant Unit of Coho Salmon." *Federal Register* 63: 42587, August 10.

———. 2000a. "Designated Critical Habitat: Critical Habitat for 19 Evolutionary Significant Units of Salmon and Steelhead in Washington, Oregon, Idaho, and California." *Federal Register* 65: 7764-01, February 16.

———. 2000b. "Endangered and Threatened Species; Final Rule Governing Take of 14 Threatened Salmon and Steelhead Evolutionarily Significant Units (ESUs)." *Federal Register* 65: 42422, July 10.

———. 2003. "Endangered and Threatened Species: Policy for Evaluation of Conservation Efforts When Making Listing Decisions." *Federal Register* 68: 15100, March 28.

———. 2004a. "Endangered and Threatened Species: Proposed Listing Determinations for 27 ESUs of West Coast Salmonids; Proposed Rule." *Federal Register* 69: 33102, June 14.

———. 2004b. "Endangered and Threatened Species: Designation of Critical Habitat for 13 Evolutionarily Significant Units of Pacific Salmon (*Oncorhynchus* spp.) and Steelhead (*O. mykiss*) in Washington, Oregon, and Idaho." *Federal Register* 69: 74572, December 14.

———. 2005. "Endangered and Threatened Species: Policy on the Consideration of Hatchery-Origin Fish in Endangered Species Act Listing Determinations for Pacific Salmon and Steelhead." *Federal Register* 70: 37204, June 28.

———. 2006. "Endangered and Threatened Species: Withdrawal of Proposals to List and Designate Critical Habitat for the Oregon Coast Evolutionarily Significant Unit (ESU) of Coho Salmon." *Federal Register* 71: 3033, January 19.

———. 2008. "Endangered and Threatened Species: Final Threatened Listing Determination, Final Protective Regulations, and Final Designation of Critical Habitat for the Oregon Coast Evolutionarily Significant Unit of Coho Salmon." *Federal Register* 73: 7816, February 11.

———. 2009. "Code of Federal Regulations. Title 50, Part 223, Subpart B. Restrictions Applicable to Threatened Marine and Anadromous Species." Section 223.203(b).

———. 2010. "Listing Endangered and Threatened Species: Completion of a Review of the Status of the Oregon Coast Evolutionarily Significant Unit of Coho Salmon; Proposal to Promulgate Rule Classifying Species as Threatened." *Federal Register* 75: 29489, May 26.

OCSRI (Oregon Coastal Salmon Restoration Initiative). 1997. *Oregon Coastal Salmon Restoration Initiative (OCSRI).* Salem, OR: Oregon Water Enhancement Board (OWEB). www.oregon-plan.org/OPSW/archives/reports-subpage.shtml (accessed July 1, 2010)

ODFW (Oregon Department of Fish and Wildlife). 2004. "Oregon Native Fish Status Report." www.dfw.state.or.us/fish/ONFSR/report.asp (accessed July 12, 2010).

———. "Oregon Coast Coho Assessment." May 16, 2005. http://nrimp.dfw.state.or.us/OregonPlan/ (accessed July 1, 2010).

———. 2007. "Oregon Coast Coho Conservation Plan for the State of Oregon." March 16. www.oregon-plan.org/OPSW/cohoproject/coho_proj.shtml (accessed July 1, 2010).

———. 2010. "Coastal Salmonid Inventory Project." http://oregonstate.edu/Dept/ODFW/spawn/index.htm (accessed July 9, 2010).

ONRC (Oregon Natural Resources Council) v. Daley, 6 F. Supp. 2d 1139 (D. Or. 1998). Stay denied, 16 F. Supp. 2d 1256 (D. Or. 1998).

Oregon Administrative Rules. 2007. Chapter 635 Division 007. Sections 0502 to 0509.

Oregon Explorer. 2010. "Oregon Watershed Restoration Inventory Access Database." Query: "SubBasinActual" = "Coos" Or "Coquille" Or "Sixes" Or "Necanicum" Or "Nehalem" Or "Wilson-Trask-Nestucca" Or "Siletz-Yaquina" Or "Alsea" Or "Siuslaw" Or "Siltcoos" Or "Umpqua" Or "North Umpqua" Or "South Umpqua" www.oregonexplorer.info/OWRI/OWRI.aspx?Res=17100 (accessed July 8, 2010).

Oregon Legislature. 1997a. "Oregon Plan." Oregon Laws, ch. 7, SB 924 codified at ORS 541.351 to 541.415. www.oregon-plan.org (accessed July 1, 2010).

Oregon Legislature. 1997b. "Appropriations of Funds for the Oregon Plan." Oregon Laws, ch. 6, HB 5042.

OWEB (Oregon Watershed Enhancement Board). 2005. 2003–2005 Biennial Report, vol 2. www.oregon.gov/OWEB/biennialreport_v2_05.shtml (accessed July 1, 2010).

Pacific Rivers Council v. Brown. 2003. U.S. Dist. LEXIS 8139 (D. Or. 2003). Dismissed without prejudice, 2004 U.S. Dist. LEXIS 19557 (D. Or. 2004).

Rohlf, Daniel J. 2004. "Section 4 of the Endangered Species Act: Top Ten Issues for the Next Thirty Years." *Journal of Environmental Law* 34(2): 483–553.

Scheberle, Denise. 2004. *Federalism and Environmental Policy*. Washington, D.C.: Georgetown University Press.

State of Oregon and NMFS (National Marine Fisheries Service). 1997. Memorandum of Agreement Between the State of Oregon and the National Marine Fisheries Service.

Stein, Amy L. 2010. "State Fish Stocking Programs at Risk: Takings under the Endangered Species Act." *Duke Environmental Law and Policy Forum* 20(1): 63.

Stouder, Deanna J., Peter A. Bisson, and Robert J. Naiman, eds. 1997. *Pacific Salmon and Their Ecosystems*. Norwell, MA: Chapman and Hall.

Taylor III, Joseph E. 1999. *Making Salmon: An Environmental History of the Northwest Fisheries Crisis*. Seattle, WA: University of Washington Press.

Trout Unlimited v. Lohn. 2007. U.S. Dist. LEXIS 77468 (D. Or. July 13, 2007). Magistrate opinion adopted, 645 F.Supp.2d 929 (D. Or. 2007).

Trout Unlimited v. Lohn. 2007. WL 1795036 (W.D. Wash. June 13, 2007). Reversed in part, 559 F.3d 946 (9th Cir. 2009).

Wolf, Edward C., and Seth Zuckerman. 1999. *Salmon Nation: People and Fish at the Edge*. Portland, OR: Ecotrust.

Endnotes

1 See ORS 541.351(7) defining the "Oregon Plan" and ORS 541.351–.415 for the overall program.

2 The term "evolutionarily significant unit" (ESU) was developed by the National Marine Fisheries Service (NMFS) in 1991 to interpret the meaning of "distinct population segment." See "Policy on Applying the Definition of Species Under the Endangered Species Act to Pacific Salmon," 56 *Federal Register* 58613 (Nov. 20, 1990).

3 The petitioning environmental groups included Oregon Trout, Portland Audubon Society, and the Siskiyou Regional Education Project.

4 The petitioners filed a lawsuit against NMFS in June 1995 for its failure to meet the statutory deadlines for responses to listing petitions. The original complaint was dismissed after NMFS proposed to list in July 1995, but the plaintiffs filed an amended complaint in September 1995 for failure to meet the deadlines. The court ordered NMFS to make its listing decision. Because of disagreement about the sufficiency of data on Oregon coastal coho, NMFS obtained an extension of time within which to make the listing decision, but it faced a firm deadline of April 21, 1997.

5 The complete assessment is available at http://nrimp.dfw.state.or.us/OregonPlan/ (accessed July 1, 2010).

6 Following the NMFS status review and adoption of the hatchery policy, 16 West Coast ESUs (not including the Oregon coastal coho) were listed in 2005. In *Alsea Valley Alliance v. Lautenbacher* (2007) the district court upheld the listings and their consideration of hatchery fish. The court also upheld revised 4(d) regulations that allowed harvest of hatchery fish with clipped fins, but not "natural" fish from the same ESU, noting that NMFS is not required to treat natural populations and hatchery stocks equally. This ruling was upheld by the Ninth Circuit in a memorandum opinion concurrent with its 2009 ruling in *Trout Unlimited v. Lohn*.

7 The *Trout Unlimited v. Lohn* case heard in the Western District Court of Washington involved the same plaintiffs, defendants, and title as the case in the District Court of Oregon. The Washington judge dismissed the 2006 NMFS delisting of Oregon coastal coho so it could be heard in Oregon, but the Western Washington District Court heard challenges to the hatchery policy.

CHAPTER 11

Grizzly Bear Conservation in the Greater Yellowstone Ecosystem: A Case Study in the Endangered Species Act and Federalism

KAUSH ARHA AND JOHN EMMERICH

Grizzly bear conservation efforts in the Greater Yellowstone Ecosystem (GYE) represent one of the most high-profile, successful, and contentious conservation efforts under the Endangered Species Act (ESA). In 1974, a National Academy of Sciences Committee on Yellowstone Grizzlies observed the grizzly bear to be "widely recognized as a species confined to wild areas ... a symbol of wilderness ... and thus [it] attracts the enthusiastic support of those devoted to maintaining intact wilderness ecosystems." The story of grizzly bear conservation efforts has borne out the committee's assessment. An analysis of the listing and recovery of the grizzly bear in the GYE under the ESA provides an illustrative chronicle of state and federal cooperation and travails in conserving wildlife species and their ecosystems. The major agencies involved in grizzly bear conservation in the GYE include the United States Fish and Wildlife Service (USFWS), National Park Service (NPS), United States Forest Service (USFS), United States Geological Survey (USGS), Wyoming Game and Fish Department (WGF), Idaho Game and Fish Department (IGF), and Montana Department of Fish, Wildlife, and Parks (MFWP). This chapter analyzes the responsibilities and concerns of these agencies vis-à-vis grizzly bear conservation in the GYE and underlying political and economic considerations. In discussing the roles of the concerned agencies, particular emphasis is directed to two agencies, USFWS and WGF. USFWS is the lead agency coordinating grizzly bear conservation in the lower 48 states, including the GYE. WGF has jurisdiction over the majority of the GYE (outside the National Parks) and has the largest state agency role in grizzly conservation efforts in the Yellowstone region.

The analysis of grizzly bear conservation in the GYE is organized into three sections. The first section presents a brief narrative of the species' natural history and a summary of the conservation efforts to the present. The second section analyzes state and federal positions and perspectives during each phase of the listing and recovery process. The third section discusses general themes that characterized state and federal working relationships in the context of grizzly bear conservation in the GYE and are applicable to a broad array of species conservation efforts.

Natural History of Grizzly Bear

Brown bears crossed the Bering Land Bridge from Asia to North America about 50,000 years ago and spread across the western half of the North American continent. Today, two subspecies of brown bear inhabit the continent: the grizzly bear (*Ursus arctos horribilis*) on the mainland and the Kodiak bear (*Ursus arctos middendorffi*) on Kodiak, Shuyak, and Afognak islands.[1]

Distribution

The historic range of the grizzly bear extended roughly from the Mississippi River to the California coast and from Alaska down to Mexico (Figure 11-1). With the westward expansion of the American settlers, the bear's range shrank dramatically. The grizzly bear population purportedly decreased from about 50,000 bears in 1800 to less than 1,000 by 1975. The major causes for the precipitous decline included habitat deterioration, unregulated hunting, commercial trapping, livestock depredation control, and human–bear confrontations. Use of professional hunters and trappers by state and federal agencies and livestock groups, in the wake of growing settlement, greatly accelerated the extirpation of the bear from large parts of its range. Texas extirpated bears by 1890, and California saw the last of its state animal in 1922, followed by Utah in 1923, Oregon in 1931, New Mexico in 1933, and Arizona in 1935.[2] In marginal habitats, grizzly bears are characteristically vulnerable to human-caused mortality, due to their opportunistic feeding habits that draw them to food sources associated with humans, including livestock, garbage, and carrion.

The current grizzly bear distribution in the lower 48 states is limited to the mountainous ecosystems of four states—Idaho, Montana, Washington, and Wyoming—amounting to about two percent of its historic range. The two major areas of self-perpetuating grizzly bear populations in the coterminous 48 states include the GYE and the Northern Continental Divide Ecosystem (NCDE). The latter encompasses Glacier National Park and Bob Marshall Wilderness in northwestern Montana. Smaller remnant populations exist in the Cabinet-Yak Ecosystem, the Selkirk Ecosystem, and the North Cascade Ecosystem in northern Idaho and Washington. These smaller populations represent offshoots of grizzly populations in Canada.

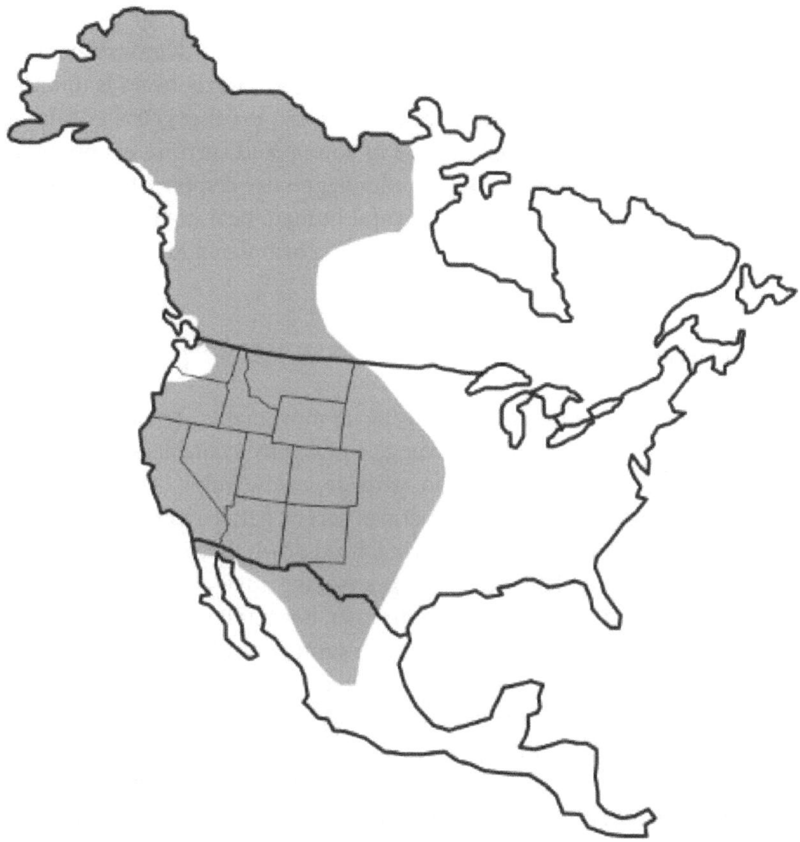

FIGURE 11-1. Historic Grizzly Bear Distribution in North America

The GYE encompasses roughly 9,500 square miles of mountainous habitat, including Yellowstone and Grand Teton National Parks; John D. Rockefeller Memorial National Parkway; parts of six National Forests; Bureau of Land Management lands; and state and private lands in Idaho, Montana, and Wyoming. National Forests comprise the largest portion (about 60 percent) of the GYE.

Physical Characteristics

Grizzly bears are distinguished from black bears by a pronounced hump at the shoulders and a dish-shaped head with a relatively raised brow. The color of a grizzly bear's fur ranges from light to dark brown, with some older bears developing a silvery shine to their pelt—the grizzled look, whence the common name of the bear is derived. In the contiguous 48 states, an average male grizzly weighs 400 to 600 pounds, and an average female bear 250 to 350 pounds. Male bears weighing 1,000 pounds are also found in the lower 48 states but tend to be rare. The life span of a grizzly can reach 30 to 40 years. Except when tending to their young or breeding, grizzly bears demonstrate a solitary pattern of behavior. Strict territoriality among grizzly bears is unknown, and defensive behavior of a grizzly bear in relation to other grizzlies is limited to specific food concentration, defense of young, and surprise encounters. In particular, a sow with cubs exhibits a reflex aggressive response to any surprise intrusion or perceived threat. The harmful human–bear confrontations most often result from a sow defending her cubs or a bear defending its food supply, with the former being more common.[3]

Habitat Characteristics

The search for food dominates a grizzly bear's movements. As an opportunistic omnivore, the bear will prey or scavenge almost any available food, including ground squirrels, ungulates, carrion, garbage, roots, fungi, insects, tree cambium, berries, nuts, and fish. The general observed pattern indicates that bears, upon emerging from their dens in spring, head down to lower elevation drainage bottoms and ungulate winter ranges to take advantage of available and emerging food sources. Through the summer, bears follow the emerging plant growth back up to high elevations. In late summer and autumn, the proportion of fruits and nuts in the bear diet increases. Individual bears can divert from the general pattern, based on local conditions.

Bears have been documented to strongly favor dense forest cover in proximity to forest openings. Most bear bed sites are found in forested habitats. Thus, forest management activities can have a significant impact on bears. Through the winter, the bears hibernate in dens, usually dug on steep terrain with deep snow accumulation. Deep snow, low ambient air temperature, and paucity of food through long winter months necessitate winter hibernation as essential to bear survival. Bears prefer remote and isolated locations for den sites. The grizzly is not known to exhibit any overt defense of its den site and often abandons the den when disturbed by human presence.

Population Characteristics

Population density of grizzly bears is determined by habitat productivity including the abundance and quality of food sources. Home range sizes of grizzly bears vary in relation to food availability, weather conditions, and interactions with other bears. Male grizzlies have larger home ranges than females, and home ranges of adult bears often overlap. Dispersing sub-adult males, due to their curiosity and use of unnatural food sources, demonstrate disproportionately high mortality as a result of human–bear conflict.

The grizzly bear has one of the lowest reproductive rates among terrestrial mammals. Consequently, brown bear populations do not lend themselves to rapid population increases. The bear's low reproductive rate argues for a conservative conservation strategy. The mating season for grizzly bears lasts from late May through mid-July, with a peak in mid-June. The estrus in females can last from days to over a month. The reproductive age of a female grizzly varies from 3.5 to 8.5 years, with an average of 5.5 years in bear populations studied in the lower 48 states. Litter size ranges from one to four, with the average being two. A reproductive interval for a female grizzly bear averages three years. Under optimal conditions, a single female grizzly could potentially add 3.5 females to the population during her lifetime. However, the existing grizzly habitat conditions in the lower 48 states, given increasing human encroachment, are less than optimal. Therefore, grizzly conservation puts a high premium on protection of females in a population.

Grizzlies have no natural predators in the wild, except other grizzlies. Adult males are known to kill juveniles and other adult bears. Overall causes of natural mortality for grizzly bears are not well understood. Human-caused mortality is a significant portion of overall mortality and can result from a variety of circumstances, including direct human–bear confrontations (hikers, hunters, etc.); human actions that can lead to food conditioning or habituation of the bears to inadequately stored food or garbage; careless livestock husbandry; erosion of grizzly bear habitat; hunting; and poaching.

Grizzly Bear Conservation in the GYE

In 1975, the grizzly bear was listed as a threatened species in the lower 48 states. Critical habitat for the bear was proposed in 1977, withdrawn in 1978, and never put forth again. The first recovery plan for the bear was written in 1982. In 1984, the Interagency Grizzly Bear Committee (IGBC) was formed to coordinate grizzly bear recovery efforts. In 1993 the grizzly bear recovery plan was revised and ever since has guided recovery efforts. The GYE bear population met its demographic recovery goals in 1998. The GYE bear population was estimated at 596 in 2008 (Haroldson 2009). The Grizzly Bear Conservation Strategy—an effort begun in 1994 to manage the recovered grizzly bear population in the GYE, after the bear was delisted—was published in March 2003. The strategy includes state grizzly management plans from Idaho, Montana, and Wyoming. All relevant state and federal agencies signed a Memorandum of Understanding to implement the Conservation Strategy on March 26, 2003, at the North American Conference for Wildlife and Natural Resources. USFWS released the draft rule for delisting the GYE grizzly bear population on November 15, 2005. The rule included the distinct population segment (DPS) analysis and a review and update of the most recent population monitoring and estimation techniques. Over 215,000 comments were received during the comment period, which ended on March 20, 2006. Final review of comments, along with recommendations for any changes to the rule, were submitted to the Director of the USFWS in Washington, D.C., on December 31, 2006. On

March 29, 2007, USFWS issued a final rule, designating GYE grizzly bears as a DPS and simultaneously removing that DPS from the list of threatened species. USFS completed the forest plan amendment process in 2006, for the six forests within the recovery area, to incorporate habitat criteria contained in the Conservation Strategy. The record of decision (ROD) on these amendments was signed at the same time as the delisting ROD. NPS also incorporated the habitat criteria into the Yellowstone National Park (YNP) and Grand Teton National Park (GTNP) management plans.

On June 4, 2007, Western Watersheds Project et al. (see *Western Watersheds Project v. Servheen*) filed a complaint in Federal District Court for the District of Idaho, challenging the Final Rule. On October 31, 2007, Wyoming was granted intervenor status in that Idaho case. On August 2, 2007, an individual plaintiff in this case filed a complaint in Federal District Court in the Northern District of Illinois raising substantially similar issues and requesting the same relief as in *Western Watersheds Project v. Servheen.* The Illinois District Court transferred this plaintiff's case to the District of Idaho for those reasons. On January 30, 2008, the federal defendants in this case moved to consolidate it with *Western Watershed Project vs. Servheen.* Wyoming did not oppose the consolidation. Final resolution of all current associated litigation was still not complete as of January 1, 2011. Table 11-1 lists the significant conservation efforts since 1974.

TABLE 11-1. Chronology of Grizzly Bear Conservation

1974	Fund for Animals petitions USFWS to list grizzly bears as endangered in the lower 48 states.
1974	National Academy of Sciences Committee on Yellowstone Grizzlies publishes report stating no evidence of Yellowstone grizzly population threat of extinction.
1975	USFWS designates the grizzly bear as threatened in lower 48 states.
1975	Interagency Grizzly Bear Study Team is formed to research grizzly bear biology.
1977	Critical habitat for grizzly bear is designated.
1978	Critical habitat for grizzly bear is withdrawn.
1982	MFWP's Don Brown writes the Grizzly Bear Recovery Plan.
1984	Interagency Grizzly Bear Committee (IGBC) replaces the Steering Committee for bear recovery efforts.
1986	Interagency Grizzly Bear Management Guidelines are formulated to direct federal land management activities in grizzly bear habitat.
1989	Chris Servheen is appointed USFWS Grizzly Bear Coordinator.
1993	Grizzly Bear Recovery Plan is thoroughly revised by the USFWS Grizzly Bear Coordinator.
1994	IGBC directs USFWS to develop Conservation Strategy to manage a recovered grizzly bear population in the GYE after it is delisted.
1995	Court orders USFWS to address specific habitat standards and genetic concerns in the Recovery Plan.
1998	IGBC undergoes an organizational review.

2000	USFWS Coordinator presents draft Conservation Strategy.
2001	A Governors' Citizen Roundtable (ID, MT, and WY) is constituted to comment on the draft Conservation Strategy.
2001	IGBC appoints a working group comprising state and federal agency personnel to incorporate the roundtable's comments into the Conservation Strategy. Idaho, Montana, and Wyoming develop state plans to manage grizzly bears.
2002	Final Conservation Strategy, with state grizzly bear management plans appended, is submitted to IGBC.
2003	Relevant state and federal agencies sign memorandum of understanding (MOU) to implement the Conservation Strategy.
2005	USFWS conducts DPS analysis for the GYE grizzly population and publishes draft delisting rule in the Federal Register November 15.
2005–2006	USFS and NPS complete amendment process of National Forest Plans and National Park Plans to incorporate habitat criteria stated in the Conservation Strategy.
2006	USFWS reviews over 215,000 comments on draft delisting rule and submits final analysis to FWS director December 31.
2007	USFWS proposes to release a Final Rule to delist the GYE grizzly population in spring 2007.
2007	USFWS issues a Final Rule designating GYE grizzly bears as a DPS and simultaneously removing that DPS from the list of threatened species on March 29.
2007	Western Watersheds Project et al., file a complaint in Federal District Court for the District of Idaho, challenging the Final Rule on June 4.
2007	Wyoming granted intervenor status in Idaho case on October 31.
2007	Second plaintiff, Robert Aland, files a complaint in Federal District Court in the Northern District of Illinois raising substantially similar issues and requesting the same relief as in *Western Watersheds Project vs. Servheen* on August 2. Plaintiff's case transferred to the District of Idaho.
2007	Third plaintiff, Greater Yellowstone Coalition, files a complaint in Federal District Court in the District of Montana challenging the Final Rule November 13.
2008	The federal defendants move to consolidate the complaints and move to transfer the Montana case to Idaho January 30. Wyoming does not oppose the consolidation. The consolidation and transfer requests are denied.
2009	On September 21, the Montana District Court issues an order vacating the USFWS Final Delisting Rule and remanding it to USFWS. On October 5 the federal defendants move that the Montana District Court amends the judgment entered September 21. Plaintiffs' counsel opposes. Court ruling on the motion does not change the effect of the September 21 judgment.
2010	By the January 2010 appeal deadline, five appeals of the September 21 Montana District Court judgment are filed, including an appeal from USFWS and State of Wyoming.
2011	As of January 1, 2011, litigation is still pending.

From Being Hunted to a Species of Concern

Grizzly bears were historically hunted with little or no restrictions throughout the West, including Idaho, Montana, and Wyoming. Through most of the 20th century they were pursued as a predator, and no distinction was made between black and grizzly bears. Montana, host to the largest population of grizzly bears in the lower 48 states, implemented regulated grizzly hunts in the 1920s (see Table 11-2). Consequently, Montana boasts better records on grizzly numbers than any other western state.

Prior to 1968, Wyoming did not restrict harvest of grizzly bears.[4] Consequently, records of grizzly bears taken by hunters in that era are less than reliable (see Table 11-3). From 1970 to 1974, the Wyoming Game and Fish Commission issued a limited number of licenses to hunt grizzly bear in Park and Teton counties (east and south of YNP, respectively) and required mandatory reporting by hunters. The number of hunting licenses ranged from 30 in 1970 to 12 in 1974. Both Montana and Wyoming suspended grizzly bear hunting in the GYE in 1974.

TABLE 11-2. Montana Grizzly Bear Management Prior to 1975

1920s	Regulated hunting seasons
1921	Abolition of baiting and use of hounds to hunt bears
1923	Listing of bears as a game species
1947	Prohibiting the killing of cubs or females with cubs

Source: USFWS 1993: 13

Through the 1950s, concern grew among some state and federal wildlife managers and resource professionals about declining grizzly bear numbers (Corsi 2004). In 1958, Robert Cooney,[6] Chief of the MFWP Game Department, queried his colleagues in Idaho and Wyoming, among others, on the status of grizzly bear populations in their respective states. Then-Director of IFG Ross Leonard responded:

> *The grizzly bear is considered to be very scarce and to inhabit the more remote and timbered sections. An estimate of approximately 50 is listed for the State. There has been no open season on this species since 1946. Accurate information regarding their population and trends is not available How many and how long we will have grizzly bear in Idaho is a matter of conjecture. The more isolated and relatively inaccessible areas are the only places that hold any promise for their survival.* (Leonard 1958)

Then-Chief of WGF Wildlife Development Division Fred Yeaman stated:

> *The grizzly bear has been holding its own in Wyoming for the past three years. We now estimate 50 to 75 in the state (not including the Yellowstone and Grand*

TABLE 11-3. Recorded Grizzly Bear Harvest in Montana and Wyoming[5]

Year	*Montana**	*Wyoming*
1949	47	2
1952	10	6
1953	40	7
1954	35	7
1955	35	4
1956	22	3
1957	23	8
1958	36	—
1959	60	4
1960	30	6
1961	60	—
1962	40	5
1963	35	—
1964	40	10
1965	30	**
1966	45	**
1967	24	21
1968	12	Closed
1969	33	Closed
1970	—	8
1971	—	6
1972	—	4
1973	—	3
1974	—	7

Source: Moody et al. 2005 and Draft Report of the Western Association of Fish and Wildlife Agencies (WAFWA) Special Grizzly Bear Committee August 7, 1970. *Montana numbers refer to the entire state, with the highest harvest occurring in the northwest region. In contrast, grizzly bears in Wyoming were harvested in the environs surrounding YNP. **WGF did not differentiate grizzly and black bear harvest information for 1965–1966. (*Note:* Total grizzly mortality was usually in excess of that harvested by hunters. Harvest data do not include grizzly deaths due to accidents, management control action by agency personnel, or bears killed on adjacent Indian reservations. For example, according to Montana records total grizzlies killed in 1967, 1968, and 1969 were 41, 28, and 46 respectively. Accuracy of the Montana and Wyoming harvest numbers cannot be substantiated. Data for 1949–1967 was obtained from Annual Big Game Inventory compiled by the USFWS and distributed as Wildlife Leaflets. Idaho has had no hunting season for grizzly bear since 1946.)

> *Teton National Parks). Last year we killed eight grizzly bear(s) during our regular open hunting seasons. Most of the grizzly bear found in Wyoming are adjacent to Yellowstone and Grand Teton National Parks. A rather large bear population supported by these National Parks assures Wyoming maintaining a future grizzly bear population.* (Yeaman 1958)

In the late 1960s, a controversial decision by NPS to abruptly end the practice of allowing grizzly bears to feed on open garbage dumps in YNP severely impacted the grizzly bear population in the GYE. Prior to 1968, YNP, as was the practice in several other parks, allowed bears—both black and grizzly—to feed at garbage dumps to supplement their food intake. This practice encouraged concentrations of bears at garbage dumps in and around the park. Grizzly viewing at open garbage dumpsites was a popular activity among visitors to YNP. In 1967, grizzly bears killed four individuals near Granite Creek Chalet in Glacier National Park, an incident notoriously referred to as the "Night of the Grizzlies." This incident, coupled with the ongoing, paradigmatic shift in NPS management to restore natural processes, led the Service to put an end to the practice of open garbage dumps in YNP. John and Frank Craighead, two of the first to study grizzlies in the wild and in YNP in particular, argued for a phased approach to getting rid of the open-pit garbage dumps. They argued that most of the grizzly bears in the GYE were found around the dumps, and when confronted with a sudden loss of a habitual food source they would cause trouble while looking for food and would be fatally dealt with. The ensuing control action might debilitate the viability of the Yellowstone grizzly population. The contrary view held that garbage-dump bears were only a subsection of the bear population in the region, and, further, that a shorter adjustment period would minimize the time during which emergency measures would have to be taken to prevent injuries to humans (NAS 1974). The latter view prevailed; the amount of garbage disposed in open dumps was sharply reduced beginning in 1968, and all dumps within the park were closed by 1971. The NAS report concluded that from 1968 to 1973, "the removal of grizzly bears from the Yellowstone ecosystem exceeded that of the 1959–67 period by 76 animals" (NAS 1974). Hunters, park officials, or other individuals acting in self-defense killed these bears. A spirited debate ensued, concerning the impact of high grizzly mortality in the wake of garbage-dump closures, on the health of the Yellowstone grizzly population. One faction contended that an overwhelming majority of Yellowstone bears inhabited the areas around the garbage dumps and argued that the bear population had suffered a debilitating blow. Another faction held a different view, contending that garbage-feeding bears represented only a subset of the Yellowstone grizzly population and that there existed a healthy population of so called "wilderness bears." In any event, the extraordinary number of grizzly deaths in the wake of open-pit garbage-dump closure in YNP is often cited as one of the major causes leading to the listing of the grizzly bear as threatened under the ESA.

Meanwhile, the Montana Fish and Game Commission adopted a Grizzly Bear Policy in 1969, with expressed intent to perpetuate and manage the species, research grizzly habitat needs, and support wildland management programs aimed at maintaining sufficient amounts of suitable grizzly habitat in the state. The Montana Commission, based on a rough calculation by Rob Cooney—estimating the grizzly population in the lower 48 states at 800 bears—strongly argued that grizzly bears were indeed rare, but not a wildlife species in danger of extinction in the region. In 1970, IFG discussed tentative plans with USFS to release grizzly bears in Idaho if suitable specimens were available from YNP. The WGF, during the early 1970s, in the wake of high grizzly mortality in YNP, initiated efforts to get a more accurate sense of grizzly numbers in the state. Outfitting and ranching interests put heavy pressure on WGF to state that grizzly bears were neither rare nor endangered in the state. The prevailing Wyoming political climate was not conducive to the state undertaking proactive steps to research and conserve the GYE grizzly bear population (Corsi 2004).

The Western Association of Game, Fish, and Conservation Commissioners adopted a resolution in July 1970 titled "Perpetuation of Grizzly Bear" (Montana Special Grizzly Bear Committee Report 1970). Subsequently, both the Western Association of Fish and Wildlife Agencies (WAFWA) and the International Association of Fish and Wildlife Agencies (IAFWA), currently recognized as the Association of Fish and Wildlife Agencies (AFWA) appointed an ad hoc Grizzly Bear Committee and passed a resolution requesting

> *... sufficient funds be appropriated by the U.S. Congress and allocated to assist the U.S. Forest Service, the Bureau of Sport Fisheries and Wildlife (predecessor to FWS) and the concerned State wildlife management agencies in the establishment of a continuing intraagency [sic] study team charged with making a comprehensive national inventory of preferred habitat, distribution and population of the Nation's grizzly bears and the identification of the influence that might detrimentally affect those populations...* (NAS 1974)

AFWA supported WAFWA's position to oppose any recommendations or actions of the Secretary of Interior to classify the grizzly bear as endangered or threatened. The National Academy of Sciences Committee evaluating Yellowstone grizzlies found the Yellowstone grizzly bear population, based on the information presented to it, to be stable, at a conservative estimate of 234 bears between 1959 and 1967 (NAS 1974). The peak in grizzly mortalities from 1968 to 1971, when garbage dumps were closed off to bears, then waned in 1972 and 1973. Importantly, the National Academy of Sciences Review Committee concluded that at the time "there is no convincing evidence that the grizzly bears in Yellowstone ecosystem are in immediate danger of extinction" (NAS 1974). Further, the NAS Committee recommended establishing a nongovernmental coordinating entity to facilitate agreements among state, federal, and private agencies to direct their efforts in preserving Yellowstone grizzlies, without ceding their basic authority. In particular, the National Academy of Sciences Committee urged the state and federal agencies to get better information on grizzly bear biology, limit human-caused removals of grizzlies to 10, and encourage a phase-out of sheep grazing permits in the ecosystem (NAS 1974).

Grizzly Bear Deemed Threatened under the ESA

On February 14, 1974, the Fund for Animals petitioned the Department of Interior to list the grizzly bear in the lower 48 states as an endangered species. USFWS proposed a rule to list the bear as threatened on January 2, 1975. The Service received 545 comments, out of which 421 favored the bears' listing as endangered and, in the alternative, opposed any sport hunting of a threatened species; 310 comments opposed depredating grizzlies to be taken on federal lands.[7]

Section 4 of the ESA establishes five criteria for determining whether a species should be listed as an endangered or threatened species (16 U.S.C. 1533[a][i]).[8] USFWS reasoned that four of the five stated criteria called for listing the grizzly bear as a threatened species in the lower 48 states. The Service concluded that the grizzly bear range was significantly curtailed (at less than two percent of its historic range) and that its remaining habitat faced significant threat of destruction or modification given the timbering practices and building of numerous forest roads in the areas where bears still reside (40 *Federal Register* 31734 [1975]). According to the Service's assessment, the prevailing cultural attitude still treated bears as vermin, and this resulted in continued loss of the animals through indiscriminate illegal killing. Disease was not found to be a major factor limiting grizzly bear numbers in contiguous states. In particular, USFWS found that gaps in the scientific information relating to grizzly bears severely hindered the prevalent management regime for grizzly bears and, as such, provided an inadequate regulatory mechanism for protection of the bears. The two additional factors contributing to the decline in grizzly numbers identified by the Service included the isolated nature of existing grizzly populations and the increasing human and livestock use of surrounding National Forests. The Service, nevertheless, concluded that threatened, and not endangered, was the appropriate designation for the grizzly bear, as most of the reduction in the bear's range had occurred in the 19th century. It is unclear what bearing the date of a species range constriction has on whether the species is in danger of extinction or under a threat of extinction. On July 23, 1975, USFWS Director Lynn Greenwalt designated the grizzly bear as a threatened animal in the lower 48 states.

Interestingly, the rule designating the grizzly bear as threatened also allowed for a limited bear hunt in the Bob Marshall Ecosystem, excluding Glacier National Park in northwestern Montana. To back its holdings, USFWS cited section 3(2) of the ESA of 1973, which provided for regulated taking "*in the extraordinary case where population pressures within a given ecosystem cannot be otherwise relieved*" (16 U.S.C. 1532[2]). USFWS stated that the bear population in the area was large enough that bears were wandering into settled areas, threatening human safety, and committing significant depredations on legally present livestock. According to the Service, live trapping and translocating were too expensive, dangerous, and ineffective in relieving the population pressure.

The Service endorsed MFWP's argument that a regulated sport hunt, where the bears are exposed to a relatively large number of humans for a short period of time, will instill an appropriate level of fear, teaching the bears to avoid humans. One may argue that if this rationale held true, then grizzly bear hunting seasons, prior to the bear's listing, would have alleviated the bear population pressure. The counterargument suggests that continued hunting was keeping the situation in check and that a cessation of hunting activities would lead to an increase of grizzly-induced damage to human enterprise in the area. It is plausible that, confronted with strong opposition from the states to the listing of the bear under the ESA, USFWS compromised by designating it a threatened rather than endangered species. Alternatively, it can be argued that if the grizzly population in NCDE could sustain a steady harvest, then it was probably not under any threat of extinction. The mood of the moment is captured by the opening paragraph of a letter written by then-Director of the Montana Fish and Game Department Wesley Woodgerd to USFWS Director Lynn Greenwalt, dated November 25, 1974:

> *Dear Lynn: In the interest of maintaining an effective and harmonious program of grizzly bear management and in order for the U.S. Department of Interior to find inadvisable and unnecessary the placing of any restrictive federal regulations on grizzly bear in the Bob Marshall ecosystem, the Montana Fish and Game Commission has developed the following program ... [limiting the hunt to 25 bears per year].*[9]

The federal action engendered a strong and vociferous reaction from the concerned states, in particular the leaders of the state wildlife agencies, who perceived their jurisdiction being limited by the federal action (Flyn 2004).

Road to Recovery for Grizzlies in GYE

The period immediately following the listing of the grizzly bear in 1975 is not marked by determined or coordinated action. The ESA was then a relatively new statute, and the road to recovery for grizzly bear populations in the GYE was far from clear, particularly as it required coordination among diverse federal regulatory and land management agencies and state wildlife agencies. Most important, grizzly bear habitat needs and population characteristics were poorly understood. Therefore, the early focus in recovery efforts was directed toward gathering more biological information on the bear.

Critical Habitat Designation

On November 5, 1976, USFWS proposed critical habitat for the grizzly bear (41 *Federal Register* 48,758 [1976]). The Service proposed to designate four zones coinciding with the distribution of the grizzly bear at that time: 1) YNP and adjacent areas; 2) northwest Montana; 3) extreme northwestern Montana and northern Idaho; and 4) extreme northern Idaho and northeastern Washington. The areas were selected for their relative inaccessibility and lack of the kinds of human developments and activities that tend to result in conflicts between bears and humans. The proposed rule determining critical habitat for the grizzly bear was withdrawn by USFWS in 1979, because the 1978 amendments to the ESA had imposed additional obligations on the Service before it could determine critical habitat (44 *Federal Register* 12,382 [1979]). Thenceforth, given the impracticability of delineating critical habitat for an opportunistic omnivore such as a grizzly bear, USFWS did not propose critical habitat designation for the bear.[10] The states also favored not designating arbitrary critical habitat for the grizzly bear.

Interagency Grizzly Bear Study Team (IGBST)

In 1975, all state and federal agencies involved in grizzly conservation agreed on the need for coordinated efforts to research bear ecology. During the same year, a cooperative agreement between USGS, NPS, USFS, FWS, IGF, MFWP, and WGF was signed to establish the Interagency Grizzly Bear Study Team (IGBST) to collect, manage, analyze, and distribute information related to grizzly bear conservation. Lack of information on basic bear ecology such as habitat use, demographic information—including population structure and trend—and dispersal rates was cited as a major cause of concern warranting a threatened status for the bear. The responsibilities of IGBST are to 1) conduct both short- and long-term research projects addressing information needs for bear management; 2) monitor grizzly bear populations; 3) monitor grizzly bear habitat trends, including impacts of human activities on bears; and 4) provide technical support to agencies responsible for bear management in the Yellowstone area (Schwartz and Moody 2009). The IGBST has produced annual reports since 1975. The NPS provided the bulk of initial funds needed to support the team's operations, and a biologist from its research staff, Dr. Richard Knight, led the team from 1975 to 1997. Since its creation, IGBST has undertaken an intensive—and rather expensive—bear research program in the GYE, which includes considerable field research involving annual trapping, radio collaring, and monitoring of grizzly bears (see Table 11-4). This is in stark contrast to the research effort devoted to gray wolf conservation in the GYE. The difference can be explained in part by the fact that gray wolf biology was well studied in Minnesota and northwestern Montana, as well as by the species' proclivity to rapid population growth and the lack of strict land-use restrictions associated with its reintroduction in the GYE.

TABLE 11-4. Annual Record of Grizzly Bear Monitoring in the GYE since 1980

Year	*# Monitored*	*# Trapped*	*Total Captures* *Research*	*Management*	*Transports*
1980	34	28	32	0	0
1981	43	36	30	35	31
1982	46	30	27	25	17
1983	26	14	0	18	13
1984	35	33	20	22	16
1985	21	4	0	5	2
1986	29	36	19	31	19
1987	30	21	15	10	8
1988	46	36	23	21	15
1989	40	15	14	3	3
1990	35	15	4	13	9
1991	42	27	28	3	4
1992	41	16	15	1	0
1993	43	21	13	8	6
1994	60	43	23	31	28
1995	71	39	26	28	22
1996	76	36	25	15	10
1997	70	24	20	8	6
1998	58	35	32	8	5
1999	65	42	31	16	13
2000	84	54	38	27	12
2001	82	63	41	32	15
2002	81	54	50	22	15
2003	80	44	40	14	11
2004	78	58	38	29	20
2005	91	63	47	27	20
2006	92	54	36	25	23
2007	86	65	54	19	8
2008	87	66	39	40	30

Source: Haroldson et al. 2009.

States followed the establishment of IGBST by creating their own state study teams. WGF appointed the Wyoming Grizzly Bear Study Team (WGBST) to assist IGBST with collection and analysis of bear data in Wyoming. WGBST was initially staffed with one permanent biologist and two seasonal employees during the field season. Montana also developed a state grizzly bear study program to monitor bear populations in the NCDE, GYE, Selkirk, and Cabinet-Yak ecosystems, with the NCDE being its high priority. The state agencies were willing to assist the federal agencies with the grizzly bear research, motivated in part by responsibility for and pride in managing their own wildlife and in part by the expectation that a coordinated conservation effort would lead to prompt recovery and delisting of the bear. WGF's commitment to grizzly bear research has increased in cost and personnel since 1975, as bears expanded their numbers and range (see Table 11-4) and more sophisticated information was required. Given that the largest percentage of the GYE falls in Wyoming, WGF has collected the most grizzly bear data for the region, particularly in the areas outside YNP.

Interagency Grizzly Bear Coordinating Committee (IGBC)

A distinguished panel of wildlife scientists, appointed by the National Academy of Sciences and AFWA in the early 1970s, had called for a coordinated interagency effort to conserve grizzly bears in the GYE. An ad hoc steering committee oversaw grizzly conservation from 1975 to 1982. In 1983, the Assistant Secretary for Natural Resources and Environment, Department of Agriculture, and the Assistant Secretary for Fish, Wildlife and Parks, Department of Interior—recognizing the need for coordinated management and research activities of federal and state agencies to achieve the recovery of the grizzly bear—established the Interagency Grizzly Bear Committee (IGBC). The membership of the IGBC included the top-line officers of the member agencies in the region: USFS regional foresters for the Northern, Rocky Mountain, and Inter-Mountain Forest regions; Regional Director of the NPS; regional directors of USFWS; and directors of state wildlife agencies from Idaho, Montana, Washington, and Wyoming. Some environmental groups have criticized IGBC for its lack of stakeholder involvement. The USFWS Grizzly Bear Coordinator acts as the chief staff advisor to the Committee.[11] IGBC created three subcommittees: the Yellowstone Ecosystem Subcommittee (YES), Northern Ecosystem Subcommittee, and Research Subcommittee. The two ecosystem subcommittees were responsible for coordinating the recovery efforts in their respective regions. The Research Subcommittee had the responsibility to oversee and prioritize the workings of the IGBST.[12]

The primary responsibility of the IGBC is to direct grizzly bear management and research in recovering the grizzly bear populations. Its major purpose is to engage top decisionmakers of the affected agencies to coordinate policy actions across state and federal agencies. The chairmanship of the IGBC rotates among the members every two years. The committee convenes a minimum of twice each year to direct and evaluate the implementation of the Grizzly Bear Recovery Plan. The recovery efforts in the GYE were overseen by the YES subcommittee, comprised of affected National Forest supervisors, National Park superintendents, and state wildlife agency representatives.

The creation of IGBC, a unique institution charged with the recovery of the grizzly bear, is a testament to the importance assigned to the grizzly bear. The committee's sense of direction, purpose, and leadership, however, fluctuated with its composition. The committee functioned best when the top regional line offices of the affected agencies were in attendance and engaged in deliberations directing research and management activities to recover grizzly bear populations. The committee did not work as well when the primary decisionmakers delegated their roles on the IGBC to rotating representatives from their respective agencies. In the latter instance, the committee attendees often looked to the USFWS Grizzly Bear Recovery Coordinator for guidance in setting the agenda and suggesting the appropriate course of action. The disproportionate influence of the Recovery Coordinator on the IGBC's workings during select periods of time was often resented by state wildlife and federal land management agencies. It is difficult to assign blame to a Recovery Coordinator for using his discretion when leadership and direction is wanting from the supposed executive body. In 1997, the IGBC contracted with a consultant to conduct an organizational review of the institution to evaluate its effectiveness and efficiency. In 1998, the committee reasserted its leadership role by incorporating most of the resulting recommendations. The structure of IGBC meetings was redesigned to give the committee better control of its own agenda and enabling it to engage in substantive policy deliberations and decisionmaking. The subcommittees were asked to develop five-year conservation goals for their respective ecosystems and report on their progress to the public biannually. Revitalization of the IGBC, welcomed as a positive development by state wildlife agencies, is often heralded as one of the major turning points in state and federal collaboration in conserving the grizzly bear (Smith 2004).

Interagency Grizzly Bear Management Guidelines

In 1986, the Interagency Grizzly Bear Management Guidelines directing federal land management activities in grizzly bear habitat were published, to comply with section 7 consultation requirements of the ESA.[13] The guidelines describe five Management Situations[14] applicable to all federal lands in the recovery area. For each Management Situation, the guidelines state a specific range of actions to 1) maintain and improve habitat; 2) minimize grizzly–human conflict potential; and 3) resolve grizzly–human conflict. The guidelines are subject to change, as new information is forthcoming from ongoing research in grizzly bear ecology.

Recovery Plan

The first Grizzly Bear Recovery Plan was written in 1982 by Don Brown, a biologist with MFWP and the designated Grizzly Bear Recovery Plan Leader.[15] By the mid-1980s, the IGBST research and monitoring efforts were beginning to provide useful information on bear habits and ecology. However, the monitoring requirements stated in the 1982 plan were found to be too ambitious and impractical (U.S. FWS 1993). In 1989, USFWS, unsatisfied with the progress of conservation efforts,[16] appointed a full-time Grizzly Bear

Coordinator. In light of a decade's worth of new field observations, the 1982 Recovery Plan was thoroughly revised by the newly appointed coordinator and distributed among the IGBC member agencies for comments. The Revised Recovery Plan was approved on September 10, 1993. Major changes from the 1982 plan to the 1993 revised plan addressed delineation of recovery zones and revised population monitoring methods, population targets, linkage zone assessment, the requirement of a conservation strategy prior to any proposal for delisting, and the long-term strategy for managing the isolated Yellowstone population.

The stated goal of the 1993 Recovery Plan was to recover grizzly bear populations in each of the designated ecosystems in the lower 48 states to the level where the bear was no longer threatened with extinction. The Recovery Plan called for two requirements before a population within an ecosystem could be delisted:

1 Attainment of the population demographic parameters for that ecosystem within the monitoring period specified, and
2 The development and completion of an interagency conservation strategy that will ensure that adequate regulatory mechanisms will continue to be present after delisting. (U.S. FWS 1993)

The 1993 Grizzly Bear Recovery Plan "defines a sequence of actions" to recover grizzly bear populations in the lower 48 states, whereupon the species can be removed from its present "threatened" listing.[17] The Recovery Plan was intended as a general document stating the broad objectives needed to recover the grizzly populations. The specific tasks to be implemented to achieve recovery were left to federal land management planning processes, including National Forest Plans, National Park management plans, and state game management processes and regulations. These agency management plans, unlike the Recovery Plan, were then subject to either the National Environmental Policy Act (NEPA) or other state review processes.

Recovery Zones

The 1993 Plan discontinued the term "occupied habitat" and demarcated specific zones where grizzly populations were to be "recovered." The plan defined a recovery zone as "*the grizzly bear ecosystem within which the population and habitat criteria for achievement of recovery will be measured.*" The plan identified six recovery zones: the Yellowstone, Northern Continental Divide, Cabinet-Yaak, Selkirk, North Cascades, and Bitterroot ecosystems. Recovery zones are further divided into administrative units referred to as Bear Management Units (BMUs). The BMUs constitute the primary units used for the purposes of grizzly bear habitat evaluation and population monitoring.

The Recovery Plan categorically declares that, although bears are expected to move and reside outside the recovery zone, only the area within the recovery zone will be managed primarily for grizzly habitat. According to the Recovery Plan, presence of bears outside the recovery zone will not lead to change in the boundary of the recovery zone, unless resources occur outside that are important to bears living primarily within the recovery zone. According to the plan, the need to expand the recovery zone would only arise when an important seasonal habitat for grizzlies was found deficient within the recovery area and is available outside the recovery zone. The state wildlife agencies and USFS strongly argued that such vague criteria amounts to a self-fulfilling prophecy, as grizzly bears expand their numbers and range, resulting in ever expanding recovery zones. Large predators such as grizzly bears are best suited to wilderness areas, where human activities and presence are at a minimum. Yet an expanding grizzly bear population will use lands well beyond designated wilderness areas around the GYE. An expanding grizzly population under the ESA's protection carries with it land-use restrictions associated with a wilderness area—in other words, an area where grizzly conservation enjoys primacy over all other management objectives. Thus, from the beginning, the agencies responsible for managing federal lands and wildlife conflicts—particularly USFS and state wildlife agencies—were keen to determine the outer boundaries of grizzly-occupied habitat. Since 1993, the scope of grizzly occupancy, or more specifically areas where grizzly bears would enjoy priority in land management decisions, has remained a major source of contention between USFWS and state wildlife agencies, in particular WGF. The current grizzly bear occupancy management plan, developed by Wyoming, established specific occupancy language for areas outside the recovery zone (primary conservation area).

Population Monitoring

The 1993 Recovery Plan forthrightly recognized the practical limits to determining with any precision total population size of a secretive, wide-ranging wildlife species occupying rugged, mountainous terrain. Striving to strike a balance between precision and cost, the 1993 Recovery Plan narrowed the complex list of population variables called for in the 1982[18] plan to three measurable parameters as indicators of population index: 1) number of females with cubs; 2) number of known human-caused mortalities; and 3) distribution of family groups throughout the ecosystem.[19] The most important population parameter tracked by the Recovery Plan involved the number of unduplicated females with cubs-of-the-year (COY), on a running six-year average accounting for two breeding cycles, based on a three-year breeding interval of a female grizzly. The second population parameter measured is the occupancy by females with young throughout the recovery zone as an indicator of habitat quality, since successful reproduction is often seen as an indicator of habitat sufficiency. The third population parameter tracked known human-caused grizzly mortality. A USFWS-sponsored workshop on grizzly bear genetics indicated that a healthy bear population could sustain a six percent human-caused mortality without population decline (Harris 1985). The Recovery Plan, by tracking the three population parameters, attempts to ensure that more bears are added to the population than taken out on a sustained basis to allow the bear populations to recover. WGF, pointing to this explicit bias towards a growing bear population in the Recovery Plan, has consistently argued for more flexible population parameters directed at stabilizing grizzly population growth—specifically, in portions of the GYE where bears are beginning to occupy populated areas with large potential for human–bear conflict, and when all biologically suitable and socially acceptable habitat is occupied in the state.

Effective Habitat

The Recovery Plan defines effective grizzly habitat as containing all components necessary for the survival of the species, including food, cover, denning habitat, solitude, and space. For a grizzly bear, effective habitat translates into large tracts of relatively undisturbed wild land. Prime grizzly habitat is characterized by diversity of habitat types and food sources, including both plants and animals, so that stochastic shortage in one food source is offset by the availability of others. The Recovery Plan calls for managing both natural and human influences on habitat effectiveness for a grizzly bear. Human activities that impact grizzly habitat the most include forestry, grazing, and recreation. The Recovery Plan identifies roads as the most imminent threat to grizzly habitat. Managing for an effective grizzly bear habitat involves controlling public access management through a rigorous road management policy, strict application of particularized forest management policies, corridors linking isolated populations, and strict regulation of sanitation and garbage disposal standards. A Cumulative Effects Model is used for each ecosystem to assess existing habitat for effectiveness and report on the trend in habitat quality. The plan recognizes corridors linking disparate grizzly populations as desirable but not essential to the recovery of the grizzly bear populations.

1993 Recovery Plan Goals for the Greater Yellowstone Ecosystem

The recovery goal in the 1993 Recovery Plan for the Yellowstone Grizzly Bear Recovery Zone calls for:

> *15 females with cubs over a running 6-year average*[20] *both inside the recovery zone and within a 10-mile area immediately surrounding the recovery zone;*

16 of 18 BMUs within the recovery zone occupied by females with young from a running 6-year sum of observations; no two adjacent BMUs shall be unoccupied; and known, human-caused mortality not to exceed 4 percent of the minimum population estimate based on the most recent 3-year sum of females with cubs. Furthermore, no more than 30 percent of this 4 percent mortality limit shall be females. These mortality limits cannot be exceeded during any 2 consecutive years for recovery to be achieved.

The final conservation strategy includes a goal of managing the GYE grizzly population at or above 500 total grizzly bears.

The population objective for the GYE was based on information gathered since 1975 on bear food habits, distribution, sex/age ratios, bear/human conflicts, home range, and density.[21] The open habitat of the GYE, as opposed to dense forested habitat in the north, facilitates a higher observation rate of females with cubs. USFWS does recognize that not all females with cubs are observed in any given year; hence, the sightings represent only the known minimum number of bears in the ecosystem. Nevertheless, USFWS did not employ any sighting efficiency correction factor or population estimate based on the number of bears observed in the initial years of monitoring. The Service consciously opted for a maximally conservative monitoring approach, given the isolation of the Yellowstone grizzly bear population. Consequently, the Yellowstone grizzly population estimate, based on actual sightings of females with cubs, systematically underestimates the actual minimum number of bears in the GYE. Based on the assumption that one-third of the adult females are with cubs in a given year, 15 females with cubs sighted on average in a year will indicate a total minimum population of 158 bears.[22] The cautious approach favored by USFWS during recovery was further apparent in the allowable mortality limit for the GYE grizzly population. The known human-caused mortality limit for the GYE was set at four percent (no more than 30 percent females) of the most recent minimum population estimate, based on the three-year sum of females with cubs.[23] USFWS chose four percent instead of six percent to account for unknown and unreported bear mortalities. In addition, GYE recovery goals require dispersal of grizzly family groups throughout the recovery zone, with the exception of two BMUs.[24] The BMUs were chosen as units of observation not because of inherent habitat value but for the sake of convenience as existing administration demarcations familiar to the management agencies. Each BMU was subjectively evaluated by the IGBST for its habitat quality and probability of observing a grizzly bear family group. Taken together, these three recovery goals for grizzly bear in the GYE represent the precautionary values of those who prepared and endorsed the Recovery Plan, based on conservative interpretation of available scientific information.

The criteria for determining and controlling nuisance grizzly bears are set out as part of the Interagency Grizzly Bear Guidelines and approved by the IGBC.[25] The provisions stated in 50 CFR 17.40 govern grizzly bear management in the lower 48 states. The regulations allow for the taking of a bear in defense of human life, removal of nuisance bears by federal or state officials, or handling of bears for research purposes by federal or state agencies, pursuant to a permit issued by the Director of USFWS.

The Recovery Plan relies primarily on state law enforcement efforts to prevent poaching or malicious killing of grizzly bear; accidental killing of grizzlies by black-bear hunters; defensive killing of grizzly bears by livestock operators, apiarists, outfitters, hunters, and resort operators while protecting person or property; and losses due to mishandling of bears during research and management actions. Handling of nuisance bears outside the National Parks is the primary responsibility of the state wildlife agencies, in conjunction with USFWS and Wildlife Services (WS). NPS is responsible for handling nuisance bears within the National Parks. After capturing a bear as part of a control action, the decision to either relocate the bear to another area or remove it from the population is to be made within 24 hours of its capture. Each National Forest has its own criteria describing conditions and locations where a nuisance grizzly bear may be relocated in that forest.

Major Judicial Decisions Influencing Grizzly Recovery

The District Court of the District of Columbia decision in *Fund for Animals v. Babbitt*, 1995, had the most impact on the grizzly recovery process and, by extension, on state and federal collaboration. In May 1994 two lawsuits were filed by The Fund for Animals and The National Audubon Society, plus 41 organizations and individuals, challenging the adequacy of the 1993 Recovery Plan. The two combined actions charged that USFWS, in developing the Recovery Plan, had failed to adequately set forth "site specific management actions" or "objective, measurable criteria." The court found the Service had incorporated site-specific management action into the Recovery Plan but that it had failed to incorporate objective, measurable recovery criteria. This failure to establish objective measurable criteria, particularly in the context of monitoring genetic diversity in grizzly populations, was deemed arbitrary and capricious by the court. The court remanded the matter to the USFWS to fix the shortcomings in the Plan. The Service addressed most of the court's concerns in a document published on August 29, 1997.[26] USFWS, cautioned by the court order, devoted particular attention to developing specific habitat criteria to be met in the GYE before the population was deemed recovered. USFS and state wildlife agencies found the late addition of specific habitat criteria beyond those mentioned in the Recovery Plan both unreasonable and unwarranted. The legal action caused significant strain on the USFWS Grizzly Bear Recovery Team's time and resources while it was in the midst of developing the Conservation Strategy called for in the Recovery Plan. As a result, the development of the Conservation Strategy was significantly delayed. The plaintiffs in the action considered the court's order a significant victory, for it "bought ten years from delisting the bear" (Honnold 2004).

Another major judicial decision was handed down in 1994 by the Ninth Circuit in *Resources Limited v. Robertson*, holding USFS in violation of section 7 of the ESA for not adequately accounting for impacts on grizzly populations in developing the 10-year forest plan for the Flathead National Forest. The court decision led to a paradigmatic shift in how national forest plans were developed in grizzly habitats. Thereupon, all National Forests in the grizzly recovery zones account for the impact of forest roads on grizzly habitat effectiveness.

In 1996, the District Court of the District of Columbia rejected an action by The Fund for Animals against USFS to compel the latter to regulate hunting on National Forest lands by banning baiting of wild game (in particular black bears) (*Fund for Animals v. Thomas*). The State of Wyoming intervened as a party-defendant and supported USFS' contention, citing *Geer v. Connecticut, 1896*, that regulation of hunting, even in National Forests, is the responsibility of the state. Soon thereafter, WGF issued state regulations banning baiting for black bear hunting in additional areas of National Forests in the GYE that are occupied by grizzly bears.

The overall impact of the litigations related to grizzly bears, according to some, turned USFWS overly cautious, whereby the overriding objective of subsequent Service actions was to survive a legal challenge. In retrospect, a more participatory recovery strategy, allowing state wildlife agencies to gain more ownership in the development of the Recovery Plan, would have enabled all concerned to view these litigations as a joint problem rather than one of individual agencies.

Recovery Achieved

The grizzly bear population in the GYE has achieved a remarkable recovery by most standards, and that fact deserves to be recognized as one of the great wildlife conservation success stories of the last 30 years. Since the mid-1980s the GYE grizzly population has grown at approximately three percent to four percent per year (Boyce et al. 2001). During the first decade and a half of the listing, grizzly bear-occupied habitat expanded by 48 percent (Schwartz et al. 2002), and the minimum grizzly population doubled in the 10 years from 1991 to 2002. The population estimate for 2008 was 596 (Haroldson 2009). The true number of grizzlies in the GYE is probably higher than the numbers estimated, since all levels of the population monitoring and estimation process are based on conservative parameters. The recovery criteria of 15 unduplicated females with COY was met in 1988, and the number of females with COY has exceeded 30 ever since 1996 (see Table 11-5). The recovery criterion of having 16 of 18 BMUs occupied, with no two adjacent units vacant, was achieved

TABLE 11-5. Number of Unduplicated Grizzly Females with Cubs-of-the-Year (COY) Observed in the GYE, 1991–2008

Unduplicated Females with COY	
Year	*Min Count*
1991	24
1992	25
1993	20
1994	20
1995	17
1996	33
1997	31
1998	35
1999	33
2000	37
2001	42
2002	52
2003	38
2004	49
2005	31
2006	47
2007	50
2008	44

Source: Haroldson 2009.

in 1999 (see Table 11-6). By 2000, all 18 BMUs in the GYE were occupied by a grizzly family group. Total and female human-caused mortality have been under thresholds nearly every year. Between 1991 and 2001,[27] 110 known and probable human-caused grizzly bear mortalities were documented in the GYE. Female grizzlies accounted for 41 percent of the mortality. Nonetheless, the grizzly population increased in number and expanded distribution. During the decade, the annual human-caused grizzly bear mortality never exceeded 22, and female mortality never exceeded 8. The GYE bear population continues to grow and expand with each passing season, and with it grows the probability of human–bear conflicts.

Since the mid-1990s, in light of growing bear numbers and expanding distribution (Figure 11-2), WGF has consistently advocated delisting of the bear and turning the management of the species back to the states to stabilize the growing bear population. WGF's aversion to an expanding grizzly population is in part explained by Wyoming statutes addressing wildlife damage. Wyoming wildlife statutes designate grizzly bear a trophy game animal (Wyoming Statute § 23-1-101[a] [xii]). The WGF, while managing the bear as trophy game, is statutorily responsible for designated types of damage caused by the bear. Wyoming Statute 23-1-901(c) states:

TABLE 11-6. BMUs Occupied by Females with Young, 1999–2008

BMU	*1999*	*2000*	*2001*	*2002*	*2003*	*2004*	*2005*	*2006*	*2007*	*2008*	*Years Occupied*
Hilgard	X	X	X	X	X	X	X	X	X	X	10
Gallatin	X	X	X	X	X	X	X	X	X	X	10
Hellroaring/Bear	X	X	X	X	X	—	X	X	—	X	8
Boulder/Slough	X	X	X	X	X	X	X	—	X	X	9
Lamar	X	X	X	X	X	X	X	X	X	X	10
Crandall/Sunlight	X	X	X	X	X	X	X	X	X	X	10
Shoshone	X	X	X	X	X	X	X	X	X	X	10
Pelican/Clear	X	X	X	X	X	X	X	X	X	X	10
Washburn	X	X	X	X	X	X	X	X	X	X	10
Firehole/Hayden	X	X	X	X	X	X	X	X	X	X	10
Madison	X	X	X	X	—	X	X	X	X	X	9
Henry's Lake	—	X	X	X	—	X	X	X	X	X	8
Plateau	X	X	X	X	X	X	X	—	X	X	9
Two Ocean/Lake	X	X	X	X	X	X	X	X	X	X	10
Thorofare	X	X	X	X	X	X	X	X	X	X	10
South Absaroka	X	X	X	X	X	X	X	X	X	X	10
Buffalo/Spread Creek	X	X	X	X	X	X	X	X	X	X	10
Bechler/Teton	X	X	X	X	X	X	X	X	X	X	10
Total	17	18	18	18	16	17	18	16	17	18	

Source: Podruzny 2009.

> *The department shall consider the claims based upon a description of the livestock or bees damaged or killed by a trophy game animal. ... Claims shall be investigated by the department and rejected or ... paid in the amount determined to be due. In the event the department fails to act within ninety (90) days, the claim, including interest based on local bank preferred rates, shall be deemed to have been allowed.*[28]

The WGF, for financial and philosophical reasons, has always tried to manage trophy game or big game numbers at biologically sustainable population objectives that also consider the level of damage caused and compensation required. The WGF is a self-funded state agency that receives less than 10 percent of its

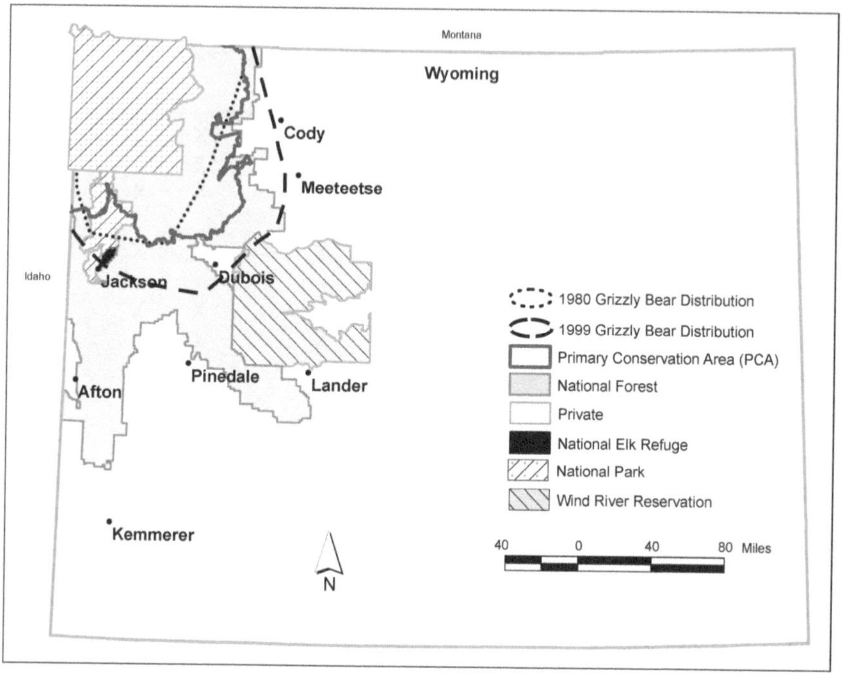

FIGURE 11-2. Comparison of Grizzly Bear Distribution in Wyoming, 1980 and 1999, Based on Radio-collared Bears

Source: Moody et al. 2005.

budget from general appropriations from the state treasury. The annual budget of the agency has ranged from $50 million to $70 million over the last decade. About 60 percent of WGF's budget comes from the sale of hunting licenses, 21 percent from federal grants including State Wildlife Grants, and Pittman-Robertson and Dingell-Johnson funds, three percent from interest on its Trust Fund, up to eight percent since 2004 from state general fund appropriations, and the remainder from other federal and state grant programs. Neither Idaho nor Montana has such onerous statutory provisions to compensate landowners for losses deriving from wildlife.

In 1990, WGF created a Grizzly Bear Conflict Management Program to address the growing number of human–grizzly conflicts. In 1990 there were 24 conflict situations, which grew to more than 200 by 2002. Between 1992 and 2005, an average of 134 grizzly–human conflicts per year occurred in Wyoming. During the same time period, grizzly–human conflicts grew approximately 13 percent a year. Most conflicts involve property damage, human injury, human-caused bear deaths, livestock depredations, apiary damage, and unnatural aggression (see Table 11-7). Livestock losses and apiary damage averaged 55 incidents per year for the period 1992 to 2005. Since 1990, about

TABLE 11-7. Incidents of Grizzly–Human Conflicts, 1992–2007 and 2008

Type of Conflict	*1992–2007 Average + S.D.*	*2008*
Human injury	4 (+/– 3)	6
Property damage	20 (+/– 12)	38
Anthropogenic foods	56 (+/– 39)	72
Gardens/orchards	6 (+/– 5)	7
Beehives	3 (+/– 4)	0
Livestock depredations	51 (+/– 18)	67
Total	140 (+56)	190

Source: Gunther et al. 2009.

2.6 bears per year have been killed by the public, in self-defense, in Wyoming, and management removals and illegal losses have amounted to about one bear per year. These incidents are predicted to grow as bears expand their range—in particular to the mountain ranges south of the GYE, where livestock use is heavy. From 1993 to 2005, damage payments associated with livestock and apiary damage have averaged about $41,441 per year, ranging from $3,228 in 1993 to $99,092 in 2000, with a steady increase over the time period of $3,011 per year. The numbers in Table 11-7 reflect only confirmed conflicts; the true number of claims investigated was much larger. The FY02 budget for field operations of the Wyoming grizzly bear conflict management program was $277,282, excluding damage payments. The total budget expenses by WGF attributed to grizzly bear management range from $800,000 to $1,000,000 per year over the last five years.

An additional reason why Wyoming and other states are concerned with growing grizzly numbers concerns the potential restrictions that can ensue on backcountry hunting opportunities. Table 11-8 describes the number of elk hunters using the GYE. The majority of human-caused mortality in Wyoming over the last decade has resulted from hunter–grizzly encounters; however, the total human-caused mortality has still allowed population growth. The states also are concerned that fear of grizzly bears might either deter hunters from hunting the backcountry or that USFS could succumb to national pressures and impose hunting restrictions on National Forest lands occupied by grizzly bears.

Scrimmage to Delist the Grizzly in the GYE

Through much of the 1990s the state wildlife agencies—WGF in particular—were increasingly frustrated with what they perceived to be delays by USFWS in initiating the delisting process for grizzly bear in the GYE. USFWS, in return, argued that even though the demographic goals had been met, a comprehensive Conservation Strategy, including particular habitat criteria, needed

TABLE 11-8. Estimated Number of Elk Hunters in the GYE Recovery Zone Plus a 10-mile Perimeter in Idaho, Montana, and Wyoming, 2000–2008

State	2000	2001	2002	2003	2004	2005	2006	2007	2008
ID	*	2,914	3,262	3,285	3,454	3,619	3,016	2,592	1,763
MT	17,329	15,407	17,908	16,489	14,320	12,365	12,211	12,635	*
WY	12,812	13,591	13,709	11,771	10,828	9,888	9,346	8,716	8,792
Total	*	31,912	34,879	31,905	28,602	25,872	24,573	23,943	*

Source: Moody et al. 2009. * = Total not available

to be developed prior to moving forward with any proposal to delist grizzly bear in the GYE.[29] In 1994, the IGBC made a proposal to move ahead with developing a Conservation Strategy as a first step in delisting the grizzly bear in the GYE.[30] After six years, the Recovery Coordinator released a Draft Conservation Strategy. In March 2000 it was collaboratively developed with state agency input for public review and comment.

Governors Appoint a Citizen Roundtable

The directors of IFG, MFWP, and WGF—increasingly frustrated by having to deal with the consequences of a listed expanding grizzly population, and dissatisfied by the progress of the Conservation Strategy—asked their respective governors to appoint a citizen roundtable to review the Draft Conservation Strategy and develop consensus recommendations constituting a tri-state response to the Draft Strategy. In response, the three governors convened a 15-member citizen roundtable. The biggest concern weighing on most state residents' minds was the uncertainty related to the management of grizzly bears outside the GYE recovery zone (referred to as the Primary Conservation Area (PCA) in the Conservation Strategy). Environmentalists were afraid that bears would be limited to the PCA, while the other interests feared increases in human–bear conflicts and severe land-use restrictions expanding with the expanding bear population.

The roundtable's recommendations were greeted by most as balanced in accommodating all interests. State wildlife agency executives regard the roundtable as one of the most significant events influencing state and federal collaboration in grizzly conservation in the GYE. The success of the roundtable was attributed to two primary factors: 1) the timing was right, as all factions—tired of fighting and delays—were ready to talk; and 2) the governors were engaged and took the exercise seriously in appointing responsible and thoughtful individuals, representing major stakeholder interests in the state.

The Governors' Roundtable, upon deliberation, made the following unanimous recommendations to the Draft Conservation Strategy:

- **Conservation Strategy and PCA**
  - Endorsed the cautious approach focusing on the PCA
  - Asked for explanation of how adaptive management will be applied and how the Conservation Strategy may be amended in the future
  - If data demonstrate flourishing bear populations outside the PCA, then more flexible management should be considered within the PCA, provided its basic objectives are being met
  - Eliminate the 10-mile buffer zone and leave monitoring and management outside the PCA to be carried out by state wildlife agencies per their grizzly management plans; agency actions described in the strategy should be based on monitoring results from within the PCA and should not include data from outside areas, unless state management plans so require
- **State Grizzly Bear Management Plans**
  - Recommended that states initiate grizzly bear management plans for areas outside the PCA. The state planning efforts should be concurrent with the revision of the draft Conservation Strategy and be harmonized with the federal planning process. In particular, the state plans should be developed, with public input, to:
    - Ensure long-term viability of grizzly bears and preclude relisting
    - Support expansion of grizzly bears beyond the PCA to areas that are biologically suitable and socially acceptable
    - Manage grizzly bears as a game animal, allowing regulated hunting when and where possible
- **Funding**
  - Short-term
    - Continue state funding for grizzly bear management and make state management plans a priority
    - Reevaluate state costs associated with bear management by including local government needs
    - Continue federal funding for bear management within the PCA
  - Long-term
    - Secure long-term funding through either an Endowment for Grizzly Management or Conservation and Reinvestment Act or other congressional appropriations
    - Develop funding for livestock depredation compensation
- **Citizen Involvement**
  - Add a nonvoting citizen member per state to the Yellowstone Grizzly Management entity responsible for coordinating the implementation of the Conservation Strategy
- **Montana Legislature**
  - Revise Montana law with regard to taking of grizzly bears for potential livestock depredation in a manner that eliminates the barrier to delisting

    - When the grizzly bear is delisted, it should be added to species in Montana State Law 87-1-115, providing enhanced protection and significant fines under Montana law
- **Additional recommendations**
    - Establish and fund an interagency working group
    - Provide citizens with a clear, understandable version of the plan and process
    - Clarify and define nuisance bear policies.

The roundtable recommendations extended political legitimacy to both state and USFWS perspectives. The IGBC and USFWS agreed to incorporate most of the suggestions made by the Governor's Roundtable. In 2001, the IGBC directed YES to develop a plan to complete the final Conservation Strategy using an accelerated state-led process. A new facilitated process to finalize the Conservation Strategy, led by a working group composed of YES members, was set in motion in early 2002. At the same time, the three states developed their state grizzly bear management plans, with the objective of incorporating them into the Conservation Strategy. MFWP and WGF funded the new facilitated process with both state and federal grant funds.

The YES working group delivered the finished Conservation Strategy to the IGBC for review in November 2002. After minor modification, the Final Grizzly Bear Conservation Strategy was signed by all relevant agency representatives on March 26, 2003.[31] The signatories to the Conservation Strategy signed a memorandum of understanding agreeing to "use their authorities to maintain and enhance the recovered status of the grizzly bear in the Greater Yellowstone Area by implementing the regulatory mechanisms, interagency cooperation, population and habitat management and monitoring, and other provisions of the Conservation Strategy. ..." (U.S. FWS 2003)

Grizzly Bear Conservation Strategy for the GYE

The Conservation Strategy adopted in 2003, like the 1993 Recovery Plan, is meant to be an "umbrella" document guiding grizzly bear conservation in the GYE, after the bear is taken off the threatened list. While the grizzly was on the threatened list, management guidelines articulated in the 1993 Recovery Plan applied. The Strategy contains population and habitat standards for bear management within the PCA (Figure 11-3). The bear management standards for the area outside the PCA are described in the individual state bear management plans that are appended to the Conservation Strategy.

The stated purpose of the Conservation Strategy and the attached state plans is to:

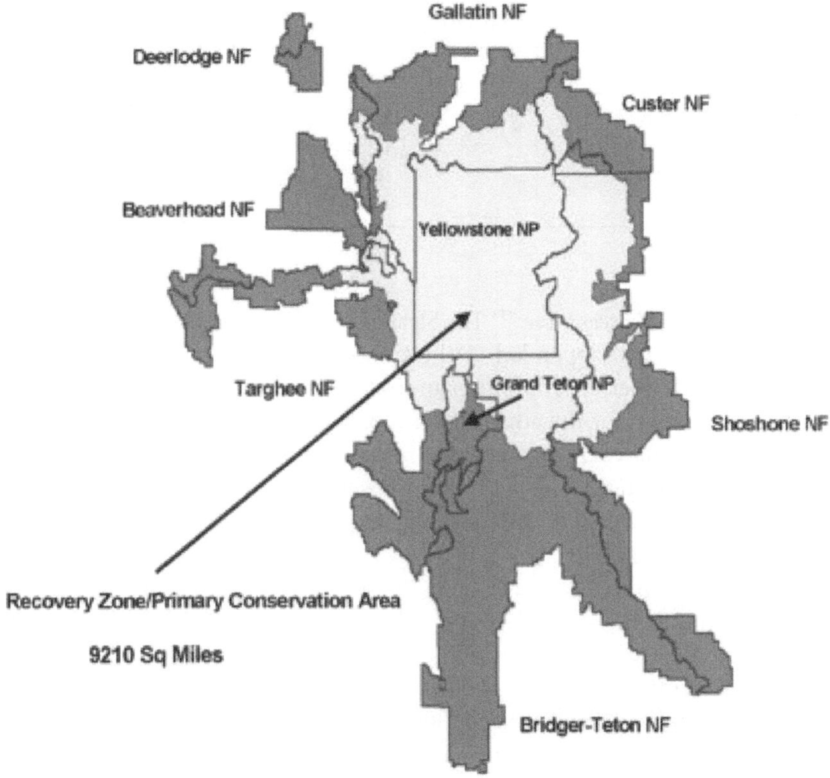

FIGURE 11-3. Recovery Zone/Primary Conservation Area (PCA) within the Greater Yellowstone Ecosystem

- Describe and summarize the coordinated efforts to manage the grizzly bear population and its habitat to ensure continued conservation in the Greater Yellowstone Area (GYA[32]);
- Specify the population, habitat, and nuisance bear standards to maintain a recovered grizzly bear population for the foreseeable future;
- Document the regulatory mechanisms and legal authorities, policies, management, and monitoring programs that exist to maintain the recovered grizzly bear population; and
- Document the commitment of the participating agencies.

Primary Conservation Area (PCA)

The management approach of the Conservation Strategy identifies a PCA composed predominantly of federal lands (see Table 11-9), replacing the recovery zone boundary and "*adjacent areas where occupancy by grizzly bears*

TABLE 11-9. Jurisdictional Lands within the PCA

Jurisdiction	*Area (square miles)*	*Percent of the PCA*
National Park Service	3,632	39.4
U.S. Forest Service	5,383	58.5
Private and other	195	2.1
Total	9,210	100

is anticipated and acceptable."[33] The Strategy anticipates that state grizzly bear management plans (appended to the Conservation Strategy), National Forest Plans, and other appropriate planning documents provide specific management direction for the adjacent areas outside the PCA. Consequently, the relevant state and federal agencies are responsible for evaluating the effectiveness of the measures described in the Conservation Strategy by conducting a thorough five-year review after delisting and implementation of the Strategy.

Population Criteria

The stated demographic goal of the Conservation Strategy is to manage the Yellowstone grizzly population for 500 or more grizzly bears. The population criteria incorporated in the Conservation Strategy reflects the compromises agreed to among the working group assigned by the IGBC to draft the document. The standards described in the Conservation Strategy apply to a much larger area than the ones prescribed in the Recovery Plan. Where the Recovery Plan focused on the recovery zone plus a 10-mile perimeter, the Conservation Strategy practically covers the entire GYA. The strategy confirmed the concerns raised by state officials in their comments on the 1993 Recovery Plan, when it was stated that the definition of a recovery zone was susceptible to an expanding grizzly bear range.

It is the specific intent of the Conservation Strategy to allow grizzlies to expand their range and numbers to occupy all biologically suitable habitats where grizzly bears are also socially acceptable. The Conservation Strategy maintained the allowable mortality threshold at the conservative level of four percent in the PCA. Female mortality is set at 30 percent of the permitted levels of adult mortalities (four percent), or at 1.2 percent of the *total* grizzly population. The WGF argued to increase the mortality threshold to six percent or higher, as it was concerned that the conservative threshold, once again, ensures a growing grizzly population in socially unacceptable areas, presenting the alarming specter of increased human–bear conflicts. The WGF did not prevail at setting a higher mortality threshold and instead settled for procedural concessions. The primary concession pertained to a provision stating that mortality thresholds would be reviewed and revised in the future based on the best science, including population trends. The standards for human-caused grizzly mortality are to be applied to all areas occupied by grizzly bears within each state's designated grizzly bear management areas (data analysis units), while the standard for spatial distribution of females with young is applied only to the PCA. Unlike the Recovery Plan, the Conservation Strategy calls for using a population monitoring methodology that allows for estimating the total rather than the minimum population of the grizzly bear in the GYA. The total population estimate—and not the minimum population estimate—is then used to calculate mortality limits for both adult grizzlies and adult female grizzlies. The strategy states that the mortality threshold will not affect the immediate management of nuisance grizzly bears, as the set limits could be exceeded in any given year. The Conservation Strategy grants WGF the flexibility to remove grizzly bears outside the Wyoming grizzly bear management area and not have them counted against the mortality threshold.

In addition, the state wildlife agencies agreed to maintain a minimum of 25 adult female grizzly bears fitted with mortality-sensing radio collars for continual monitoring to constantly assess survivorship. To maintain an adequate level of genetic diversity in the population, the Strategy requires that the grizzly bear population not be allowed to fall below 400 bears. If the genetic condition of the Yellowstone population worsens, and no movements from other grizzly populations are detected by 2020, then the Strategy recommends translocating two grizzly bears per generation from other populations into the GYA.

In the end, the states decided to go along with the population parameters stated in the Conservation Strategy with the hope of adopting more flexible management criteria based on monitoring and additional science when the states had authority over bear management rather than taking a rigid stance at the time and delaying delisting further. Without delisting the states were confronted with an ever-increasing grizzly bear population protected by ESA.

Based on a thorough review of grizzly population data, most recent science, and trends completed by the IGBST in 2005, the draft delisting rule included revised annual mortality thresholds of nine percent of the total population estimate of adult females and 15 percent of the adult males. The review also pointed out that the new thresholds are probably still conservative and will need to be reviewed as new data is collected.

Habitat Criteria

The Conservation Strategy calls for grizzly habitat conditions on federal lands in the GYA to be maintained at or improved from the conditions existing in 1998. In 1998, grizzly bear demographic goals for the GYA were accomplished for the first time. Habitat conditions were determined for each BMU in the GYA.[34] The particular habitat standards described in the Conservation Strategy include:

- "Maintenance of secure habitat at 1998 levels in each BMU subunit through management of motorized access route building and density), with short-term deviations allowed under specific conditions"[35] (U.S. FWS 2003)
- "The number of commercial livestock allotments and number of permitted domesticated sheep will not exceed 1998 levels inside the PCA"[36] (U.S. FWS 2003)
- Management of developed sites at 1998 levels within each BMU.

The Conservation Strategy calls for monitoring several habitat parameters,[37] including four major food sources in the GYA: winter ungulate carcasses, spawning cutthroat trout, army cutworm moths, and whitebark pine cones. These food sources constitute the highest digestible energy available to the grizzly bear in the GYA. The availability of high-quality food sources exerts a strong influence on grizzly bear fecundity and survival. Characteristically, the four food sources are susceptible to wide annual fluctuations. When these food sources are abundant, there is a discernible drop in grizzly–human conflicts in the GYA. Yellowstone grizzlies are unique among grizzly populations in North America in relying on wild ungulates as a major food source.[38] Usually grizzly bears in Yellowstone feed on winter-killed ungulates from March to May—the availability of winter ungulate carcasses fluctuates with winter severity. The competition with the growing wolf population in the GYA and consequent changes in ungulate population could have an effect on carcass availability to grizzly bears. Grizzly bears are also known to feed on spawning cutthroat trout in at least 36 different stream tributaries of Yellowstone Lake. In 2002, between 69 and 85 bears were documented to frequent cutthroat spawning streams. The native cutthroat trout populations in YNP, however, are presently vulnerable to the voracious appetite of the introduced lake trout. Moth aggregations are also an important food source for grizzlies in Yellowstone and contain the highest caloric content per gram of any other bear food. Moths are usually found on remote talus slopes during late summer and early fall, and when available in good quantities, bears may feed on them exclusively. On a good moth site, as many as 55 bears have been observed feeding at the same time, with high numbers of grizzlies using the sites in a given year (see Table 11-10). Whitebark pine cones, due to their high fat content, are an important fall food source for bears getting ready for winter hibernation. In a good production year (defined as more than 22 cones per tree), the bears may feed almost exclusively on pine seeds in late fall. Pine blister rust and pine bark beetles, however, are affecting stands of whitebark pine in GYA and present growing concern.

Yellowstone Grizzly Coordinating Committee (YGCC)

With the delisting of the grizzly bear on March 29, 2007, the responsibility of coordinating grizzly bear management among various state and federal

TABLE 11-10. Availability and Use of Moth Sites in GYA, 1986–2008

Year	*Confirmed Moth Sites*	*Moth Sites Used by Grizzly Bears*	*Number of Grizzlies Located/Observed*
1986	3	2	10
1987	5	4	15
1988	5	3	43
1989	10	9	51
1990	14	11	84
1991	17	14	176
1992	19	13	107
1993	19	2	2
1994	22	12	29
1995	25	11	42
1996	26	14	86
1997	28	19	95
1998	30	22	182
1999	30	17	177
2000	30	14	127
2001	31	18	141
2002	31	23	272
2003	32	26	167
2004	32	21	132
2005	33	20	190
2006	34	19	174
2007	35	24	174
2008	35	26	213

Source: Bjornlie and Haroldson 2009.

agencies transferred from the Yellowstone Ecosystem Grizzly Bear Subcommittee (YES) under the IGBC[39] to the new Yellowstone Grizzly Coordinating Committee (YGCC). The YGCC began meeting within a few months of delisting and meets twice annually. The YGCC has no authority beyond the specific actions agreed to by the signatories of the Conservation Strategy, and as such, does not in any way supersede any management agency's authority. Its meetings are open to the public. The YGCC membership is described in Table 11-11 and includes two significant additions over YES—representatives from local and tribal governments, as recommended by the Governors' Roundtable.

The YGCC coordinates the implementation of the Final Conservation Strategy for the Grizzly Bear in the GYA as well as the functions of three teams, including the IGBST, the Habitat Modeling Team in charge of the Cumulative Effects Model (CEM), and the Information and Education Team (I&E). The YGCC facilitates the sharing of information and coordination of management actions and informs the IGBC about the GYA grizzly bear population in an annual report provided to IGBC at their winter meeting. The YGCC will also identify shortfalls in needs and pursue the financial resources and budgets to

TABLE 11-11. YGCC Membership

Federal	NPS: Yellowstone and Teton USFS: Beaverhead-Deerlodge NF, Bridger-Teton NF, Caribou-Targhee NF, Custer NF, Gallatin, and Shoshone NF BLM: one representative each from ID, MT, and WY USGS: one representative from Biological Resources
State wildlife agencies	Idaho, Montana, and Wyoming
Local government	One representative from each state
Tribal	One representative from the Shoshone Bannock and Eastern Shoshone Tribes.

successfully implement the coordinated Conservation Strategy. If the YGCC observes deviations from population or habitat standards stipulated in the Conservation Strategy, it can authorize a biology and monitoring review. Any YGCC member agency can also request such a review. The review might be commenced to investigate a failure to meet particular demographic or habitat objectives, or a departure from any agency monitoring effort described in the Conservation Strategy. The review is required to consider and establish a scientific basis for possible changes in management, due to changed conditions in the ecosystem. The IGBST is responsible for conducting the review and submitting it to the YGCC for consideration. If the YGCC determines that desired population and habitat standards agreed upon in the Conservation Strategy are not being met and cannot be met, then the YGCC can petition USFWS to list the grizzly bear under the ESA.[40] USFWS, in undertaking a status review, will consider the entire GYA grizzly bear population. Moreover, any person armed with credible information may petition USFWS for a status review, and if the Service determines that the grizzly bear is in severe and immediate need of protection, it may utilize the emergency provisions of ESA section 4(b)(7).[41]

The Conservation Strategy represents a cooperative management plan among three state wildlife agencies and several federal land management agencies. With the bear off the threatened list, USFWS retains no regulatory role in conservation of grizzly bears in the GYA. It is understood that USFWS will maintain the role of an advisor but have no say in management decisions during the five-year post-delisting monitoring period. State agencies still have some concern about the potential bureaucratic and cumbersome structure in place to manage the recovered population. Furthermore, each state wildlife agency represents only one vote on the YGCC, and at the present time, there is no sunset provision for the committee. Accordingly, states are still concerned about how much influence the YGCC will have, permanently, on a state's authority to manage its native resident wildlife population of grizzly bears.

Present State of Affairs

The state wildlife and federal land management agencies assumed their regulatory responsibilities to manage the delisted grizzly population and habitat in a coordinated manner through the YGCC on April 29, 2007. Despite coordination and responsible management by the three states, on September 21, 2009 the Montana District Court issued an order vacating the USFWS Final Rule and remanding it back to the USFWS on two counts: 1) the regulatory mechanisms to protect the grizzly once it is delisted are inadequate, and 2) USFWS did not adequately consider the impacts of global warming and other factors on whitebark pine nuts. Five appeals to the Judge's order have been filed, including one from the USFWS and State of Wyoming.

A rather remarkable, exhaustive, complex, collaborative effort by all affected state and federal agencies to recover grizzly bears in the GYE and implement an agreed upon conservation plan has now entered into what appears to be a lingering legal morass. The threat of such legal actions upon species recovery stands to undermine the practical application of species recovery under ESA and the willingness of states to engage in recovery efforts.

From Threatened to Recovery to Litigation: Observations

The conservation of the Yellowstone grizzly bear population arguably represents a fine example of state and federal collaboration in recovering a wildlife population, which, 30 years ago, was deemed in threat of extinction. However, the nature of the collaboration, which was never comfortable, is best described as complex, and it fluctuated over time. Furthermore, the experience of state and federal agencies with grizzly bear conservation influenced the interaction between the two at a much broader level. A review of the grizzly bear recovery efforts in the GYA illuminates broad themes, discussed below, that have general applicability to state and federal collaboration for species conservation.

Historic Range and Present Dilemma

Prior to the listing of the grizzly bear as threatened in 1975, little was known about grizzly populations and their habitat requirements in the lower 48 states. Through the course of the twentieth century, grizzly bears were exterminated in several western states. By the 1970s only two major grizzly populations—restricted to the wild spaces around Yellowstone and Glacier National Parks—remained in the lower 48 states. At the same time, the grizzly populations north in British Columbia were described as healthy (Hatter 1958). Montana records and harvest reports indicated a stable grizzly bear population in the NCDE since the 1950s. In 1974, while USFWS was reviewing the petition to list the grizzly bear under the ESA, the National Academy of Sciences report found no convincing evidence of immediate danger of extinction for GYA grizzly bears. Thus, in the mid-1970s, USFWS was presented with a wildlife species that was restricted to less than two percent of its historic range; the two remnant populations, however, were arguably not threatened with extinction. This oft-repeated scenario presents a case of the historic range present occupancy dilemma that demands careful treatment. In administering the ESA, USFWS frequently confronts this historic range present occupancy problem, as demonstrated in the case of the Karner blue butterfly, black-tailed prairie dog, sage grouse, and others.

USFWS should enact policy guidelines to address the historic range present occupancy quandary in a consistent manner. The reduction in historic range is often considered under the habitat destruction element of section 4 of the ESA in assessing whether a species warrants a protective designation under the Act. The Act does not explicitly prescribe that reduction in historic range be a major factor in a listing decision. Most wide-ranging wildlife species have lost a significant amount of their historic range, due to human encroachment and settlement over the course of the last two centuries. Several historically wide-ranging species—such as the grizzly bear, gray wolf, Karner blue butterfly, and black-tailed prairie dogs—can be extirpated from more than 90 percent of their historic range and still maintain pockets of viable populations in presently occupied habitats.

The Service might find it helpful to consider its conservation goals when addressing wildlife species that demonstrate the historic range present occupancy dilemma. The conservation goals for such species might differ based on their ecology and associated social factors. The conservation objectives could rest on a spectrum between just ensuring long-term viability of existing populations and facilitating re-colonization of the species in parts of its historic range. The former goal might call for strong and rigid regulatory protection, while the latter could be better served by a more nuanced regulatory regime. The Service has tried the model of protecting existing populations with rigid regulations and introduced experimental populations under a more flexible regime. The failed attempt to introduce grizzly bears in the Bitterroot Mountains of Montana and Idaho, in large measure, can be described as a rebuke of that approach. The momentary regulatory distinction failed to engender the requisite trust and credibility called for by the highly charged situation.

Furthermore, a policy that placed a disproportionate regulatory burden on states that still support remnant populations of a species that has lost most of its historic range is both ecologically and politically unsound. Such misguided policies can dampen the most robust and sincere state efforts to proactively address species conservation. The Service's willingness to designate geographical exclusions in listing decisions would offer a welcome incentive to states with rare species that are imperiled in other regions of the country to develop appropriate conservation strategies. Similarly, a USFWS policy providing for geographical exclusions in delisting a species would extend a strong incentive for concerted action on the part of states to recover the imperiled population. The Service is well positioned to innovatively use the flexibility granted to the Secretary of the Interior to protect threatened species and assign distinct population segments, or significant portions of a species range, to better address the historic range present occupancy problem in a consistent manner.[42]

Grizzly Recovery: Need for Participatory Planning and Management

Recovery of a listed species is best achieved with the willing participation of state and federal wildlife agencies on a balanced plane, with each contributing according to their respective competitive advantage. The cooperation between WGF and the IGBST on grizzly research in the GYA represents a good example of enthusiastic and close collaboration. Interestingly, the executive leadership of WGF often exhibited a curious level of dissonance over the direction of certain elements of recommended management. If close state–federal collaboration is desired, the state agency must have at least an equal role in drafting the Recovery Plan and policy-level decisionmakers must be engaged. An equal role in developing a Recovery Plan is significantly different from an opportunity to comment on the draft of a plan. In response to public dissatisfaction with some components of the Conservation Strategy, the state wildlife agencies invoked the offices of their respective governors to obtain citizen comment on the proposed strategy. The IGBC then appointed a working group of state and federal agency personnel to incorporate citizen comments into the Conservation Strategy. The grizzly recovery effort in the GYA, at least since 1993, suffered from various levels of resentment and lack of trust that derived from a perceived uneven relationship.

Recovery goals are best defined in clear, measurable terms to ensure the clarity, credibility, and consistency essential to the success of any participatory endeavor. These recovery goals, to the extent possible, should be developed through a participatory process involving the concerned agencies and publics. Some wildlife species do not lend themselves to specific recovery goals early in the process, as requisite biological information on the species could be lacking. In such situations, a participatory process involving state and federal agencies and the public can play a constructive role in helping all involved agree on recovery goal modifications in light of new information as it arrives. In the absence of such a process, any future changes, however legitimate, are often viewed by the states as arbitrary and unwarranted, leading to lack of trust and credibility among the agencies and publics involved. Such was the case when USFWS—signed by the court order in *Fund for Animals v. Babbitt, 1995*, questioning the adequacy of the 1993 Recovery Plan—decided to add specific habitat criteria as a condition for recovery of the Yellowstone grizzly population.

Ever-moving recovery goals for large predators, or wildlife species that carry with them significant land-use restrictions, are particularly egregious to state agencies that are responsible for addressing the impacts of an expanding population without having any say over the recovery process. Local hostility in rural areas to restrictions associated with threatened or endangered species can be highly charged. Such hostility is usually directed to the state wildlife agency, as it is more accessible, whether that agency has any regulatory authority or discretion in the matter or not. Therefore, state and local governments often advocate for a "no surprises" policy to apply to cooperative agreements between state and federal public agencies.

Ideally, a Recovery Plan should incorporate all factors that would lead to recovery and delisting of a concerned species. A Recovery Plan is best developed by a Recovery Team constituted of all affected agencies and, if need be, submitted to an independent and professional review. When a species is listed, the Service should develop specific deadlines for developing and updating a Recovery Plan. The plan should clearly state jurisdictional responsibilities and specific population and habitat objectives, supported by documented and credible justifications. If possible, the plan should clearly articulate interim management strategies and a range of recovery approaches that can be modified as more information becomes available. In any case, the plan should specifically account for progressive steps towards recovery and delisting, along with associated regulatory flexibility as those steps are met. The progress on a Recovery Plan should be publicly reported. Furthermore, a Recovery Plan should be allowed to function as an umbrella document, whereby management actions in conformity with the plan carry the presumption of approval, including section 7 consultations, approval of habitat conservation plans (HCPs) and safe harbor agreements (SHAs), and so on.

State wildlife agencies are more willing and able to contribute to the recovery of a threatened or endangered species if needed funds are made available to do so. Traditional section 6 grants are notoriously deficient in meeting the conservation costs of listed species. WGF, on average, received $10,000 each year for its cooperation with grizzly research and management in the GYA, amounting to less than two percent of its annual cost related to grizzly bears. A stable source of funding tied to accomplishing mutually agreed conservation tasks would further facilitate state and federal collaboration in species conservation.

Grizzly Recovery: Research Based Management

Yellowstone grizzlies have the distinction of being the most studied grizzly population in the world. Through the 1970s, there was widespread recognition of the need to further research grizzly bear biology in the GYA to better inform appropriate conservation measures. The newly created IGBST took to the task with purpose and has never wavered in its enthusiasm to study the bear and its habitat. It was the responsibility of the IGBC to pose and prioritize research questions needed to inform the management decisions of the various agencies involved. In absence of consistent and sustained direction from the IGBC, research topics were decided by lead researchers in the IGBST and the Recovery Coordinator. In such situations, a possibility for conflict of interest arises, and "more research is needed" becomes the standard response to any new management proposal. State wildlife agencies involved in the recovery of Yellowstone grizzly bears often complained about the "more research is needed" moniker as a wasteful, self-fulfilling prophecy. The states argued that the marginal benefits of some research for improving management decisions were slight and, in absence of persuasive arguments to the contrary, perceived the "more research is needed" response as a strategy to delaying the delisting process.[43] Furthermore, disproportionate focus and investment on researching one species takes away limited resources from addressing basic research needs of other species.

Line officers of state and federal agencies with policymaking authority, after assessing available scientific information, should determine the appropriate recovery criteria. The appropriate role for agency biologists or staff officers is to present the available scientific information, accompanied by their recommendations. The recovery criteria for Yellowstone grizzlies reflects a conservative approach beyond what available scientific information supported, for example, the four percent mortality threshold based on a minimum, not total, population estimate. Such judgment calls carry more credibility and acceptance when made by the line officers whose perspective allows for a richer, nuanced understanding of the trade-offs involved in the decision. In contrast, a single species biologist—while perhaps most knowledgeable about the biology of the species—lacks the benefit of observing "the big picture."

The Role of the IGBC

Effective species conservation increasingly relies on coordinated action across multiple state and federal jurisdictions. The recovery of the Yellowstone grizzly population was achieved by coordinated action among federal land management agencies, primarily USFS and NPS, and the state wildlife agencies of Wyoming, Montana, and Idaho. The IGBC, a unique and rather elaborate structure, was created to coordinate grizzly recovery across jurisdictions. The concept of a supra-coordinating body such as the IGBC, in theory, was a sensible strategy. In practice, however, the committee did not consistently execute its expected role. The IGBC model has rarely been replicated; it was consciously rejected for being too cumbersome and bureaucratic in managing wolf reintroduction and recovery in the Yellowstone region. The primacy of YNP, and the interest in grizzly bears in part, explain the creation and existence of the high-profile IGBC.

However, there are important lessons to be learned by reviewing the role of the IGBC in recovering grizzly populations in the GYA. The IGBC provided institutional strength to the grizzly bear recovery efforts. The committee did provide a valuable platform to coordinate recovery efforts in a consistent manner across a wide and disparate landscape. The more effective elements of the IGBC may offer innovative institutional strategies to address increasingly complex issues surrounding species conservation.

With the growing number of species added to the threatened and endangered list, a single species conservation approach becomes increasingly untenable. Several of the key species of concern occupy an entire ecotype—for example, sage grouse in the sagebrush/grassland type, black-tailed prairie dog in the short-grass prairie type, etc. The pressing need to coordinate conservation of increasing numbers of imperiled species will only grow with time and human development. A standing Species Conservation Committee (SCC) for every state may serve a valuable role in coordinating state and federal collaboration toward more effective species conservation.

An SCC can be composed of all major state and federal land management and wildlife agencies, including USFS, BLM, NPS, USFWS, the governor's planning office, and state wildlife and land management departments. The member agencies should be represented in the committee by the appropriate line officers with the authority to make decisions. The committee can be created by either an MOU, as used for IGBC, or under the auspices of a section 6 cooperative agreement. The state wildlife department or governor's planning office represent the appropriate entities to support the workings of the committee. The committee could be an effective institution to evaluate petitions to list a particular species, in light of available information on the species and any conservation efforts to address its limiting factors. The state wildlife agency can play a leadership role on the committee in presenting best available information on the species status and suggesting appropriate conservation measures to address the challenges faced by the species. Similarly, in the context of the SCC, the state wildlife agency can request, and be granted, greater responsibility in designating critical habitat and conducting section 7 consultations. Further, the state wildlife agency is best placed to take the lead in facilitating SHAs and habitat conservation plans for listed species in the state. Thus, the SCC can offer an effective medium to states, if adequate resources are available and there is willingness to accept greater responsibility in species conservation. The committee offers an opportunity to the state wildlife agency to present its findings and recommended course of action to all concerned agencies and address their concerns, related to all species of interest, in a collective manner. Consequently, the committee structure encourages species conservation at the scale of an eco-region, focusing on a cluster of species rather than a single species. States with comprehensive state conservation plans will be at an advantage in persuading other committee members to follow their suggested course of action. States may use their state conservation plans to suggest broad habitat-based, landscape-level conservation agreements to conserve a host of species, including endangered, threatened, candidate, and common.

The SCC, as an entity, could facilitate coordination of jurisdictional authority between the state and federal agencies involved. By so doing, the committee can nurture a true partnership among member agencies committed to achieving the conservation goals underlying state and federal statutes. The SCC will be well placed to coordinate the application of not only state and federal endangered species acts, but also a host of other relevant statutes such as the National Forest Management Act (NFMA), the Federal Land Policy and Management Act (FLPMA), NEPA, etc. USFWS should retain its authority as the final arbiter of a species status. However, by actively participating in the SCC, USFWS might be able to persuade the states to take the necessary, proactive measures to preclude the listing of a given species. Additionally, the Service, by participating in the SCC, might find itself better prepared to deal with legal challenges under the ESA or the Administrative Procedures Act (APA).

The SCC, as an entity, could provide the collaborative vehicle for identifying shared support for appropriate species conservation efforts within a state and record the agreed-upon strategy under a legally binding State Conservation Agreement, as suggested by AFWA. Subsequently, it should submit state proposals for federal funding to implement the state conservation strategy. The Service may request that Congress make more funds available for traditional section 6 cooperative agreements. The committee should also publish periodic reports informing all interested publics on the progress being made in implementing the state species conservation strategy. One of the biggest benefits of the SCC might be to facilitate maturation of state wildlife departments in addressing conservation needs of all wildlife and plant species in the state.

The Role of Governors

The Governors' Roundtable played a crucial role in the acceptance of the Conservation Strategy by the people of Idaho, Montana, and Wyoming. The roundtable's recommendations, in many ways, were more balanced and reasonable than those presented by the state and federal wildlife agencies. Most important, the roundtable recommendations legitimized the Conservation Strategy, lending credibility to both state and federal considerations. The recommendations ameliorated some of the skepticism that wildlife professionals have for the layman's judgment in species conservation matters. The majority of the residents surveyed in the three states supported a recovered grizzly bear population in the GYA. The roundtable approach was suggested and is used most frequently by Montana. It is an effective tool to obtain the acceptance by the state residents of a controversial conservation measure. However, it is a tool to be used sparingly, due to the considerable political and economic investment involved. The success of a roundtable is closely related to the importance given to the process by the governor's office and the quality of individuals appointed to it.

Grizzly Conservation: Use of Threatened and Endangered Designation

The Service should more closely adhere to the congressional intent of providing two disparate levels of species protection, one greater than the other. The Service should establish a default rule of listing all warranted and not precluded species as threatened, unless they are found in immediate and imminent danger of extinction. The Service should utilize the threatened designation wherever appropriate and engage the states in developing appropriate section 4(d) rules and regulations to recover the species. The conservation needs of wide-ranging wildlife species—proposed for listing due to a precipitous decline in their historic range, but still boasting remnant viable populations—are best addressed through a separate category below a threatened designation or a warranted but precluded assessment. However, the Service should further strengthen and specify strict regulatory protections for species found in immediate danger of extinction.

Prompt Delisting When Appropriate

Species that have met the recovery goals should be taken off the list to ensure effective administration of the ESA. It is undisputed that there should be a greater urgency to granting regulatory protection to an imperiled species than to withdrawing protection from a species that no longer needs it. However, success, which is represented by a recovered population and the reward of delisting, carries a psychological benefit for all individuals involved that should not be underestimated. Moreover, delisting represents an accomplishment, and as such it engenders public support for a system that produces results and is able to demonstrate success. More important in the context of federalism, it builds much needed credibility and trust among the state and federal agencies by adhering to agreed-upon goals and procedures.

It is imperative that a Recovery Plan incorporates incremental conservation goals. When these goals are met, concurrent actions should be taken to downgrade the species protection to an eventual delisting. The Service should make greater use of downgrading species protection for distinct geographical populations and, by so doing, reward those states that have accomplished recovery goals—incremental or otherwise. The emphasis on sole consideration of best available scientific information should be consistently applied in both listing and delisting decisions. If the Service chooses to allocate burden of proof differently in the two cases, it should make that consideration transparent.

Grizzly Conservation: Spillover Effect into Gray Wolf Conservation

The benefits or fallouts from a joint state and federal conservation effort of such prominence as grizzly bear recovery in the GYA can affect the working relationship between the two agencies on a much broader level. The Wyoming Game and Fish Commission (WGFC) was soured by paying almost the entire cost of grizzly management in Wyoming while the grizzly bears were listed and by the fact that delisting took much longer than anticipated, when the commission believed the science clearly indicated the grizzly population in the GYA was recovered. As a result, the Commission turned down an invitation to participate in gray wolf management when invited to do so by the Service. The gray wolf was introduced to YNP and central Idaho in 1995 and 1996. The Recovery Plan asked for state management of the species after the translocated wolves had established themselves. All three states—Montana, Idaho, and Wyoming—agreed to decline the invitation initially, although Montana and Idaho later assumed management responsibility, under provisions of a 10(j) rule, prior to delisting of the wolves. Wyoming did not want to accept the cost and responsibility of managing another large carnivore, capable of causing even more damage to private property than the grizzly bear, with limited state authority, as long as the species remained listed. The WGFC did not want to accept management of the wolf until it was delisted—and then only in that portion of the state where gray wolves were classified as a trophy game animal—due to the unsuitable nature of habitat for wolves outside Northwest Wyoming and statutes requiring the WGFC to pay for damages caused by certain wildlife, such as wolves, that are not classified as predatory animals. The agriculture community also wanted maximum opportunity to manage wolf predation in the majority of the state. At the time wolves were released in YNP, Wyoming statutes classified the wolf as a predatory animal. In Wyoming, wildlife classified as predatory animals can be taken with few restrictions, and the WGFC is not liable for any damages they cause. Delisting in Wyoming required a reclassification of the wolf, thus making the WGFC liable for livestock damage caused by wolves classified as trophy game animals.

In the absence of state participation, USFWS was forced to manage the recovery program on its own. The approach taken by the Service in the case of wolf recovery was remarkably different from that of the grizzly bear. There was no high-level interagency coordinating entity for gray wolf conservation, and there was no interagency study team. The whole process was run on a significantly leaner budget in resources and manpower. One of the main reasons that allowed for the leaner approach is the biology of the gray wolf, which, unlike the grizzly bear, is a prolific breeder. In less than 10 years, wolves achieved the recovery goal of 30 packs distributed over the Northern Rocky Mountain (NRM) recovery area DPS. As of the end of 2009, there were 115 confirmed breeding pairs and 27 breeding pairs in Wyoming alone, including YNP. Total population estimate for NRM in 2009 was over 1,700 wolves.

The delisting process for the NRM gray wolf followed a similar approach as that of the grizzly bear, in that a state wolf management plan from each of the three states was required by USFWS before it proposed a status change in the NRM recovery area. The Wyoming Gray Wolf Management Plan calls for managing the recovered wolf population in the state under dual classification—gray wolves around the GYA are designated trophy game, and everywhere else in the state, the wolf is classified a predatory animal. The Wyoming legislature passed House Bill 299 in 2003, incorporating this dual classification into state law. The Wyoming Gray Wolf Management Plan, based on provisions contained in H.B. 229, was deemed adequate for sustaining a population of gray wolves in the NRM recovery area, in combination with the Idaho and Montana Plans, by 10 out of 11 nationally known wolf biologists asked by the Service to review the plan (the eleventh reviewer never responded). However, the designation of predatory animal for wolves in a portion of Wyoming, and the USFWS determination that Wyoming's plan had inadequate regulatory mechanisms to ensure a recovered wolf population would be sustained in Wyoming, caused USFWS to reject the Wyoming Plan and not move the delisting process forward. The USFWS contended that the Wyoming Plan needed to be revised to provide stronger assurances that 15 packs could be maintained in the state, in particular given the dual status designation. In April 2004, the State of Wyoming filed an appeal in U.S. Federal District Court, charging that USFWS's decision not to initiate the delisting process for the gray wolf in the NRM recovery area amounted to an arbitrary and capricious determination contrary to the law.

The April 2004 appeal was denied, based on the court ruling that the USFWS rejection of Wyoming's plan was not a final decision. In response, Wyoming filed a comprehensive delisting petition in July 2005. The 12-month finding by USFWS determined that delisting was not warranted, since Wyoming's plan did not provide adequate regulatory assurances that 15 packs could be maintained in the state. The State of Wyoming filed a lawsuit in October 2006 challenging this USFWS decision. Based on negotiations between the state and USFWS to resolve the impasse, the 2007 Wyoming state legislature passed H.B. 213, providing added flexibility for the state to establish a larger trophy game area for wolves in northwest Wyoming and provisions to address the pending lawsuit if a final revised 10j rule or memorandum of agreement with the State of Wyoming—allowing wolf control to benefit wild ungulates—and a final delisting rule were published no later than February 28, 2008. As a result of the passage of H.B. 213, the WGFC adopted a revised wolf management plan on November 16, 2007, which retained most components of the original plan but allowed for a larger trophy game area designation in northwest Wyoming. On December 12, 2007, USFWS accepted the revised plan as adequate and published a final delisting rule in the Federal Register on February 27, 2008. The final rule was slated to go into effect March 28, 2008, transferring management authority for gray wolves to Idaho, portions of Montana, and Wyoming, pending the outcome of any unfinished litigation. A notice of intent to sue USFWS for its delisting decision was filed February 27, 2008, by Earthjustice, on behalf of 12 plaintiffs. On July 18, 2008, U.S. District Judge Donald Molloy, District of Montana, enjoined the USFWS delisting rule finding that the USFWS has "acted arbitrarily in delisting the wolf despite a lack of evidence of genetic exchange between subpopulations; and it acted arbitrarily and capriciously when it approved Wyoming's 2007 plan despite the State's failure to commit to managing for 15 breeding pairs and the plan's malleable trophy game area.... In both instances the USFWS altered its earlier position without providing a reasoned decision for the change based on identified new information." The Judge remanded the rule to the USFWS. On April 2, 2009, USFWS issued a new delisting rule with the same DPS for the NRM population, delisting wolves within the DPS except wolves in the entire state of Wyoming. Legal action has been initiated in response to the April 2, 2009 rule by multiple parties on both sides of the delisting issue.

Conclusion

By the 1970s, the grizzly bear was poisoned, shot, trapped, and eliminated from all western states but a few. The grizzly populations in Montana and Wyoming had survived, largely because of available swaths of wild spaces and less so because of any affirmative action taken by the states. In the absence of the ESA, it is most likely that the prevalent political climate in Montana and Wyoming would have precluded the states from appropriately addressing the conservation needs of a large predator. The 1975 listing of the grizzly bear as threatened transformed it from being an elusive marauder of the mountains to the most heavily studied and protected large carnivore on the continent. Given the high profile of Yellowstone and Glacier National Parks and the grizzly bear itself, the full force of the ESA was enlisted to recover the species. The grizzly bear conservation efforts in the GYA represent perhaps the most elaborate research and management structure devoted to any one species in the nation. The result has been a stellar biological success. However, few if any wildlife professionals would recommend replicating the enormous effort invested as a necessary condition to recovering similarly placed species. The model for gray wolf recovery in the Northern Rockies—even after accounting for differences in the biology of the species concerned—presents an effective and less resource-intensive way of conserving species. Nonetheless, the wolf recovery efforts benefited greatly from the experience related to the conservation of the grizzly bear.

The ESA has been good to the grizzly bear in the GYA; the bears have thrived in the region since the late 1970s. One will never know, however, the precise role the ESA played in grizzly recovery in the GYA. The two main ingredients of grizzly recovery were reduction of human-caused mortality and availability of wilderness habitats. In retrospect, it could be argued that a cooperative management regime between the three state wildlife agencies, USFS, and NPS would have achieved a similar result. The proposed Conservation Strategy reflects such a cooperative arrangement. It is plausible that, were a petition to list the grizzly bear in the GYA proposed today, in the context of the limited information available in the 1970s, the concerned agencies might choose a more flexible and cooperative arrangement to conserve and recover the population. Were that to be the case, then perhaps one of the lasting legacies of the grizzly bear recovery in the GYA, under the ESA, could be the considerable ground covered by state and federal agencies to appreciate and commit their resources to preserving a species of concern before it was found under threat of extinction.

In the end, a major contribution of grizzly bear conservation in the GYA might lie in the paradigmatic shift it facilitated in the culture of state and federal wildlife and land management agencies. The grizzly bear reasserted the value of wilderness areas and the need to conserve large predators that play a key ecological role in the system. USFS was persuaded to direct high priority to the grizzly bear and wilderness values in planning its management activities in the national forests. The state wildlife agencies were persuaded to specifically manage for the grizzly bear and, by so doing, broaden their conservation responsibilities from game to all wildlife species in the state. USFWS learned the benefits of a streamlined participatory approach to species conservation by its handling of the gray wolf recovery in the region. The success of the grizzly bear conservation efforts in the GYA testifies to the need for close collaboration between state and federal agencies. The challenge ahead lies in achieving lasting success in species conservation at significantly less cost, both in money, goodwill and litigation. The nation's ability to meet this challenge will determine the efficacy of its efforts to conserve its imperiled species. In the final analysis, recovery of the grizzly bear, and the rich experience of coordinating species conservation across multiple jurisdictions, constitutes the legacy of grizzly recovery in the GYA.

References

Bjornlie, D. and M. A. Haroldson. 2009. "Grizzly Bear Use of Insect Aggregation Sites Documented from Aerial Telemetry and Observations." In *Yellowstone Grizzly Bear Investigations: Annual Report of the Interagency Grizzly Bear Study Team, 2008*, C. C. Schwartz, M. A. Haroldson, and K. West, eds., 32. Bozeman, Montana: U.S. Geological Survey.

Boyce M. S., B. M. Blanchard, R. R. Knight, and C. Servheen. 2001. "Population Viability for Grizzly Bears: A Critical Review." *Management Monograph Series Number 4.* International Association for Bear Research.

Corsi, Rex. 2004. Personal communication with former Wyoming Chief of Wildlife Division.

Dood, Arnold. 2004. Personal communication with Endangered Species Coordinator, Montana Department of Fish, Wildlife and Parks.

Gunther, K. A., B. Aber, M. T. Bruscino, S. L. Cain, K. Frey, M. A. Haroldson, and C. Schwartz. 2009. "Grizzly Bear–Human Conflicts in the Greater Yellowstone Ecosystem." In *Yellowstone Grizzly Bear Investigations: Annual Report of the Interagency Grizzly Bear Study Team, 2008*, C. C. Schwartz, M. A. Haroldson, and K. West, eds., 40. Bozeman, Montana: U.S. Geological Survey.

Haroldson, M. A. 2009. "Assessing Trend and Estimating Population Size from Counts of Unduplicated Females." In *Yellowstone Grizzly Bear Investigations: Annual Report of the Interagency Grizzly Bear Study Team, 2008*, C. C. Schwartz, M. A. Haroldson, and K. West, eds., Bozeman, Montana: U.S. Geological Survey. p. 9.

Haroldson, M. A., C. Dickinson, and D. Bjornlie. 2009. "Bear Monitoring and Population Trend." In *Yellowstone Grizzly Bear Investigations: Annual Report of the Interagency Grizzly Bear Study Team, 2008*, C. C. Schwartz, M. A. Haroldson, and K. West, eds., 9. Bozeman, Montana: U.S. Geological Survey.

Harris, R. 1985. *Results of a Workshop on Grizzly Bear Population Genetics*. Missoula, MT: U.S. Fish and Wildlife Service.

Hatter, J. 1958. Official letter from Mr. J. Hatter, Chief Game Biologist, Fish and Game Branch, British Columbia Department of Recreation and Conservation, to Robert F. Cooney, Chief, Game Management, Montana Department of the Fish and Game. July 22.

Honnold, D. 2004. Personal communication with lead attorney for Earthjustice.

Hunter, Gilbert N. 1958. Official letter from Gilbert N. Hunter, State Game Warden, Colorado Department of Game and Fish, to Robert F. Cooney, Chief, Game Management, Montana Department of the Fish and Game. July 23.

Leonard, Ross. 1958. Official letter from Mr. Ross Leonard, Director, Idaho Department of Fish and Game, to Robert F. Cooney, Chief, Game Management, Montana Department of the Fish and Game. July 23.

McArthur, K. L. 1979. "The Behavior of Grizzly Bears in Relation to People in the Glacier National Park—A Literature Review." *Montana Special Grizzly Bear Committee Report*. 1970. Draft copy, August 7.

Montana Special Grizzly Bear Committee Report. 1970.

Moody, D. S., D. Hammer, M. Bruscino, D. Bjornlie, R. Grogan, and B. Debolt. 2005. *Wyoming Grizzly Bear Management Plan*. Cheyenne, WY: Wyoming Game and Fish Department.

Moody, D. S., K. Frey, and D. Meints. 2009. "Trends in Elk Hunter Numbers within the Primary Conservation Area Plus the 10-Mile Perimeter Area." In *Yellowstone Grizzly Bear Investigations: Annual Report of the Interagency Grizzly Bear Study Team, 2008*, C. C. Schwartz, M. A. Haroldson, and K. West, eds.,39. Bozeman, Montana: U.S. Geological Survey.

NAS (National Academy of Sciences). 1974. *Report of Committee on the Yellowstone Grizzlies*. Division of Biological Sciences, Assembly of Life Sciences, National Research Council, 61. July.

Podruzny, S. 2009. "Occupancy of Bear Management Units (BMU) by Females with Young." In *Yellowstone Grizzly Bear Investigations: Annual Report of the Interagency Grizzly Bear Study Team, 2008*, C. C. Schwartz, M. A. Haroldson, and K. West, eds., 15. Bozeman, Montana: U.S. Geological Survey.

Salter, Robert L. 1974. *International Association's Ad hoc Committee on Grizzly Bear Report.*

Schwartz, C. C., M. A. Haroldson, K. Gunther, and D. Moody. 2002. "Distribution of Grizzly Bears in the Greater Yellowstone Ecosystem, 1990–2000." *Ursus* 13: 203–212.

Schwartz, C. C. and D. S. Moody. 2009. "History and Purpose of the Study Team." In *Yellowstone Grizzly Bear Investigations: Annual Report of the Interagency Grizzly Bear Study Team, 2008*, C. C. Schwartz, M. A. Haroldson, and K. West, eds., 2. Bozeman, Montana: U.S. Geological Survey.

Smith, Chris. 2004. Personal communication with Chris Smith, Chief of Staff, Montana Department of Fish, Wildlife and Parks.

U.S. FWS (U.S. Fish and Wildlife Service). 2003. *Final Conservation Strategy for the Grizzly Bear in the Yellowstone Ecosystem.* Missoula, MT: U.S. Fish and Wildlife Service.

Waits, L. P., D. Paetkau, C. Strobeck, and R. H. Ward. 1998. "A Comparison of Genetic Diversity in North American Brown Bears." *Ursus* 10: 307–314.

Western Watershed Project vs. Servheen.

Yeaman, Fred. 1958. Official letter from Fred R. Yeaman, Chief, Wildlife Development Division, Wyoming Game and Fish Commission, to Mr. Robert F. Cooney, Chief, Game Management, Montana Department of Fish and Game. July 23.

Endnotes

1 The natural history information of the grizzly bear is derived from the Grizzly Bear Recovery Plan, USFWS 1993; Final Conservation Strategy for the Grizzly Bear in the Yellowstone Ecosystem, USFWS March 2003; and annual reports of the Interagency Grizzly Bear Study Team and Wyoming Trophy Game Section.

2 There were unsubstantiated sightings of grizzly bears in the San Juan Mountains of Colorado until the 1970s.

3 While females with young comprise less than 20 percent of the total grizzly bear population, they perpetrated about 79 percent of human injuries during the period between 1970 and 1973 (McArthur 1979).

4 The 1899 Game and Fish Laws of Wyoming do not mention grizzly bears or their management. The 1903 Wyoming State Game Warden report alludes it to be a misdemeanor to hunt, kill, or trap grizzlies on National Forest Reserves in the state, except during open game seasons. In 1937 grizzly bears, in conjunction with black bears, were designated game animals in most National Forests and as predators in the remainder of the state. Usually an elk or deer license holder could take one bear of either species.

5

6 In the 1940s, Rob Cooney of the Montana Dept. of Fish and Game surveyed grizzly bear numbers in the North and South Forks of the Flathead River in northwest Montana. On finding diminished numbers of bear in the area, the department announced a moratorium on grizzly bear hunting season in the South Fork area, which remained in place from 1941 to 1954.

7 Information concerning the residency of the letter writers was not available. Consequently, one cannot assess how many of the letters originated in states with existing grizzly populations, in particular Montana and Wyoming.

8 ESA section 4.

9 In 1992 USFWS, in response to a court-ordered injunction, removed regulations set forth in 50 CFR 17.40 (I)(E) that allowed for the special grizzly hunt in the Bob Marshall ecosystem. The court found no rational finding of grizzly bear population pressure existing in NCDE in 1986 to justify the regulation (*The Fund for Animals v. Turner* 1991). In the court's interpretation, Congress limited the hunting of threatened or endangered species to extraordinary cases of population pressure, not whenever it is found to be a sound conservation tool. The court was persuaded by the holding in a Minnesota case, *Sierra Club v. Clark*, 1984, rejecting the rationale of reducing the level of wolf–human contact as a valid basis for authorizing a regulated take under the ESA. As a consequence of this court decision, USFWS withdrew a proposed regulation to allow state-administered hunting in the GYE to control nuisance bears.

10 For species listed after the 1978 amendments to the ESA, USFWS is required to designate critical habitat, unless it finds that the benefits of non-designation outweigh the benefits of designation (16 U.S.C. § 1533[b][2]). For species listed before 1978, such as the grizzly bear, "critical habitat may be established for those species … as set forth in subparagraph (A) of this paragraph" (16 U.S.C. § 1532[5][B]); the subparagraph defines critical habitat. USFWS has maintained that the decision to designate critical habitat for a species listed prior to the 1978 amendments is discretionary, and federal courts have agreed with the agency's interpretation (*Fund for Animals v. Babbitt* 1995).

11 Dr. Christopher Servheen has occupied the position since 1989.

12 The Research Committee was disbanded in 1998 after an organizational review of the IGBC.

13 Section 7 of ESA.

14 Management Situation (MS)1 applies to areas that contain grizzly population centers or habitat essential to the survival and recovery of the species; consequently, grizzly bear needs are given highest management priority. MS2 applies to areas other than grizzly population centers, where grizzlies may occur; here, grizzly conservation is an important but not the primary consideration for land-use decisions in the area. MS3 includes areas such as established campgrounds, highways, towns, and so on, where grizzly presence is possible but infrequent, and management focus is on minimizing grizzly–human conflict. MS4 includes areas where grizzlies do not occur now, but habitat and human conditions make the area potentially suitable for grizzly occupancy; consequently, management considerations include grizzly habitat maintenance and improvement. MS5 refers to areas where grizzlies do not occur and habitat is unsuitable; as the area lacks recovery values, grizzly considerations do not detract from other resources-related decisions. Here, a grizzly involved in a grizzly–human conflict will be controlled.

15 MFWP personnel contend that the grizzly population in northwestern Montana met the recovery criteria stated in the 1982 plan at the time it was published (Dood 2004).

16 In the early 1980s, an NPS internal memo by Roland Wauer insinuated that fewer than 200 grizzlies remained in YNP, and the number of adult females may have dropped to as low as 30. This memo elicited a flurry of activity, including the appointment of a full-time grizzly bear coordinator by USFWS. Through the 1980s, the state and federal agency personnel often held opposing views on the

population trend of GYE grizzlies. Some federal biologists, relying on emerging population modeling techniques, contended that the bear population in GYE was stagnant or declining. During the same time period, state field personnel were reporting grizzly sightings in areas where bears were not observed before, indicating an expanding population.

17 The broad steps outlined by the Recovery Plan included: a) identifying grizzly bear population goals that represent species recovery in measurable and quantifiable terms for the ecosystems where the grizzly bear has suitable habitat; b) providing a population monitoring approach that will allow determination of recovered levels; c) identifying population and habitat limiting factors that account for current populations existing at levels requiring threatened status under the Act; d) identifying management measures needed to remove population and habitat limiting factors so that populations will increase and sustain themselves at levels identified as the recovery goals; and e) establishing recovered populations in each of the ecosystems where habitat is available to sustain a grizzly bear population.

18 The 1982 plan called for specific population targets, including monitoring reproductive rate, average litter size, reproductive intervals, and annual total mortality, which are nearly impossible to collect, given the rugged terrain the bears inhabit.

19 MFWP employs its own monitoring system to track grizzly bear numbers. Its monitoring regime is outlined in the Montana Grizzly Bear EIS and Grizzly Bear Management Program.

20 The running six-year average is based on a three-year reproductive cycle, allowing at least two years when each adult female grizzly bear could be reported with cubs. Based on these numbers, it is assumed that, on average, 33 percent of adult females (at least five years old) will be with cubs each year.

21 Research data relied on in the Recovery Plan indicate average density of grizzly bears in the GYE of about 29 square miles/bear (76 square kilometers/bear).

22 When the age structure of the GYE grizzly population indicated that adult female grizzlies of cub-rearing age represent 28.40 percent of the population, the population estimate would be 15 × 3 = 45, which is 28.40 percent of 158.

23 Example: 1990–1992 (female with cubs) 24 + 24 + 23 = 71 − 4 (known adult female mortalities for the same period) = 67, which is 28.4 percent of 236. Allowable adult mortality: 236 × .04 = 9; allowable adult female mortality: 9 × .30 = 3.

24 The Plateau BMU and Henry's Lake BMU on the west-southwest border of the recovery area were initially deemed to contain low habitat value and, as such, a lower probability of grizzly bear sightings.

25 A bear is deemed a nuisance if any of these three conditions apply: a) The bear causes significant depredation to lawfully present livestock or uses unnatural food materials … which have been reasonably secured from the bear, resulting in conditioning of the bear or significant loss of property; b) The bear has displayed aggressive (not defensive) behavior toward humans, which constitutes a demonstrable immediate or potential threat to human safety; or a minor human injury resulted from a human/bear encounter; or c) The bear has had an encounter with people resulting in a substantial human injury or loss of human life.

26 Genetic variability in GYE grizzlies was deemed not depressed when compared to other brown bear populations. (See Table 11-12.)

TABLE 11-12. Genetic Variability within Healthy North American Brown Bear Populations

Population	*Alleles*	*Diversity (%)*	*Sample Size*
Kodiak Island Alaska	2.1	26.5	34
Kluane National Park, Canada	7.4	76.2	24
East Slope, Alberta, Canada	6.4	65.6	30
NCDE, Montana	6.8	70.3	35
Greater Yellowstone	4.4	55.5	46

Source: Waits et al. 1998.

27 IGBST has recorded about 307 known and probable human-caused grizzly mortalities in the GYA between 1975 and 2001.

28 The claimant has an option to appeal the Department's assessment of the claim to the Commission and, if unsatisfied with the Commission's ruling, may file for arbitration.

29 The USFWS has the authority to remove listed species from the threatened list only when species have recovered sufficiently so that the protections of the ESA no longer are needed (16 U.S.C. § 1533[c][2][B][i]). To initiate a delisting process, USFWS is required to publish a notice in the Federal Register of a proposed regulation explaining why delisting is appropriate in light of the same five factors that the agency considered in listing the species (16 U.S.C. § 1533[a], [b], [c]). To ensure that adequate regulatory mechanisms were in place to manage a delisted grizzly bear population, USFWS required that a comprehensive Conservation Strategy be developed and endorsed by relevant federal land management and state wildlife agencies, prior to USFWS proposing a status change for the bear.

30 That year the three demographic recovery criteria were attained for the first time, but not for three years running, as was required.

31 These included USFS, NPS, USFWS, BLM, MFWP, IFG, and WGF.

32 It may be confusing to change from Greater Yellowstone Ecosystem (GYE) to Greater Yellowstone Area (GYA), but that change in terminology reflects the preference of the IGBC.

33 The Conservation Strategy aspires to the following vision for grizzly bear conservation within and outside the PCA: The PCA will be a secure area for grizzly bears, with population and habitat conditions maintained to ensure a recovered population is maintained for the foreseeable future and to allow bears to continue to expand outside the PCA. Outside the PCA, grizzly bears will be allowed to expand into biologically suitable and socially acceptable areas. Outside of the PCA, the objective is to maintain existing resource management and recreational uses and to allow agencies to respond to demonstrated problems with appropriate management actions. Outside the PCA, the key to successful management of grizzly bears lies in bears utilizing lands that are not managed solely for bears but in which their needs are considered along with other uses. Public information and education efforts are to be expanded, and quick, responsive management should be provided to deal with grizzly bear conflicts. Grizzly bears are to be managed as a game animal, with regulated hunting allowed when and where appropriate.

34 The GYA PCA's 18 BMUs are divided into 40 subunits.

35 The strategy defines secure habitat as an area greater than 10 acres in size and 500 meters from an open or gated motorized access route or reoccurring helicopter flight line.

36 It is the stated objective of the agencies that signed the Conservation Strategy to phase out existing sheep allotments, as opportunity arises, with willing permittees. Historically, interactions between livestock and grizzly bears have led to relocation or removal of the latter. Grizzly bears have the potential to cause tremendous damage to sheep herds (for example, in 2001 a grizzly ran a herd of sheep—more than 70—over a cliff in the GYA. In 1998, there were 73 cattle allotments and 15 sheep allotments (totaling 17,279 sheep) within the PCA.

37 Additional habitat parameters include: open and total motorized access route density in each BMU; number of elk hunters inside the PCA; number of grizzly mortalities on private lands in the GYA; and development of a protocol to monitor private land status and condition in the GYA.

38 Among Yellowstone grizzlies, meat constitutes on average 79 percent of an adult male grizzly's diet and 45 percent of the diet of a female grizzly. In contrast, 95 percent of the bear diet in Glacier National Park is comprised of vegetation.

39 The IGBC will still persist as the main interagency entity coordinating grizzly bear recovery in other ecosystems.

40 The YGCC will strive for consensus, but if needed, decisions of the YGCC will be determined by majority vote.

41 ESA section 4(b)(7) states:

(7) Neither paragraph (4), (5), or (6) of this subsection nor section 553 of title 5, United States Code, shall apply to any regulation issued by the Secretary in regard to any emergency posing a significant risk to the well-being of any species of fish and wildlife or plants, but only if—

(A) at the time of publication of the regulation in the Federal Register the Secretary publishes therein detailed reasons why such regulation is necessary; and

(B) in the case such regulation applies to resident species of fish or wildlife, or plants, the Secretary gives actual notice of such regulation to the State agency in each State in which such species is believed to occur.

Such regulation shall, at the discretion of the Secretary, take effect immediately upon the publication of the regulation in the Federal Register. Any regulation promulgated under the authority of this paragraph shall cease to have force and effect at the close of the 240-day period following the date of publication unless, during such 240-day period, the rulemaking procedures which would apply to such regulation without regard to this paragraph are complied with. If at any time after issuing an emergency regulation the Secretary determines, on the basis of the best appropriate data available to him, that substantial evidence does not exist to warrant such regulation, he shall withdraw it.

42 Similar consistent treatment is also needed to address the conservation issues arising when treating the core and peripheral ranges of a listed species. The natural ecological processes determine in large measure the ebb and flow of a species' occupancy in its peripheral ranges. The approach described above might apply equally to resolving the core-periphery quandary as well.

43 Ironically, the states were calling for more research in the years immediately preceding the grizzly listings, which was widely regarded as a tactic to delay and divert the process.

PART V

Concluding Thoughts

CHAPTER 12

Toward Greater State and Local Commitment

KAUSH ARHA AND BARTON H. THOMPSON, JR.

More can be done to encourage effective state conservation of imperiled species. As implemented by the national government and interpreted by federal courts, the Endangered Species Act (ESA) unnecessarily restricts the role of the states in achieving the Act's ultimate goal—conserving those species listed as endangered or threatened. In the process, the ESA also fails to maximize the incentive that states have to recover and protect listed species. The national government could promote diversity protection both a) by giving states greater authority over those species that are threatened but not yet endangered or that have made significant progress toward recovery, and b) by working more effectively with states to set recovery goals that, if met through state and other efforts, would remove species from national purview. The national government could also provide for state involvement in recovery efforts, give clearer guidance on when state or local actions will preclude the need for a federal listing, make greater use of section 6, establish new state–federal coordinating bodies, and increase funding for state conservation efforts. Far from undermining the ESA, such actions could strengthen biodiversity policy in the United States and better reflect the goals of Congress when it passed the ESA in 1973.

The Arguments for Greater State Involvement

Based on the structure of other national environmental laws, one might have expected that the ESA would have established a central role for the states in designing and implementing plans for protecting the nation's biodiversity—a role more akin to states' authority under the Coastal Zone Management Act (CZMA) than their discretion under the Clean Air Act (CAA) and Clean Water Act (CWA). States have long assumed major responsibility for the protection of fish and wildlife within their borders, although the focus historically has been on the conservation of game species rather than the protection of imperiled species (George et al. 1998). Judicial decisions at both the federal and state levels have recognized the special interest of the states in conserving their fish and wildlife (Defenders of Wildlife 1995; Bean and Rowland 1997). A significant number of state constitutions, moreover, provide explicitly for the conservation of fish, wildlife, and other natural resources, although none talks specifically about the importance of protecting biodiversity (Thompson 1996).

All 56 states and territories also have valuable administrative infrastructure for protecting and restoring wildlife. Fish and game departments and independent commissions constitute some of the oldest, most firmly established, and well-funded state administrative agencies (Davis et al. 2005). In some states, fish and game commissions are of sufficient public importance to be constitutionally mandated (Thompson 1996). Fish and game departments and commissions arose in the early 20th century in response to concerns over the depletion of fish and game stocks. Although few state agencies were concerned with more general biodiversity protection at the time that Congress passed the ESA in 1973, such agencies enjoy crucial advantages in the development of restoration plans for imperiled species and in the enforcement of regulations prohibiting individuals from catching, injuring, or killing such species. State fish and game agencies, for example, enjoy significant biological expertise on local species and conditions, and they employ the front line of fish and game wardens and other officials charged with ensuring against the unlawful catch or killing of protected species (Niles and Korth 2005). State officials also have particularly useful relations with farmers, ranchers, and other owners of the "working landscape," a landscape of special importance to biodiversity but where the national government has had limited managerial success (Thompson 2005).

State wildlife agencies generally have a far larger number of biologists on their staffs who can perform on-the-ground conservation work than the national government employs. The two national agencies responsible for administering the ESA—the United States Fish and Wildlife Service (USFWS) in the Department of the Interior and the National Marine Fisheries Service (NMFS) in the Department of Commerce—are both leanly staffed regulatory agencies (Vaughan 1995). Many of the national agency personnel, moreover, are deskbound. States, by contrast, often have scores or hundreds of biologists working in the field. In 2005, the average state wildlife diversity program employed 25 biologists, although the number per state varied tremendously from one biologist in Rhode Island to 374 in California (Niles and Korth 2005).

States also historically have spent significant sums on fish and wildlife protection—a trend which continues today. Given the number of agencies and programs devoting at least some resources to the protection of imperiled species at both the state and national level, it is impossible to confidently estimate the relative amounts of money spent by the states and the national government on biodiversity protection. One survey, however, estimates that state wildlife agencies spent approximately $300 million on wildlife diversity in 2003 (Niles and Korth 2005). The national government's expenditures for the ESA in fiscal year 2003, by contrast, were no more than approximately $240 million.[1] Since federal funding may have supported approximately a quarter (or $75 million) of state expenditures in 2003, the best estimate is that state funding of wildlife diversity efforts ($225 million) was approximately equal in that year to federal funding ($240 million).[2] Other studies have concluded that state funding has generally been significantly greater than federal funding (Richie and Holmes 1999). Whatever the exact figures, the national government can significantly leverage its own resources by promoting state protection of imperiled species.

States, moreover, can appeal to their local populations for multiple forms of support for the protection of biodiversity in ways that the national government cannot. State agencies such as fish and wildlife departments and agricultural programs have often built significant working relationships with ranchers, farmers, and other local property owners over decades. As a result, landowners often trust the state agencies, and the state agencies may be able to obtain critical cooperation from such landowners that frequently is lacking in nationally driven programs.[3] State wildlife agencies also can appeal to local interest and pride in imperiled native species, as well as economic self-interest, to tap new sources of revenue generally unavailable to the national government. One of the fastest growing sources of revenue for state biodiversity efforts, for example, is the sale of vehicle license plates featuring important local species (Richie and Holmes 1999). Florida generated almost $20 million in just seven years from the sale of license plates showing the Florida panther and manatee (George et al. 1998). Some states have also begun to raise private contributions toward their local conservation efforts (Richie and Holmes 1999). Other uniquely state sources of revenue used to protect biodiversity include state lottery funds, state sales taxes, and revenues from the sale of hunting and fishing licenses (Defenders of Wildlife 1995; Richie and Holmes 1999). To generate critical support for protection efforts and expenditures, local governments have sometimes appealed to the fact that such efforts protect not only the species but also a variety of valuable ecosystem services (Defenders of Wildlife 1995; Behan 2005).

Effective protection of imperiled species, moreover, requires the integration of recovery plans with land-use and water management decisions (Tarlock 2005). The greatest threat to imperiled species in the United States is the loss of land and water habitats. Land-use and water decisions, in turn, have historically been the responsibility of state and local government (Defenders of Wildlife 1995; Sax et al. 2006). For instance, local zoning authorities have the expertise and authority to determine how individual parcels of land should be used (U.S. National Commission on Urban Problems 1968). Similarly, state water agencies control how much water can be withdrawn from rivers and other bodies of surface water, as well as the points of diversion and the uses of the water (Sax et al. 2006). Absent a national takeover of these functions, effective implementation of the ESA requires the active cooperation of such state and local agencies.

States themselves, moreover, often own lands of significant habitat value to federally listed and other imperiled species (Defenders of Wildlife 1995). This is particularly true in the western United States where the states own large swaths of land, frequently in the form of school trust lands, parks, and wildlife refuges.[4] States in other portions of the nation, however, also often own critical habitats for imperiled species. In Florida, for example, the state and local governments own almost 60 percent of all publicly managed lands, much of which is critical to species protection (Swain 2005).

Active state involvement in the protection and restoration of biodiversity also allows for critical policy diversity. The art of species protection and restoration is still in its infancy and often plagued by uncertainty. One method of managing such uncertainty is to build and use a portfolio approach, effectively spreading the risk that any one approach will fail and testing a variety of approaches for efficacy. (Portfolios of stock, in a similar fashion, reduce the risks of investing all of one's funds in a single stock.) Cooperative federalism can help to encourage and test new and alternative methods of protection and restoration, providing exactly such a portfolio (Defenders of Wildlife 1995). State efforts in other areas, such as housing and community development, have often produced valuable innovations that other states (as well as the national government) have later adopted (George et al. 1998). Diversity in regulatory and managerial approaches is of further value in the biodiversity context because of the need to tailor solutions to the unique local conditions and factors that often determine the effectiveness of protection and restoration measures (Davis et al. 2005; Niles and Korth 2005).

Not all arguments for cooperative federalism are as strong in the context of the ESA as for other national environmental statutes. States, for example, had not historically played any significant role in protecting imperiled species at the time Congress passed the ESA (Defenders of Wildlife 1995). This contrasts sharply with the traditional pollution context, where a number of states had adopted often significant statutory protections before Congress proposed national legislation.[5] Although every state had a fish and game department or commission at the time of the ESA's passage, these agencies, as noted above, were focused on the conservation of fish and game for recreational or commercial purposes and typically did not have either the mission or expertise to preserve imperiled species that happened to be nongame or predator species. While approximately a third of the states had adopted legislation protecting imperiled species from direct takings by hunters, fishers, or collectors, no state had passed laws providing for broad protection within their borders of species facing extinction from habitat degradation. Nor had state laws protected imperiled species from state projects and expenditures (George et al. 1998). Thus, states had not been engaged in protecting biodiversity prior to enactment of the ESA to the same extent as they had sought to reduce air and water pollution prior to the enactment of the CWA and CAA. The other arguments for cooperative federalism, however, still provide strong and convincing support for actively involving state and local governments in the protection of biodiversity.

Current State and Local Involvement

Over the last thirty years states have accepted substantial and growing responsibilities in protecting biodiversity and conserving imperiled species. As demonstrated by the conservation of the Karner blue butterfly, grizzly bear, coho salmon, and red-cockaded woodpecker, as well as by state programs such as California's Natural Community Conservation Program (NCCP), states have increasingly brought to bear their valuable advantages in conserving threatened and endangered species within their respective borders. States can bring critically important resources, expertise, and relationships to the conservation of the nation's wealth of biodiversity. Just as states have helped advance the nation's goals for clean air and clean water, states also can assist in meeting the ESA's aims. Without active state involvement, efforts to protect the nation's imperiled species will be a much steeper hill to climb.

Since the enactment of the ESA in 1973, federal wildlife agencies have played a formative role in shaping the future course of state wildlife agencies in pursuing species conservation—and vice-versa. USFWS's administration of the ESA has prodded state agencies to evolve from game management agencies to wildlife agencies that address the conservation needs of all wildlife species, including imperiled nongame species. California's NCCP and Oregon's Plan for Salmon and Watershed Restoration are two of the prominent consequences. Today, moreover, all 56 states and territories have species conservation plans in place, encouraged by grants from the Department of the Interior. On the other hand, the realities of having to implement species conservation measures on the ground have persuaded the USFWS to develop new regulatory policies that allow for enhanced state roles in species conservation efforts. Habitat Conservation Plans (HCPs), Safe Harbor Agreements (SHAs), Candidate Conservation Agreements with Assurances (CCAAs), and the Policy for Evaluating Conservation Efforts when Making Listing Decisions (PECE) are prime examples of such policies. USFWS has also taken steps toward providing financial assistance in the way of grants to states to mitigate costs associated with species conservation and encourage state species conservation planning.

Potential Federalism Reforms

Although states are playing an increasing role in protecting the nation's biodiversity, often with the encouragement and approval of the departments of Interior and Commerce, the national government can take a variety of steps to further encourage valuable state involvement. Under the shadow of changing climate, USFWS and NMFS need—more than ever—the assistance of state wildlife agencies to monitor sensitive species, implement strategies to recover threatened and endangered species, and achieve species conservation on private lands. As J. B. Ruhl discusses in Chapter 3, without greater state involvement, growing aversion to centralized federal regulation and the limited resources available to the federal agencies will continue to stunt efforts at species conservation and recovery. Jurisdictional skirmishes between state and federal agencies also will continue to divert critical time and resources. State governments and landowners are likely to continue to show at least a measure of antipathy towards the ESA, impeding important efforts at species conservation.

Further progress in species conservation is thus difficult to imagine without active state participation. Most of the cooperative state–federal successes described in previous chapters were enabled or facilitated by unique circumstances and the work of innovative and dedicated individuals.

This chapter builds on the information presented earlier to suggest a range of general, broadly applicable regulatory, institutional, and financial actions that together could greatly assist federal and state collaboration in species conservation. Based on discussions at the Stanford conference that was the genesis for this book, seven steps would appear to be particularly valuable. Interior and Commerce can take most of these steps without additional legislative authority, although explicit Congressional authorization and support would be useful for all of the proposals and would be essential for the final recommended action of providing additional federal funding for state programs.

Listings

First, Interior and Commerce should develop more useful guidance for when state conservation efforts will eliminate the need for federal listing of species. The USFWS developed its PECE policy with close state collaboration to establish clear criteria to evaluate the certainty and effectiveness of state conservation efforts as part of federal-listing determinations. PECE, however, has seen little use since its publication in 2004, in part because PECE is insufficiently clear and in part because it may set too high a standard that is inconsistent with similar federal programs. PECE lists nine factors for determining whether state efforts are certain to be implemented and another six for determining effectiveness. Candidate Conservation Agreements (CCAs) appear to use a lower bar for determining when local actions are sufficient to avoid federal listings, both raising questions regarding the higher standards established under PECE and effectively undermining the value of direct state actions and PECE.

Clear, consistent and pragmatic guidance for state conservation efforts that may preclude a federal listing of species will go a long way in facilitating state–federal partnerships in species conservation. The cases of coastal coho in Oregon and Atlantic salmon in Maine, as well as the judicial decisions described by Sobeck and Weiland in Chapter 4, highlight the judicial and political complexities that arise from the lack of a common understanding of what level of state conservation efforts is appropriate for eliminating the need for a federal listing of an imperiled species. Several western states and the USFWS became embroiled in determining this very question in the fight over listing sage grouse. Resolution of this issue is critical because it is more effective to conserve a species prior to the species reaching a stage requiring federal protection under the ESA.

Threatened Species

Second, Interior and Commerce should provide states with a greater opportunity to protect and conserve species that are listed as threatened under the national ESA. States often can play the major role in conserving threatened species and regulating state or private actions that might further imperil the species.

The interest of the national government increases both normatively and legally as a species becomes more imperiled. For instance, where a species has not yet qualified as threatened or endangered, the ESA provides for no national regulation.[6] Once a species qualifies as threatened, the national government has a substantial interest in ensuring its protection, but the level of peril may not justify fully nationalizing protective measures. States may be able to adequately protect such species, subject to national oversight. Where a species is endangered, however, the national interest in ensuring the species' preservation may mandate that the national government itself implement conservation measures.

The ESA arguably allows for exactly this type of transition from full state authority over the conservation of unlisted species, to joint state–federal conservation for threatened species, to full national authority for the protection of endangered species. As noted, the ESA provides no authority to Interior or Commerce to regulate unlisted species. Although Congress may have intended that the national government would delegate authority for the conservation of many listed species to states under section 6, the courts, as discussed earlier, have effectively limited the relevance of this provision. Interior and Commerce, however, can use section 4(d) to give the states greater authority over threatened species. Section 4(d) allows Interior and Commerce to issue whatever regulations they "deem necessary and advisable to provide for the conservation of [threatened] species" (16 U.S.C. § 1533(d)), arguably permitting the national agencies to delegate regulation to the states. In the case of endangered species, by contrast, all of the prohibitions of the ESA apply fully, and the national government must exercise its regulatory authority.

Interior and Commerce can therefore use the classification of threatened species to encourage states to volunteer for greater leadership in species conservation. To provide maximum flexibility for state conservation, the national government should list species as threatened before listing them as endangered except in cases where extinction is imminent. The national government should also use section 4(d) to allow willing and capable states to exercise primary authority for the conservation and recovery of these threatened species. To ensure effective state conservation of threatened species, the national government could enter into specific agreements for each species between relevant states and the national government that would define the ecological criteria by which to measure conservation programs for such species, as well as the deadlines for meeting them. If a state fails to achieve measurable progress for a threatened species as specified under the agreement for that species, the national government could reclaim primary authority over the conservation and recovery of that species.

Interior and Commerce should consider providing for the reverse federal evolution as a species recovers. Although the ESA contemplates downlisting of species from endangered to threatened as the species recover,[7] neither Interior nor Commerce has made significant use of this option. In the first 30 years of the ESA, the USFWS downlisted only 22 species (Scott et al. 2005). Complete delisting of an endangered species raises concerns whether state and other protective measures that will remain in effect after delisting will be adequate to ensure that the species does not go back into decline. Interior and Commerce might be able to reduce the risk from delisting, and thus increase the chances of successful recovery, by reclassifying species from endangered to threatened as they improve and adopting section 4(d) rules that delegate principal conservation authority to the states, subject again to continuing national oversight. The national government can evaluate the effectiveness of state protections during the period that the species is listed as threatened in deciding whether to ultimately delist the species entirely.

State Involvement in Recovery Efforts

Third, even where the national government takes the lead in the conservation and recovery of a species, Interior and Commerce should provide willing and competent states greater authority to participate in those efforts. States should be able to participate in and play a substantial role in section 7 consultations, the development of special rules under sections 4(d) and 10(j), and the issuance of incidental take permits. In addition, states should be encouraged to forge conservation agreements with third parties and submit them for the approval of the secretaries of Interior and Commerce.

Over the last thirty years, states have been critically involved in the recovery of federally listed species under the ESA. From the Karner blue butterfly to the grizzly bear, there are few, if any, examples of recovering federally listed species—requiring considerable on-ground conservation actions—absent strong state contributions to their recovery. In light of state efforts in the recovery of listed species, a consensus emerged among many of the participants at the Stanford conference that facilitating greater state and nonfederal actor involvement in species conservation can further contribute to recovery.

Criteria for Downlisting and Delisting

Fourth, Interior and Commerce can increase the willingness of states to undertake significant recovery measures by setting out clearer and more certain downlisting and delisting criteria as part of species recovery plans. A major motivation for some states to participate in the implementation of national recovery plans is the potential to recover state authority over the species and eliminate national intervention. In many cases, however, downlisting and delisting criteria are insufficiently specific. In other cases, Interior and Commerce change the criteria as species come close to achieving them, without providing clear explanation to the states as to why. The national government changed recovery plans two or three times for most of the species discussed in prior chapters. States also have criticized Interior and Commerce for inconsistency in their evaluations of whether the criteria for downlisting or delisting have been met.[8] Vague and changing criteria, as well as inconsistent implementation, can lead to state confusion and frustration, undermining the willingness of states to participate actively with the national government in implementing recovery plans. Ill-defined and moving recovery targets are the bane of a sustained state–federal partnership built on credibility and trust. To states, it is akin to encouraging an individual to train for an Olympic marathon and when she apparently wins the race, informing her that the finish line has been moved.

Interior and Commerce can also increase the willingness of individual states to pursue recovery efforts by allowing species to be downlisted or delisted in some states ahead of their downlisting or delisting in other states. If a state effectively implements its conservation measures, as Wisconsin has done for the Karner blue butterfly, the species' status in that state should be adjusted. By rewarding those states that adopt and implement successful conservation measures, and not punishing those states for lack of progress elsewhere, Interior and Commerce can encourage states to pursue aggressive conservation of interstate species.

Species Conservation Committees

Fifth, Interior and Commerce should develop state-based "Species Conservation Committees" (SCCs) that would actively coordinate conservation measures between the national and state governments. Effective state–national conservation partnerships require close collaboration and coordination. Yet as discussed earlier, states sometimes do not even know what the USFWS is doing within their borders. An SCC would enable the regular meeting of national, state, and local officials working on biodiversity conservation within any specific state and thus enhance coordination and credibility between state and federal agencies. SCCs could particularly help provide for coordination in the listing of species, an area where coordination is especially lacking. SCCs also could help ensure that scientific efforts, such as the mapping of wildlife habitats undertaken by individual states, ultimately benefit conservation efforts by neighboring states and the national government (Niles and Korth 2005). Most importantly, however, SCCs could build trust among the various agencies involved, reduce transaction costs in interagency work, and facilitate a more strategic multi-species eco-regional approach to species conservation.

Effective conservation of species—particularly those that inhabit or are dependent on large landscapes—often requires close collaboration among a variety of state and federal agencies and a broad range of stakeholders. Recovery efforts for coastal coho in Oregon and grizzly bears in Wyoming, as well as ongoing efforts to conserve sage grouse over most of the western states, demonstrate this need. The need for collaboration is heightened because different state and federal agencies often have discrete management responsibility for species, land, and water. For example, any meaningful effort to conserve salmon species in Oregon requires close coordination among the state agencies for wildlife, forestry, agriculture, and water, and relevant federal agencies including the departments of the Interior, Commerce, Agriculture, and Energy. Yet, it would be impractical and imprudent to establish separate interagency policy-level committees, in addition to science-based species recovery teams, for the conservation and recovery of each species of concern such as salmon, northern spotted owls, grizzly bears, and gray wolves. In most states, a working SCC could better define and structure existing informal procedures to coordinate species conservation among the various state and federal agencies and other stakeholders. The primary goals of a SCC would be coordination and information-sharing concerning species conservation within a state.

Section 6 Cooperative Agreements

Sixth, Interior and Commerce should use section 6 of the ESA to facilitate greater cooperative federalism in species conservation. Section 6, which authorizes cooperative agreements between states and the national government, offers an ample and largely untapped opportunity to foster strong and sustainable state–federal partnerships for conserving our nation's species. Although section 6 would appear to provide significant discretion in the use and structuring of cooperative agreements, section 6 until recently has served as a mere formality for providing federal funding for state conservation efforts.

The national government should use section 6 in the future to encourage and empower willing states to take leading roles in species conservation through state programs that are functionally equivalent to the federal ESA and that complement national regulatory efforts. Cooperative agreements under section 6 can take one of two forms: (1) a state can agree to accept responsibility for virtually all tasks related to species conservation under the ESA (as Arizona has done), or (2) a state can choose to accept responsibility more selectively under the ESA (as Idaho has done). In either case, the national government should judge the adequacy of state programs for the purpose of section 6 cooperative agreements based on a "functionally equivalent" standard, with an imperative for sustained improvement in species' ecological condition. States, acting as creative laboratories, should be the harbingers of innovation and initiative in designing robust cooperative agreements under section 6. Interior and Commerce should embrace and encourage this process.

Funding

Finally, Congress should provide the funding necessary to support effective state conservation efforts. As discussed earlier, states have brought considerable financial resources, perhaps more than the national government, to biodiversity conservation. But current state funding represents only a small fraction of the resources needed to conserve the species now listed under the ESA (George et al. 1998). States spent approximately $300 million on biodiversity conservation in 2003 (of which about 25 percent came from the national government). The International Association of Fish and Wildlife Agencies (IAFWA) in 1999, however, estimated that states needed $1 billion or more to adequately implement state wildlife diversity programs (Richie and Holmes 1999). In those states with limited local resources and funding sources, the low level of national funding presents an insurmountable barrier to greater state participation in conservation efforts (George et al. 1998). Even in those states that dedicate significant levels of their own funds, funding is often unpredictable, making it difficult for the state agencies to adequately plan and implement long-term programs (Richie and Holmes 1999). Given the national interest in preservation of federally listed species, Congress should provide additional funding to those states willing to undertake more significant conservation efforts.

Additional national funding is also justified by the uneven financial burden that biodiversity conservation places on individual states. The cost of recovering some high-profile species can be astronomical. The desert tortoise budget since 1999 has been more than $100 million. The captive breeding program for California condors has cost in excess of $40 million since its inception (leading some to call the condors "million dollar birds"). These can be high costs for individual states to bear. Given the uneven distribution of federally listed species across the United States and the national interest in their protection, greater national funding is needed.

Not only is additional national funding needed, but a dedicated source of funding is also critical to ensuring that states can enhance their roles in achieving species conservation. Annual appropriations and competitive grants are not conducive to multi-year programmatic conservation efforts because of the inherent financial uncertainty involved. A source of assured base funds would provide states with needed leverage to fully avail themselves of available competitive grants. The states should complement federal funding by developing a core base of funding at the state level.

Conclusion

A major thread in many proposals to reform the ESA in Congress has been the importance of increasing state involvement in the protection and restoration of federally listed species. This is not surprising. As discussed earlier, states can bring critically important resources, expertise, and relationships to the conservation of the nation's wealth of biodiversity. Just as states have helped advance the nation's goals for clean air and clean water, states can help effectuate the purposes of the ESA. As illustrated by the case studies described earlier, without active state involvement, efforts to protect the nation's imperiled species are often an uphill fight. A study of the ESA on its thirtieth anniversary, based on papers presented at a three-day conference in Santa Barbara, California, in 2003, reached similar conclusions on the general importance of federalism in the implementation of the ESA:[9]

> *It was telling that in their keynote remarks in Santa Barbara, Idaho governor Dirk Kempthorne [subsequently the United States Secretary of Interior], former secretary of the interior Bruce Babbitt, and assistant secretary of the interior Craig Manson all called for more use of cooperative agreements with the states. Since 1972, the role of states has expanded continuously but still falls short of state roles under similar cooperative federalism provisions in the Clean Water Act.* (Davis et al. 2005)

The importance of active national involvement in the protection of imperiled species should not be forgotten in the search for a greater and more effective state role. One of the clearest lessons that emerges from the chapters in this book is the critical role that the national ESA has played both in elevating the plight of threatened and endangered species in the national consciousness and in spurring state action. States unfortunately often do not have the political will to take the necessary steps to protect biodiversity absent the ESA (George et al. 1998). The Act has spurred state agencies across the nation to action by substantially enhancing the bite of national biodiversity conservation laws and by providing national funds for species recovery and management. From the Aleutian goose to the Florida manatee, states did not begin a sustained or sincere commitment of resources to the species' recovery and management until after the national government listed the species under the ESA. At the threat of listing the sage grouse most Rocky Mountain and Great Basin states sprang into action to develop comprehensive sage grouse conservation strategies. Without the ESA, all of these species would likely have been worse off, and the California condor and desert tortoise might well have gone extinct. The ESA is particularly important in the protection of the many endangered and threatened species that cross state lines (George et al. 1998).

Cooperative federalism is not about devolution of authority from the national to the state governments, but about the creation of effective mechanisms for gaining the benefits of joint regulation and recovery efforts. Consider, for example, regulations of land use designed to protect species. National agencies, because they are far removed from the communities in which the land is located, often err in favor of conservation. Local agencies, however, often tend to favor local property owners. Neither national nor local regulation by itself is likely to be effective. A collaborative process among national, state, and local agencies, by contrast, is more likely to respond to the multiple considerations involved in land-based conservation efforts. This is a major lesson from the design of marine sanctuaries and from boating restrictions to protect the Florida manatee.

Interior and Commerce have used their discretion under the ESA to begin building more effective and dynamic partnerships among the national, state, and local governments in biodiversity conservation. The state and local roles, however, remain too constrained. While continuing to maintain national oversight of conservation efforts, the national government can increase the states' efficacious involvement through the seven measures detailed above: 1) providing clearer guidance on when state conservation efforts will relieve the need to list local species, 2) allowing states greater primacy in the protection of threatened species, 3) allowing greater state involvement in recovery efforts, 4) increasing the incentive for states to invest in the recovery of species through clearer and more stable recovery and delisting criteria, 5) establishing state-based SCCs, 6) making greater use of section 6 cooperative agreements and 7) increasing national funding for state conservation efforts. Through these and related steps, all levels of government in partnership will move closer to the crucial goal of restoring and protecting biodiversity in the United States.

References

Bean, Michael J. and Melanie J. Rowland. 1997. *The Evolution of National Wildlife Law*. 3rd ed. Westport, Conn.: Praeger.

Behan, Maeveen. 2005. "County Conservation Planning." In *The Endangered Species Act at Thirty*, vol. 1, Dale D. Goble, J. Michael Scott, and Frank W. Davis eds., 156–160. Washington, D.C.: Island Press.

Davis, Frank W., Dale D. Goble, and J. Michael Scott. 2005. "Renewing the Conservation Commitment." In *The Endangered Species Act at Thirty*, vol. 1, Dale D. Goble, J. Michael Scott, and Frank W. Davis eds., 296–306. Washington, D.C.: Island Press.

Defenders of Wildlife. 1995. *Saving Biodiversity: A Status Report on State Laws, Policies, and Programs*. Washington, D.C.: Defenders of Wildlife.

Elliott, Donald E. 1985. "Toward a Theory of Statutory Evolution: The Federalization of Environmental Law." *Journal of Law, Economics, and Organization* 1: 313–360.

George, Susan, William J. Snopes III, and Michael Senatore. 1998. *State Endangered Species Acts: Past, Present, and Future*. Washington, D.C.: Defenders of Wildlife.

Niles, Lawrence and Kimberly Korth. 2005. "State Wildlife Diversity Programs." In *The Endangered Species Act at Thirty*, vol. 1, Dale D. Goble, J. Michael Scott, and Frank W. Davis eds., 141–155. Washington, D.C.: Island Press.

Percival, Robert V., Christopher H. Schroeder, Alan S. Miller, and James P. Leape. 2006. *Environmental Regulation: Law, Science, and Policy*. 5th ed. New York, NY: Aspen Publishers.

Richie, Deborah and Jennifer Holmes. 1999. *State Wildlife Diversity Program Funding: A 1998 Survey*. Washington, D.C.: International Association of Fish and Wildlife Agencies.

Salzman, James and Barton H. Thompson, Jr. 2010. *Environmental Law and Policy*. 3rd ed. New York, NY: Foundation Press.

Sax, Joseph L., Barton H. Thompson, Jr., John D. Leshy, and Robert H. Abrams. 2006. *Legal Control of Natural Resources: Cases and Materials*. 4th ed. St. Paul, MN: Thomson/West.

Scott, J. Michael, Dale D. Goble, Leona K. Svancara, and Anna Pidgorna. 2005. "By the Numbers." In *The Endangered Species Act at Thirty*, vol. 1, Dale D. Goble, J. Michael Scott, and Frank W. Davis eds., 16–35. Washington, D.C.: Island Press.

Swain, Hilary. 2005. "A Reality Check from Florida." In *The Endangered Species Act at Thirty*, vol. 1, Dale D. Goble, J. Michael Scott, and Frank W. Davis eds., 137–140. Washington, D.C.: Island Press.

Tarlock, A. Dan. 2005. "The Dynamic Urban Landscape." In *The Endangered Species Act at Thirty*, vol. 1, Dale D. Goble, J. Michael Scott, and Frank W. Davis eds., 127–136. Washington, D.C.: Island Press.

Thompson, Barton H., Jr. 1996. "Environmental Policy and State Constitutions: The Potential State Role of Substantive Guidance." *Rutgers Law Journal* 27: 863–908.

———. 2005. "Managing the Working Landscape." In *The Endangered Species Act at Thirty*, vol. 1, Dale D. Goble, J. Michael Scott, and Frank W. Davis eds., 101–126. Washington, D.C.: Island Press.

U.S. Department of the Interior. 2007. *Fish and Wildlife Service, FY 2007 Budget Justification: Endangered Species*. Washington, D.C.: Department of the Interior.

U.S. National Commission on Urban Problems. 1968. *Building the American City*. Washington, D.C.: U.S. Government Printing Office.

Vaughan, Ray. 1995. "State of Extinction: The Case of the Alabama Sturgeon and Ways Opponents of the Endangered Species Act Thwart Protection for Rare Species." *Alabama Law Review*: 569–598.

Endnotes

1 According to Scott et al. (2005), national expenditures for all activities under the ESA in 2003, when approximately 1,260 species were listed, was about $170,000

per species. According to the Department of the Interior, national expenditures on endangered species protection in 2006 were only $140 million (U.S. Department of the Interior 2007).

2 The most recent comprehensive study of the sources of state expenditures was in 1998. At that time, federal funds constituted 24.5 percent of all the monies that states spent on wildlife diversity. It should be noted, however, that state funding for biodiversity protection varies dramatically from state to state, with a few states providing the great bulk of biodiversity expenditures (Richie and Holmes 1999).

3 Efforts by the state agency to reach out to local hunting organizations and explain the dangers of lead bullets to wild California condors similarly helped gain stakeholder support for restrictions on lead bullets.

4 Arizona, for example, manages almost 10 million acres of land, Utah manages four million, and Washington three million (Defenders of Wildlife 1995).

5 Oregon, for example, created the first air quality regulatory agency in 1952. California adopted the first controls on automobile emissions in 1959 (Percival et al. 2006). Indeed, the proliferation of state air pollution laws led automobile manufacturers to support federal legislation in order to avoid being subjected to a growing number of conflicting state laws (Elliott 1985).

6 Under the ESA, a species is "endangered" if it "is in danger of extinction throughout all or a significant portion of its range" (16 U.S.C. § 1532(6)). A species is threatened, by contrast, when it "is likely to become an endangered species within the foreseeable future throughout all or a significant portion of its range" (16 U.S.C. § 1532(20)). The ESA provides no protection to species that are not yet listed, even if those species are candidates for protection (Salzman and Thompson 2010).

7 Section 4(c)(2) of the ESA provides for the regular review of species to determine whether species should be "changed in status from an endangered species to a threatened species" (16 U.S.C. § 1533(c)(2)).

8 In the case of Aleutian geese, for example, USFWS delisted the species even though only one of three recovery targets had been met. Although the total population had recovered to the desired level, there were not 50 pairs of geese nesting in all three geographic areas of its historic range and the government had not acquired the desired amount of secure wintering grounds. By contrast, the national government made no effort to downlist or delist the manatee even after manatee populations had recovered in three out of the five management regions in Florida.

9 The study also warned, however, that states have often been "reticent to establish regulatory programs that are as protective as the ESA, a basic requirement of cooperative federalism." While noting that the exact role that states can play "is a matter of continuing debate," the study found agreement on several reforms. First, the national government should consult states earlier in the ESA process and give the states a "greater role in recovery planning and implementation." Second, state and national wildlife agencies should more freely share expertise, data, and information. Finally, Congress should increase its funding of state efforts (Davis et al. 2005).

Index

Administrative Procedure(s) Act 56, 293
Air Force Integrated Natural Resource Management Plan 73
alevins, definition 191
Alsea Valley Alliance v. Evans 237
American Fisheries Society 61
amphibians
  Barton Springs salamander 62–65, 78
  Chiricahua leopard frog 183
  Houston toad 183
anadromous, definition 191
Andrus v. Allard 40
anti-take provision 7
aquaculture
  effluent limitation guidelines 212
  Maine 188
Aquatic Conservation Strategy of Northwest Forest Plan 224
Arizona
  Game and Fish Department 110
  Yuma County 67
Association of Fish and Wildlife Agencies 261
Atkins, Charles G.
  fisheries commissioner, Maine, 1874 193
Atlantic salmon. *See* salmon, Atlantic
Atlantic Salmon Authority 203
  inception of, 1995 202
Atlantic Salmon Federation 204
Atlantic Salmon of Maine, Inc. 212
Atlantic Salmon Task Force 202
Atlantic Sea Run Commission
  history 194
Attwater's prairie chicken 183
Audubon Society 133, 151
  grizzly bear litigation 272
  Wisconsin 153
Babbitt v. Sweet Home Chapter of Communities for a Greater Oregon 45
Babbitt, Bruce 199, 318
  ESA reforms 42, 58, 104
  ESA reforms, California 116
  litigation 45, 272, 289
  role in California NCCP 134
  support of dam removal, Maine 213
  withdrawal of proposed listing of Atlantic salmon 204
Baird, Spencer
  director, U.S. Fish Commission, history 193
Balcones Canyonlands HCP 104
bald eagle 28, 99
  existence values of 29
Baldacci, John E.
  Maine governor 211
Barton Springs salamander 65, 78
  Conservation Agreement and Strategy 63
  listing history 63
  violation of ESA 62
Baum, Edward
  salmon biologist, Maine 189, 215
Bay Delta Conservation Plan 132
Bear Management Units 268
bear, grizzly. *See* grizzly bear
Beattie, Molly
  Fish and Wildlife Service Director 204
Bennett v. Spear 56
biodiversity, protecting 3, 4, 10, 115
  federal funding 309, 317
  license plate revenue 309
  recommendations for federal government 307

sources of revenue 309
state funding 309
state programs 11–12, 115
Biodiversity Legal Foundation 199
black-tailed prairie dog 112
decentralized management of 30
blue mussel farms in Maine 195
blueberry, effect of cultivation on Atlantic salmon 194
Bob Marshall
Ecosystem 262
Wilderness 252
Breaux, John
Louisiana Representative 99
Brown, Don
author of Grizzly Bear Recovery Plan 267
Buck, Richard A.
role of in intercept fisheries 197
bull trout, devolution of 29
Bureau of Reclamation, Ten Point Plan 62
Bureau of Sport Fisheries, history 193
Bureau of Sport Fisheries and Wildlife 261
Bush administration, failings of 43
butterfly, Karner blue. *See* Karner blue butterfly

Cabinet-Yak Ecosystem, 252, 265
California
biodiversity protection 115
Building Industry Association 124
coastal sage scrub 125
Department of Fish and Game 62, 107, 115, 117, 119
Department of Parks and Recreation 107
Department of Transportation 107
Endangered Species Act 70, 117
Environmental Quality Act 117
gnatcatcher 59, 108, 116, 126, 134
Imperial County 67
incidental take permits 119
legal definition of take 103
Native Plant Protection Act 70
Natural Communities Conservation Program 311
Natural Community Conservation Planning Act. *See* Natural Community Conservation Planning Act, California
Proposition 12, 1999 132
Proposition 84, 2006 132
regional conservation plans, 1996–2010 127
Resources Agency 126
Riverside County, 2003 HCP 107
Riverside County Habitat Conservation Agency 124
San Bruno Mountain HCP 116
urban expansion, history 116
Water Quality Control Board 62
Candidate Conservation Agreement
for slickspot peppergrass 73
purpose of 108
with Assurances Policy 71–72, 73, 109
Carter, Gene
Maine District Court Judge 211
caselaw relevant to ESA 59
CCAA Policy. *See* Candidate Conservation Agreement with Assurances Policy
CDFG. *See* California, Department of Fish and Game
centralization
increase of 24
regulatory 24
CEQA. *See* California, Environmental Quality Act
Certificate of Inclusion 169
CESA. *See* California, Endangered Species Act
Cheeca Lodge Schaus swallowtail butterfly 183
Chinook salmon. *See* salmon, Chinook
Chiricahua leopard frog 183
Clean Air Act 4, 307
Clean Water Act 4, 13, 36, 38, 43, 221, 307
compared to CZMA 43
federal 188
clear-cutting, effect on Atlantic salmon 194

Clearwater River 110
Coachella Valley
fringe-toed lizard 121
MSHCP 138
coastal sage scrub 125
Scientific Review Panel 127
Coastal Salmon Restoration and Production Task Force 226
Coastal Salmon Restoration Initiative
effectiveness 231
funding 227
history 226
memorandum of agreement 227
purpose of 221
Coastal Zone Management Act 7, 13, 36, 38, 43, 307
compared with Clean Water Act 43
Code of Practice on the Introduction and Transfer of Marine Organisms
history 199
Cohen, William
Maine senator 197, 201
coho salmon, Oregon coastal
Coho Assessment 232–233
conservation history 223
conservation lessons 240
critical habitat designation 234
ESA listing of 230
life cycle 221
listing history 64, 207
no list decision 227, 228
population decline, history 223
recovery, history 234
relisting 237, 239
restoration program 226
restoration themes 242
risk factors 234
threatened listing, history 225
withdrawal of proposed listing 239
Columbia River, litigation regarding management of 57
commerce, interstate, regulatory power over 27
Committee on Merchant Marine and Fisheries 90
Committee on the Atlantic Salmon Emergency 197
comprehensive cooperative agreements 98
conservation efforts
challenges to state wildlife agencies 163
factors used 72
federal funding 317
grizzly bear 251, 255–256, 291
private contributions 309
recommendations for improved federal role 312–118
rewarding success of 160
role of states, recommendation 307
Conservation Management Training Program 140
conservation planning
limiting of private property rights 117
of privately owned land 116
Consolidated Papers, Inc 151
Constitution, governmental power 21
control, localized 27
Convention on International Trade in Endangered Species 90
Convention on Nature Protection and Wildlife Preservation 106
cooperative agreements 9, 10
impediments to 96
Cooperative Endangered Species Conservation Fund 101, 104, 105
cooperative federalism 35
and CZMA 44
history 37
limitations of 41
costs
principal-agent 23
transaction 22, 26, 27, 29
Cove Brook
tributary of Penobscot River 209
covered species 117
Craighead, Frank
early grizzly bear researcher 260
Craighead, John
early grizzly bear researcher 260
critical habitat, designation of 5
Crosby, Bing
involvement in Atlantic salmon plight 197

dams, Maine, effect on salmon populations 189
Defenders of Wildlife 29, 133, 153
  litigation 67, 207
delisting, by state 29
Delta Waterfowl 28
Department of Commerce
  authority regarding takes 7
  authority to implement recovery plans 6
  authority to list species 5
  jurisdiction 7
Department of Interior
  authority of Secretary 91
  authority regarding takes 7
  authority to implement recovery plans 6
  authority to list species 5
  jurisdiction 7
  violation of ESA 63
Department of Marine Resources 195
desert pupfish 183
devolution, levels of 21
Dingell, John
  Michigan Representative 91
Dingell-Johnson federal aid program 100
distinct and significant population segment 188
distinct population segment, history 199, 201
Douglas County v. Balsiger 235
Down East rivers, Maine 187, 197
DPS. *See* distinct and significant population segment
dual federalism 37
Ducks Unlimited 28
Dudleya, short-leaved 79
  listing history, 70

eagle, bald 28, 99
  existence values of 29
Earthjustice
  litigation regarding delisting of gray wolf 297
East Contra Costa County NCCP 139
economies, scale 27, 28
Edwards Dam, Maine, removal of 213
Elliott, Jeffrey
  New Hampshire resident 199
Endangered and Threatened Species Act, Wisconsin 149
Endangered Species Act
  1977 amendments 97
  1982 amendments 99
  1988 amendments 100
  addition of plants, 1978 97
  administration of, recommendation 244
  advantage of listing 243
  authority 44
  California, history 116
  citizen suit provisions 57
  criticism of 45
  delisting of species, recommendation 294
  downlisting of species 314
  evolution of federalism under 89–114
  funding 99
  importance of 319
  inception of 4
  limitations of 36, 44, 307
  listing of Atlantic salmon 187
  number of listed species 101
  overview of litigation 56–59
  reform of 30, 80, 318
  relationship of NCCP to 122
  section 4 38, 229. *See also* section 4 of ESA
  section 6 5, 9–10, 35, 41, 89–114, 163, 307. *See also* section 6 of ESA
  section 7 38, 39, 244, 267, 273. *See also* section 7 of ESA
  section 9 39, 42. *See also* section 9 of ESA
  section 10 7, 39, 47, 151. *See also* section 10 of ESA
Environment and Public Works Committee, 100
Environmental Careers Organization 140
Environmental Defense Fund
  role in red-cockaded woodpecker conservation 173
environmental federalism
  comparative assessment 35–52

economic assessment of 21–33
Environmental Protection Agency, regulatory function 43
ESA. *See* Endangered Species Act
evolutionarily significant units 65

falcon, peregrine 99
Federal Land Policy and Management Act 293
federalism
cooperative, anatomy of 39
cooperative, inception 4
environmental, economics of 22
incoherent, costs of 44
reforms, potential 311–318
tools of cooperative 37–38
Ferge, Les
Wisconsin lepidopterist 149
Fish and Wildlife Foundation, National
grant funding to California 129
Fish and Wildlife Service 3, 5, 35, 55, 106, 308
authority of 41
cooperation with states 40
five-point policy 136
funding to red-cockaded woodpecker safe harbor programs 178
grant funding to California 129
involvement in grizzly bear conservation 251
listing of Karner blue butterfly, 1992 148
Listing Priority Guidance 59
nationwide grant program 131
regulations, no surprises 58
regulations, regarding section 6 93
responsibilities of 174
Fish Commission, United States, history 193
fish farming, effluent limitation guidelines 212
five-point policy of USFWS 136
Fjord Seafood 212
Flathead National Forest 273
Flathead River 29
flat-tailed horned lizard 66, 79
Forest Practice Advisory Committee 226
Forest Service Land Resource Management Plans 56
free-rider 23
problem with 26
Friends of the Wild Swan 77
fry, definition 191
Fund for Animals 262
grizzly bear litigation 272
v. Babbitt 272, 289
v. Thomas 273

Garcia, Terry
Commerce Under Secretary for Oceans 204
Under Secretary of Commerce 228
Geer v. Connecticut, 1896 273
Geographic Information System development 132
Georgia, Red Hills region 173
Georgia Pacific Corporation 151
Glacier National Park 252, 260
gnatcatcher, California 59, 108, 116, 126, 134
Goodling, George
Pennsylvania representative 92
gopher tortoise 183
government, national
responsibilities of 5
role of 4
Governors' Roundtable 293
recommendations regarding grizzly bears 278
Gowdy, Curt
involvement in Atlantic salmon plight 197
Grand Teton National Park 253
Granite Creek Chalet 260
grants, Congressional, to states 104
gray wolf
conservation, Yellowstone 264
delisting 296
management, Wyoming 295
population recovery 296
reintroduction to Yellowstone 29
greater sage grouse 112
listing history 74, 313
Greater Yellowstone Ecosystem. *See* Yellowstone

Greenwalt, Lynn
  FWS Director 262
Greer v. Connecticut 40
grisle, definition 192
grizzly bear
  allowances for taking of 272
  conflict situations, history 276
  conservation, delisting 255
  conservation, efforts 311
  conservation, history 255–258
  conservation, protection of females 255
  conservation, strategy 280
  conservation, success of 255, 287
  controlling nuisance animals 271
  critical habitat designation 263
  cubs-of-the-year 273
  damage by, economics of 277
  decline of population 252
  delisting 277, 284
  demographic goal 282
  designations, recommendation 294
  diet 254
  differences from black bears 253
  distribution, Wyoming, 1980–1999 276
  effective habitat, definition 270
  estrus cycle, length 255
  food sources 284
  habitat 254, 262, 270
  historic distribution and range 252
  history, in Western states 252
  human confrontation 254
  hunting of 263
  impact of litigations 273
  Interagency Grizzly Bear Study Team 264
  Interagency Grizzly Bear Coordinating Committee 266–267
  judicial decisions influencing recovery 272–273
  legality of hunting, history 258
  listing history 262
  management funding 279
  management guidelines 267
  management plans 279
  management, recommendations 291
  monitoring, history 265
  mortalities, recent history 274
  mortality due to humans 255
  natural history 252–253
  population density 254
  population monitoring 269
  population recovery 273
  predators, lack of 255
  present situation 287–289
  Primary Conservation Area 281
  radio collars 283
  recovery plan 267–269
  recovery, recommendations 289–291
  reproductive age, female 255
  reproductive rate 254
  research, recommendations 291
  species of concern 258–261
  threatened listing 262–263
Grizzly Bear Conflict Management Program 276
Grizzly Bear Conservation Strategy 255
Grizzly Bear Policy, history 260
Grizzly Bear Recovery Plan, history 266–267
Gulf of Maine Distinct Population Segment 208
GYE. *See* Yellowstone

habitat acquisition 12
habitat conservation plan (HCP) 7, 58
  compared to NCCPs 124
  compared to safe harbor agreement 166–167
  compatibility with NCCPs 123
  history 199
  Karner blue butterfly 151, 154, 159, 166, 167
  process 8
  red-cockaded woodpecker 166
  regional 8
  statewide, history 171
  Wisconsin, accomplishments of 154 155
  Wisconsin, effectiveness 159
Habitat Conservation Plan Land Acquisition Grants, purpose of 104
habitat modification
  prohibition against 11

regulation of 44
Hales, David
  Acting Assistant Interior Secretary 99
harm, definition 57
hatchery-based restocking, Atlantic salmon 193
Hawaiian goose 9
HCP. *See* habitat conservation plan
House Committee on Merchant Marine and Fisheries 98, 100
Houston toad 183
Hughes, Royston
  Assistant Interior Secretary 95

Idaho
  Bitterroot Mountains 288
  Forestry Program, voluntary 110
  Game and Fish Department 251
  native plants 73
Imperial County, California 67
incidental take permits 7
  California 119
  Karner blue butterfly 149
Independent Multidisciplinary Science Team, Oregon 226
Interagency Cooperative Policy 106
Interagency Grizzly Bear Coordinating Committee 255, 266
  role of 291
Interagency Grizzly Bear Management Guidelines 267
Interagency Grizzly Bear Study Team 264–266
Interior Department, responsibility of Secretary 60
International Association of Fish and Wildlife Agencies 28, 96, 261, 317
International Council for Exploration of the Seas 198
International Paper Company 151

John D. Rockefeller Memorial National Parkway 253
Joint Explanatory Statement of the Conference Committee 92

kangaroo rat, Stephens' 108
Karner blue butterfly 108, 147–164
  causes of endangerment 148
  conservation efforts 311
  delisting possibility 159
  habitat 147, 154
  history 148
  largest occurrence of 157
  population decline 148, 157
Karner Blue High Potential Range 150
Kempthorne, Dirk
  Idaho governor 318
  Interior Secretary 318
Kennebec River, Maine 192
King, Angus S.
  Governor of Maine 187, 200
kit fox, San Joaquin 183
Kitzhaber, John
  Oregon governor 65, 225
Klamath Mountains ESU of steelhead 69, 79
Kleppe v. New Mexico 40
Knight, Richard
  IGBST Director 264
Kulongoski, Ted
  Oregon governor 232

Lacey Act 27
Lake Matthews, California, NCCP 125
Landowner Incentive Program 11
landowners, incentives offered to 31
landownership, size of 27
Leonard, Ross
  Director of IFG 258
limited authorities agreements 98
Listing Priority Guidance, USFWS 59
listing, of species
  determination process 62
  history 58
  litigation 58
  protection not required 60
litigation
  assessment of risk 75
  limitations of 59
  overview of trends 55
lizard
  Coachella Valley fringe-toed 121
  flat-tailed horned 66, 79

localized control 27
longleaf pine habitat in Sandhills region 169
lupine, wild 147, 155
Lusk Company 124
Lycaeides melissa samuelis. See Karner blue butterfly

Maine
  Aquaculture Association 206
  aquaculture, history 195
  Atlantic Salmon Authority 203
  Atlantic salmon commission 189, 195
  Atlantic Salmon Task Force 202
  blue mussel farms 195
  blueberry cultivation's effect on salmon 194
  Cobscook Bay, salmon farming area 211
  conflict with federal officials 187
  conservation plan, history 203
  Down East rivers 187, 197
  ESA litigation 188
  forest industry, opposition to listing of salmon 200
  Gulf of, Distinct Population Segment 208
  Kennebac River 192
  Land and Water Resources Council 206
  opposition to salmon listing 187, 202
  Penobscot Indian Nation 213
  Penobscot River 189, 190, 213
  Penobscot River restoration agreement 213
  Recovery Plan for the Gulf of Maine DPS of Atlantic Salmon 213
  removal of dams 213
  rivers having wild salmon runs 196
  salmon farming industry 188
  salmon farming, history 195
  statewide salmon restoration and management plan 197
  v. Norton 211
management
  agreements 9
  endangered species, centralization of 25
managerial federalism 10
Manson, Craig
  Assistant Interior Secretary 318
memorandums of agreement
  for red-cockaded woodpecker 180
Merchant Marine and Fisheries Committee 91
Migratory Bird Treaty Act 27, 40
migratory species 28
Missouri River
  litigation regarding management of 57
Missouri v. Holland 40
Mitchell, George
  Maine senator 197
Molloy, Donald
  U.S. District Judge, Montana 297
monitoring agents
  costs of 26
monopoly distortion 24
Montana
  Bitterroot Mountains 288
  Department of Fish, Wildlife, and Parks 31, 110, 251
  Fish and Game Commission 260
  grizzly bear harvest, history 259
  law regarding grizzly bear takes 279
  legal definition of take 103
  Special Grizzly Bear Committee 261
Mountain Lion Foundation 124
Multiple Species Conservation Program 70

National Academy of Sciences
  committee evaluation of grizzly bears in YNP 261
  conservation of Atlantic salmon report, 2004 213
  grizzly bear report, 1974 288
National Association of Homebuilders 235
  v. Evans 235
National Environmental Policy Act 119, 169, 268
National Flood Insurance Program 56
National Forest Management Act 293
National Forest Plan 69
National Marine Fisheries Service 55, 102, 225, 308

Northwest Fisheries Science Center 233
Policy for Evaluation of Conservation Efforts 232
tension with Oregon 229
National Oceanic and Atmospheric Administration (NOAA)
authority of 41
cooperation with states 40
Fisheries Division 35, 106
National Pollutant Discharge Elimination System 43
National Wildlife Federation 153
Natural Community Conservation Planning Act, California 8
approved plans 131
area statistics 129
budget problems 136
challenges to 138–140
champions, definition 138
collaboration with FWS and NMFS 134–135
compared to habitat conservation plan 124
compared to project permitting 120
conservation, definition 123
desired result 117
effect on state and federal wildlife agencies 135–137
evolution of 133
grant funding 129–133
historical performance 127
history 115–143
land use planning 116
legal challenges to 125
major changes to, 2000-2002 133
origin of 115–116
performance 127–129
planning, benefit of 122
process 118
project-by-project review and permitting, drawbacks of 119
requirements 123
suggested future features 141
supporters and opponents of 123
Natural Resources Defense Council v. Department of the Interior 59
Nature Conservancy 23, 26, 27, 151, 182
nene (Hawaiian goose) 9
New England Fishery Management Council 198
New York v. United States 40
Night of the Grizzlies 260
Nixon, Richard M.
signer of ESA 93
NMFS. *See* National Marine Fisheries Service
NOAA. *See* National Oceanic and Atmospheric Administration
North Atlantic Salmon Conservation Organization 198
North Carolina
red-cockaded woodpecker safe harbor program 169–171
Sandhills Physiographic Region 166–176
Wildlife Resources Commission 170
North Cascade Ecosystem 252
Northern Ecosystem Subcommittee of IGBC 266
northern spotted owl, listing history 58, 224
Northwest Forest Plan 66, 230
history 224
Northwest Power Planning Council 224
Norton, Gail
Interior Secretary 189

Obama Administration 43
ONRC v. Daley 229, 242
Oregon
Alsea Valley Alliance 237
Coastal Coho Conservation Plan 239
Coastal Salmon Restoration Initiative. *See* Coastal Salmon Restoration Initiative
conservation timeline of coho salmon 221, 245
Coho Assessment 232–233
coho salmon. *See* salmon, Oregon coastal coho
Department of Fish and Wildlife 225
Department of Forestry, conflict with NMFS 229

Forest Practices Act 66
Governor's Watershed Enhancement Board 224
Independent Multidisciplinary Science Team 226
Native Fish Conservation Policy 233
Natural Resources Council, challenge of no list decision 229
Oregon Plan, purpose of 221
Plan for Salmon and Watershed Restoration 311
protection from take liability 236
Secure Rural School and Community Self Determination Act 230
timber industry, conflict with NMFS 229
Watershed Enhancement Board 231–232
Oregon chub 183
Oregon coastal coho salmon. *See* coho salmon, Oregon coastal
Oregon Plan
implementation of 230
success of 243–247
Oslo resolution, history 198
Outer Continental Shelf and Lands Act 57
owl, northern spotted, listing history 58, 224

Pacific Coast Federation of Fishermen's Associations 235
Pacific Coastal Salmon Recovery Fund, federal 231
Pacific Rivers Council v. Brown 237
Pacific Rivers Council, litigation 236
Pacific salmon. *See* coho salmon, Oregon coastal
Pacific Salmon Task Force, history 225
Pacific Salmonid Biological Review Team 237
parr, definition 191
PECE. *See* Policy for Evaluation of Conservation Efforts When Making Listing Decisions
PECE process 238
pelican, brown 99
Pelly Amendment, to Fisherman's Protection Act 197
Penobscot Indian Nation, Maine 213
Penobscot River, Maine 189, 190, 213
Cove brook tributary 209
restoration agreement 213
peppergrass, slickspot 73
peregrine falcon 99
Peregrine Fund 28
permits, take, incidental 7, 43
Picoides borealis. See red-cockaded woodpecker
pine, longleaf, habitat in Sandhills region 169
Pittman-Robertson federal aid program 100
poaching
grizzly bear 255
salmon, Maine 208
Policy for Evaluation of Conservation Efforts 5
Policy for Evaluation of Conservation Efforts When Making Listing Decisions 71, 73
evaluation of conservation, Oregon coho salmon 238
limitations of 312
state involvement 312
policy, interagency 106
Poway Property Owners Group v. The City of Poway 125
Poway, California, NCCP 125
Powell, John Wesley 24
prairie dog, black-tailed 112
decentralized management of 30
Primary Conservation Area
grizzly bear, jurisdictional lands 281
principal-agent costs 23
prohibition, take 41
project agreements, definition 94
project-by-project review and permitting under NCCP
drawbacks of 119

Ramsey v. Kantor 77
Rangewide Strategy Working Group 66
Recovery Land Acquisition Grant

program, funding 105
Recovery Plan
Gulf of Maine DPS of Atlantic Salmon 213
grizzly bear, goal, 1993 270
roles of Interior and Commerce in developing 6
recovery zone, definition 268
red-cockaded woodpecker 9, 165–186, 311
Conservation Coalitions 171
conservation on private land 165
Interagency Working Group 170
memorandums of agreement 180
roles of scientific advisory committees 172
roles of steering committees 172
safe harbor agreements 108
safe harbor agreements, accomplishments of 170, 179–180
safe harbor agreements, funding for 174, 178
safe harbor program, implementation of 169
safe harbor agreements, status of by state 170
redd, definition 191
regulation, decentralized 28–30
regulatory agencies, environmental 24
relations, federal-state 37, 46
Research Subcommittee of IGBC 266
Resource Conservation and Recovery Act 4
Resources Limited v. Robertson 273
Restoration of Atlantic Salmon in America, Inc. 197
RESTORE The North Woods 199
Rivers and Harbors Act 212
Riverside County, California
2003 HCP 107
Habitat Conservation Agency 124
Role of State Agencies in Endangered Species Activities 6

Sacramento River, California
Chinook salmon 61
safe harbor agreement/plan
accomplishments of, red-cockaded woodpecker 179–180
benefits of 181, 183
challenges of 165
compared to habitat conservation plan 166–167
funding for red-cockaded woodpecker 174, 178
implementation of, red-cockaded woodpecker 169
purpose of 108
management agreements 169
safe harbor program 8, 11, 165
state involvement in 171–173
red-cockaded woodpecker 167–186
safe harbor program 169
first in nation 173
success of 176
sage scrub, coastal 125
salamander, Barton Springs 62–65, 78
Salmo salar. *See* salmon, Atlantic
Salmon, Atlantic
aquaculture, history 195
aquaculture, impact of 198
ban on use of nonnative strains 211
Conservation Agreement 205
conservation before 1993 193
decline of 187, 192
diseases, infectious anemia 211
effect of sawmill debris on 192
escape of penned salmon, 2001 210
farming of, history 195
genetic risk to 209
life cycle of 191
listing history 187, 188, 199, 202, 209, 210, 216
migration routes 191
natural history 189–192
nonnative species 188
ocean fishery, history 197
prelisting recovery plan 196
recovery plan 196, 211, 213–214
risk factors for 213
river-specific stocking program, history 196
threats to 201

withdrawal of proposed listing 204
salmon, Chinook
listing history 61
Sacramento River 61
winter-run 78
salmon, Oregon coastal coho 79, 207, 311
Assessment 232–233
conservation history 223
conservation lessons 240
conservation timeline 221, 245
critical habitat designation 234
delisting 237–238
ESA listing of 64, 229
life cycle 221
listing history 64
management of 28
Oregon Plan 221–234
no list decision 227–228
population decline, history 223
recovery, history 234
recovery, lessons 240
relisting 238–239
restoration program 226
restoration themes 242
risk factors 234
section 4 take limits 236
threatened listing, history 225
withdrawal of proposed listing 239
Salmon Habitat and River Enhancement 200
salmon restoration and management plan, Maine 197
Salmon River 110
San Bernardino Valley Audubon Society v. Metropolitan Water District of Southern California 125
San Bruno Mountain HCP 116
San Diego County, California
Resource Protection Ordinance 70
MSCP 125, 139
Multiple Species Conservation Program 127
San Joaquin kit fox 183
Sandhills Physiographic Region 166
Private Land Initiative 170
scale economies 27, 28
Scalia, Antonin
Supreme Court justice 45
section 4 of ESA
cases, state-federal relationship 61–70
criteria for listing 262
governing of listing process 58
litigation involving 55
recommendations 244, 314
role of states 105
state-federal relations 60
section 6 of ESA
1997 extension 104
amendments 198, 100–101
amendments, 1976 95
amendments, 1977 and 1978 97–98
amendments, 1982 99–100
authority of states 95–96, 313
cooperative agreements, creation of 94, 317
criteria 9
funding 96, 104–105, 160
history 11
implementation of cooperative agreements 102
implementation of ESA 9
judicial interpretation of 103
legislative history 89–93
limitations of 42, 47
narrow interpretation of 93
projects eligible for funding 103
regulatory interpretation of provisions 93
shift of emphasis 94
to enhance cooperative federalism 316
section 7 of ESA
litigation involving 56, 77
regulations 106
section 9 of ESA
litigation involving 57
prohibitions 57
prosecution of violators 57
section 10 of ESA
authority 107, 108
enlarged role of 109
HCP process 122
litigation regarding permits 58

role of states 106–107
Secure Rural School and Community Self Determination Act 230
Selkirk Conservation Alliance v. Forsgren 77
Selkirk Ecosystem 252, 265
Senate Committee on Environment and Public Works 99
Senate Natural Resources Committee 136
shifting mosaic strategy 155
short-leaved Dudleya 79
Sierra Club 151
litigation 58
slickspot peppergrass 73
smolt, definition 192
snail darter 3
Snake River Basin Adjudication 110
Snake River Water Rights Act of 2004 110
Snowe, Olympia
Maine senator 210
Solid Waste Agency of Northern Cook County v. Army Corps of Engineers 45
Southern California Coastal Sage Scrub NCCP 126
Southwest Center for Biological Diversity
litigation 70
v. Berg et. al. 125
v. Bureau of Reclamation 77
species
covered, under HCPs 107
covered, definition 117
determination of status 60
Species and Habitat Conservation Agreement
Wisconsin, Karner blue butterfly 152
Species Conservation Committee
recommendation for development of 315
statewide, recommendation 292
Species Recovery Land Acquisition Grants 104
spotted owl, northern 183
listing history 58
State Conservation Agreement 293
State Wildlife Grant program 11
states, role of
constitutional power of 22
encouraging involvement 312
greater commitment 307–319
importance of active involvement 310
in protecting biodiversity 307
in takes 111
increased authority 314
limitations of 36
recovery plans 315
regional endangered species lists 12
responsibility of 36, 311
Steelhead Supplement to Oregon Plan 69
steelhead trout
Klamath Mountains, listing history 69
Stephens' kangaroo rat 108
Stevens, Ted
Alaska Senator 90
Swan River 29
Swan View 47
Coalition, Inc. v. Turner 42, 103

take
definition 57, 103
grizzly bear, allowances 272
incidental permit 57, 119, 149
incidental statements 57
incidental, of red-cockaded woodpecker 169
prohibitions 7–9, 41, 42, 93, 97
Oregon, history 236
Tellico Dam 3
Ten Point Plan 62
Tennessee Valley Authority v. Hill 56
total maximum daily load allocations 224
transaction costs 22
demand for endangered species 26
for habitat inputs 26
of gray wolf management 29
transaction-cost approach 27
trout
bull, devolution of 29
Klamath Mountains steelhead 79
listing history 69
Umpqua River 66

Westslope cutthroat 110
Trout Unlimited
litigation 75, 207, 238, 239
role of in Atlantic salmon fisheries 197
v. Lohn 75, 238, 239
Tunney, John
California Senator 90
Turner, Ted
land owner 23, 26
litigation 42, 103

U.S.-Danish Fisheries Agreement
history 197
Umpqua River cutthroat trout 66
United States Fish and Wildlife Service. *See* Fish and Wildlife Service
United States v. Glenn-Colusa Irrigation District 103
USFWS. *See* Fish and Wildlife Service

vireo, least Bell's 108

Wallop, Malcolm
Wyoming Senator 97
WDNR. *See* Wisconsin, Department of Natural Resources
Western Association of Fish and Wildlife Agencies 261
Western Association of Game, Fish, and Conservation Commissioners 261
Western Governors' Association 46
Western Riverside MSHCP 139
Western Watersheds Project v. Servheen 256
Westslope cutthroat trout 110
wetlands, conservation law 45
Williams, Ted
involvement in Atlantic salmon plight 197
Wisconsin
Department of Natural Resources 108, 147
Department of Transportation 156
Division of Forestry, funding 160
Endangered and Threatened Species Act 149
habitat conservation plan, effectiveness 159
Initiative 153–155
Karner blue butterfly 108, 147–164. *See also* Karner blue butterfly
Woodland Owners Association 151
wolf, gray. *See* gray wolf
Woodgerd, Wesley
MFGD Director 263
woodpecker, red-cockaded. *See* red-cockaded woodpecker
Wyoming
Game and Fish Commission 295
Game and Fish Commission, proposed grizzly delisting 274
Game and Fish Department 110, 251
Gray Wolf Management Plan 295–296
grizzly bear conservation 251
grizzly bear distribution, 1980–1999 276
grizzly bear harvest, history 259
Grizzly Bear Study Team 265
proposed grizzly delisting 290

Yeaman, Fred
Chief of WGF Wildlife Development 258
Yellowstone
elk hunters 277
grizzly bear conservation 251
grizzly bear conservation, success of 287
grizzly bear demographic goal 282
grizzly bear monitoring, history 265
grizzly bear population recovery 273
grizzly bear recovery 263
prohibition of grizzly bear garbage dump feeding 260
Yellowstone Ecosystem Grizzly Bear Subcommittee 285
Yellowstone Ecosystem Subcommittee of IGBC 266
Yellowstone Grizzly Coordinating Committee 285
Yellowstone National Park 29
Yuma County, Arizona 67